Surveys and Tutorials in the Applied Mathematical Sciences

Volume 15

Featuring short books of approximately 80-200pp, Surveys and Tutorials in the Applied Mathematical Sciences (STAMS) focuses on emerging topics, with an emphasis on emerging mathematical and computational techniques that are proving relevant in the physical, biological sciences and social sciences. STAMS also includes expository texts describing innovative applications or recent developments in more classical mathematical and computational methods.

This series is aimed at graduate students and researchers across the mathematical sciences. Contributions are intended to be accessible to a broad audience, featuring clear exposition, a lively tutorial style, and pointers to the literature for further study. In some cases a volume can serve as a preliminary version of a fuller and more comprehensive book.

Gábor Orosz · Tamás G. Molnár

Dynamics and Control of Connected Vehicles

Springer

Gábor Orosz
Department of Mechanical Engineering
University of Michigan
Ann Arbor, MI, USA

Tamás G. Molnár
Department of Mechanical Engineering
Wichita State University
Wichita, KS, USA

ISSN 2199-4765 ISSN 2199-4773 (electronic)
Surveys and Tutorials in the Applied Mathematical Sciences
ISBN 978-3-031-94554-0 ISBN 978-3-031-94555-7 (eBook)
https://doi.org/10.1007/978-3-031-94555-7

Mathematics Subject Classification: 37N35, 93C43, 93D15, 93D25

This Springer imprint is published by the registered company Springer Nature Switzerland AG
The registered company address is: Gewerbestrasse 11, 6330 Cham, Switzerland

To our families

Foreword

I personally have a broad interest in nonlinear dynamics and its engineering applications, especially in systems where time delay has an essential role. When I have the first look at a new book, I often wonder whether there is a chapter, a section in it which might be interesting or relevant for my research. Honestly speaking, I rarely consider that a book should be read in full, either by myself, or in a reading group of Ph.D. students. This book of *Gábor Orosz* and *Tamás G. Molnár* belongs to the rare cases, and not just because I consider them as former student and "grandstudent", respectively.

One obvious reason, why this book is suggested for reading it, is its topic. In the age when microprocessors appear in all our machines from a hammer to a rocket, the dynamics of these systems present some unconventional, often counter intuitive characteristics due to the interplay of the Newtonian mechanical systems moving in continuous time and their computer control taking place in discrete time. Even when these systems are understood, the collaboration of the human operators and the computer-controlled machines results in further unorthodox dynamics, while the mathematical modeling of the human decision making still involves uncertainties, sometimes even mysteries. Nowadays, traffic dynamics is like a system of several human-machine systems with increasing level of intelligence at the machine side, where the intelligence is partly induced by the communication of the vehicles, that is, by their virtual connectivity. The beauty of this topic is that we all can experience and test the appearance of the various traffic scenarios predicted by means of the introduced mathematical models in the subsequent chapters of the book.

Another reason for suggesting this book to be read in full is its manageable length. It is compiled with a clear and simple structure with balanced and properly weighed chapters and sections. The text is easy to read, and the figures provide straight and colorful information transmission. The dense knowledge transfer is made possible with the use of a relatively high level of mathematical abstraction, which is mainly required by the relevance of nonlinearities and the presence of time delays in the models. There are two ways the unprepared reader is supported in this regard. On the one hand, the mathematical models are continuously, tirelessly and relentlessly linked to their physical meanings, which are, as mentioned above, emerging traffic configurations experienced along our roads and highways. On the other hand, in case the mathematical background needs to be revisited or studied by those who are less familiar with dynamic systems theory, the Appendices provide a concise and correct overview of the basic principles also using simple non-trivial examples.

The book is written in a compact form with a unique, rigorous, still practice-oriented engineering approach. All researchers have their preferred style between theory and practice. In case of this book, they are likely to find their specific point of view since the authors use a broad spectrum of methods to introduce and to discuss the specific traffic models of connected vehicles. The predictions are bright: this research direction promises improved traffic management, decreasing number of accidents, enhanced environment protection that are all in the forefront of actual economic and societal problems to be solved. Clearly, the leading author has decades long experience in traffic dynamics, which made it possible to construct a strained introduction to the research field that is recommended not just to those interested in the future traffic dynamics of connected vehicles, but also to those interested in the control of time delay systems in general, or just in a relevant application of time delay systems.

March 2025

Gábor Stépán
Professor of Applied Mechanics
Budapest University of Technology
Budapest, Hungary

Preface

Since the invention of the wheel, humans have been constructing vehicles to transport people and goods. Many ancient civilizations relied on extensive road networks that fueled their prosperity and growth. Such developments were accelerated at the beginning of the twentieth century by the rise of automobiles, and nowadays, more than a billion vehicles travel along the millions of miles of highways tracing the surface of the Globe.

While motorized road vehicles have been operated by human drivers for a century, during the last few decades they have been becoming more and more automated. Moreover, wireless communication enables vehicles to exchange information with each other and the infrastructure. These technologies carry a large potential to make road transportation safer and more efficient. Achieving this, however, requires a profound understanding of how vehicles move when driven by human drivers or automated systems and how connectivity can benefit different vehicle classes and the transportation system as a whole. These needs motivated the authors to organize their knowledge on this subject into a book, which can serve students, researchers and practitioners working on this exciting, interesting and ever-developing field.

The content of this book is based on the research carried out by the authors during the last 10–15 years, and on the graduate course (of the same title as the book) they taught at the California Institute of Technology, at the University of Michigan, at the Budapest University of Technology, and at the Shanghai Jiao Tong University. Concepts from multiple disciplines, including physics, applied mathematics and engineering, are used in this monograph. Readers are expected to have some background in modeling physical systems, solving differential equations and understanding integral transformations, as these are needed to embrace the topics covered in the book. Nevertheless, no expert knowledge is necessary on any of these areas. Some background knowledge on stability analysis and time delay systems is placed in the Appendices. While later chapters build on the knowledge imparted in the previous ones, each chapter is designed to be readable on its own. Some sections, namely, Sects. 2.4, 3.5, 4.5, 4.6, and 4.7, require somewhat deeper background knowledge, but these may be skipped without braking continuity.

The layout of the book is as follows.

Chapter 1 briefly reviews the history of vehicular traffic and describes the efforts made in introducing vehicle automation and connectivity over the last few decades. By no means we aim for a complete review of this large area but rather use this chapter to motivate the reader by providing some background for the forthcoming technical chapters.

Chapter 2 focuses on the modeling and analysis of vehicular traffic including human-driven vehicles. After describing car-following models, we establish some fundamental concepts, like plant stability and string stability, and link these to the stability of ring configuration. Then we extend these models and the analysis techniques for the case when the human reaction time delay is incorporated. First order lag approximation is also discussed, and at the end, the role of nonlinearities in human-driven traffic is highlighted and the notion of bistability is explained.

Chapter 3 focuses on vehicle automation, in particular, on the automation of the longitudinal dynamics. The design principles of cruise control and adaptive cruise control are laid out and the importance of the effects of time delays in the control loop are discussed. Then the impact of automated vehicles on the traffic flow is presented under different penetration levels. Finally, the effects of sampling and digital control are discussed while highlighting the corresponding time-varying time delays.

Chapter 4 introduces vehicle-to-everything communication and its advantages in controlling connected automated vehicles. Here we present the concept of connected cruise control. After discussing the use of high-quality motion signals, like acceleration, we provide the general mathematical tools that can be used to analyze connected vehicle networks. In particular, the concept of head-to-tail string stability is discussed in detail since this facilitates the analysis and control design in mixed transportation systems containing human-driven and connected automated vehicles. The role of nonlinearities in connected vehicle systems is highlighted, and the chapter is finished by discussing performance measures like energy efficiency and robustness against uncertainties in human behavior.

Chapter 5 collects the experimental results obtained with full-size connected and automated vehicles on test tracks and public roads. The results are organized according to the complexity of the experiments which increases with the number of vehicles involved. The benefits of connectivity in terms of safety, energy efficiency and mobility are highlighted.

The numerical results and the figures presented in this book can be reproduced using the Matlab codes publicly available at the following repository: https://github.com/molnartamasg/dynamics-and-control-of-connected-vehicles.

We greatly acknowledge the help and support of our colleagues during the preparation of this monograph. First and foremost, we thank Gábor Stépán and Tamás Insperger from Budapest for their continuous encouragement and for facilitating many international exchanges between Budapest and Michigan. We greatly appreciate the support of Richard Murray and Aaron Ames from Caltech, where the first version of the course, that served as the foundation of this book, was taught. We thank Huei Peng and Henry Liu, the directors of the Mcity test facility at Michigan, for their continuing support of our experimental work. Our extensive collaborations with Harvey Bell from Michigan, Dénes Takács from Budapest, Ardalan Vahidi from Clemson, Mike Hopka, Devesh Upadhyay, Eric Tseng and Gaurav Pandey from Ford, Sergei Avedisov, Gaurav Bansal and Onur Altintas from Toyota, László Virág, András Váradi and András Wippelhauser from Commsignia, and Chaozhe He from Navistar were instrumental for putting our research into a broader context. At last but not at least, the contributions of the more than dozen Ph.D. students from the University of Michigan and the contributions of the half a dozen Ph.D. students from the Budapest University of Technology are gratefully acknowledged.

Ann Arbor, MI, USA
January 2025

Gábor Orosz
Tamás G. Molnár

Contents

Acronyms

ACC	Adaptive cruise control
ADAS	Advanced driver assistance system
AHS	Automated highway system
ARZ	Aw-Rascle-Zhang
AV	Automated vehicle
CACC	Cooperative adaptive cruise control
CAT	Connected automated truck
CAV	Connected automated vehicle
CC	Cruise control
CCC	Connected cruise control
CHV	Connected human-driven vehicle
CV2X	Cellular vehicle-to-everything
DDE	Delay differential equation
DSRC	Dedicated short range communication
ECU	Electronic control unit
FFT	Fast Fourier transform
GPS	Global positioning system
HTSS	Head-to-tail string stability
HV	Human-driven vehicle
IC	Internal combustion
IDM	Intelligent driver model
LTE	Long term evolution
LWR	Lighthill-Whitham-Richards
NUE	Number of unstable characteristic exponents
OBU	On-board unit
ODE	Ordinary differential equation
OVM	Optimal velocity model
P	Proportional
PD	Period doubling
PDE	Partial differential equation
PI	Proportional-integral

PW	Payne-Whitham
ReLU	Rectified linear unit
SE	Stable equilibrium
SLC	Stable limit cycle
UE	Unstable equilibrium
ULC	Unstable limit cycle
US	United States
V2I	Vehicle-to-infrastructure
V2P	Vehicle-to-pedestrian
V2V	Vehicle-to-vehicle
V2X	Vehicle-to-everything
WiFi	Wireless fidelity
ZOH	Zero-order hold

Chapter 1
Introduction

In this introductory chapter we provide a brief history of vehicular traffic and point out the continuing challenges societies have been facing while developing and managing such systems. We also describe recent advances in road vehicle automation and connectivity and discuss how such technologies may be utilized to solve (or at least mitigate) the above issues. By no means we plan to cover all aspects of road transportation but rather to motivate the technical material presented in the forthcoming chapters.

1.1 Modeling Vehicular Traffic

One of the first inventions of early human civilizations was the wheel with axle which enabled the transportation of people and freight in an efficient manner. Early vehicles were towed by humans and domesticated animals. Still, archaeological evidence [171] shows that high density and speed often led to accidents and congestion in ancient settlements. In the early twentieth century, motorized vehicles, which were capable of traveling with much higher speed, amplified such issues and by the 1930s researchers were quantifying the problems experimentally. These included the early measurements of driver reaction time [72] and traffic capacity [71].

During the 1950s, the US started to build its (interstate) highway system aiming for uninterrupted flow of traffic. The higher speed limit set on these motorways enabled people to cover larger distances and live farther away from work, which, in turn, led to the development of large, highly-populated urban areas. It was soon discovered that the flow along the highways is not always smooth which led to the construction of early traffic flow models. These so-called continuum models are based on the Eulerian view and utilize partial differential equations (PDEs) to describe how traffic density and speed (and consequently their product, the flux) change in space and time along the highway. Such models include the Lighthill-Whitham-Richards

G. Orosz and T. G. Molnár, *Dynamics and Control of Connected Vehicles*,
Surveys and Tutorials in the Applied Mathematical Sciences 15,
https://doi.org/10.1007/978-3-031-94555-7_1

(LWR) model [121, 182], the Payne-Whitham (PW) model [162, 239], and the more recent Aw-Rascle-Zhang (ARZ) model [10, 250].

At the same time, other researchers (many of whom worked in the booming automotive industry) focused on the interactions between individual vehicles and developed so-called car-following models. These consider the Lagrangian view and describe the longitudinal motion of automobiles using ordinary differential equations (ODEs) and delay differential equations (DDEs) [29, 48, 59, 60, 89, 146, 165]. It is interesting to note that the theory of DDEs did not really exist at that time, as that was developed in the forthcoming decades [19, 80, 111, 112, 197]. Also, only linear analysis was possible, as the theory of nonlinear differential equations has not yet been developed and the computational power of computers (which often occupied a whole building) did not allow extensive simulations of the developed models.

From the 1970s the development of microprocessors led to a large increase in computational speed and also the miniaturization of computers. By the 1990s researchers had their own desktop computers which enabled them to simulate the mathematical models (even with nonlinearities and time delays), which led to a large boom of different models. Initially, cellular automaton models, in which both time and space were discretized [143], became popular as these were capable of reproducing the spatiotemporal patterns observed in traffic (at least qualitatively) and had very little computational demand. However, as such models failed to capture many details in driver behavior, they were replaced by car-following models and such models will be discussed in the next chapter in detail. These include the optimal velocity model (OVM) [15] and the intelligent driver model (IDM) [108, 205], both of which are generalizable by incorporating driver reaction time; see [14, 36, 37] for the OVM and [107, 207] for the IDM. A larger variety of car-following models can be found in [88, 144, 206, 243], and some of these have been implemented in agent-based simulation software like SUMO.

Recent modeling efforts include linking car-following models to continuum models. For example, the simplified car-following model [147], called the Newell model, can be solved explicitly and these solutions also satisfy the LWR model when the latter is rewritten into the Lagrangian framework [115, 139]. This connection also becomes relevant when processing trajectory data [114] and when doing data-based traffic predictions [134, 263]. Similar ideas can also be applied to more complex car-following models like the OVM [21, 104] and driver reaction time delay can also be incorporated [139]. In addition, machine learning techniques are also entering traffic modeling at a rapid pace due to their ability of capturing complex nonlinearities [54, 101, 227, 242]. There exist many extensions for multi-lane traffic and highway junctions but we omit the discussion of those for the sake of brevity.

During the first decade of the twenty first century, efforts were made to analyze the dynamics of car-following models both at the linear and nonlinear levels. As will be shown in Chap. 2, most car-following models admit an equilibrium where vehicles follow each other with the same constant speed. Then linear stability analysis can be utilized to reveal the onset of instability where the system diverges from the equilibrium [65, 240, 241]. The arising oscillations can grow into stop-and-go waves which travel upstream with a characteristic speed of 15–20 [km/h] [105, 106, 159]. The formation of such traffic jams was also validated experimentally on a closed track [199]. To understand the mechanisms behind such phenomena, nonlinear analysis of the system is required [58]. Once reaction time delay is included in car-following, drivers may respond differently to small perturbations and large perturbations, and consequently, stop-and-go waves may also appear when the equilibrium is linearly stable [154, 157, 158, 160]. Such bistable behavior will also be discussed in detail in Chap. 2. Due to the nonlinear phenomena, identifying the car-following dynamics from data is a challenging task [7, 228] especially when one also wants to identify the driver reaction time delay [64, 100, 101].

1.2 Automation of Vehicular Traffic

As early as in the late 1930s, engineers were conceptualizing about automated highways [66]. However, the technology of the era prohibited real deployments. Even at the end of the 1950s, the distance between consecutive vehicles was measured by stretching a wire using a spring-loaded reel [29]. These limitations vanished by the 1980s with the arrival of range sensors (like radar and laser) and the concept of automated highway systems (AHS) came to reality. This culminated in an experimental demonstration with eight automated vehicles driving autonomously on a highway section between Los Angeles and San Diego [180]. According to today's categorization these were automated vehicles (AVs) with level 2 automation [195], as they controlled the intervehicle distance (sensed by radar) and followed the lane center (by utilizing magnetic markers). In fact, the vehicles also communicated with each other using dedicated radios, so they can be considered as early representations of connected automated vehicles (CAVs) [196].

These efforts led to the revival of dynamical analysis of vehicle chains and new control algorithms for the longitudinal control of vehicles. The notion of string stability was established [188, 202] to characterize whether an automated vehicle can attenuate the velocity fluctuations of the vehicle ahead. Adaptive cruise control (ACC) [97, 203] was born, allowing AVs to obtain motion information about the vehicle ahead via range sensors (radar, lidar, camera) and respond to it by controlling the throttle and the brake [116, 210]. The dynamics of chains of ACC driven vehicles have also been analyzed and it was highlighted that string stable ACC designs can benefit the traffic flow [38, 109]. The effects of nonlinearities in the dynamics and control design have been investigated [155, 246] and it was shown that time delays, that arise in the control loop from sensing, computation and actuation, have detrimental effect on string stability [173, 245]. More recent efforts use different predictor strategies to compensate for such delays [18, 135, 224, 244].

The auto industry also started to deploy ACC on commercial vehicles as part of advanced driver assistance system (ADAS) packages, and by now, it is available for literally all brands. As the control algorithms are considered proprietary, research efforts have been made to identify them from experimental data using online parameter estimation [229, 230]. Somewhat surprisingly, recent experiments show that most commercial ACC algorithms are in fact string unstable [34, 76, 77]. This is due to other considerations (mainly driver and passenger comfort) during the design process as well as due to large time delays arising in the control loops [44, 45]. After all, ACC is still considered to be a comfort feature, and while adding more sensors can potentially help the situation awareness of AVs, processing and fusing such information takes considerable time which increases the time delay in the control loop significantly.

It is important to note that even though most new vehicles are equipped with ACC (including vehicles with higher levels of automation), drivers may choose not to activate it. Consequently, the actual penetration of AVs in the traffic is still quite low; see Fig. 1.1a where an AV is inserted into a chain of human-driven vehicles (HVs). It was demonstrated experimentally on a closed track that (by keeping large enough distances) even a single string stable AV can help to mitigate velocity fluctuations and smooth the traffic flow [198]. However, when AVs keep similar distances as human drivers (in terms of time headways), experiments show that to compensate for the string unstable behavior of human drivers, either large penetration of AVs is necessary [176] or AVs need to utilize large-scale traffic information [82] which may not always available. The design of ACC algorithms and their impact on the traffic flow will be discussed in Chap. 3. We remark that vehicles with higher levels of automation also require controlling the lateral and yaw motions via steering control which needs to be integrated with the longitudinal control [67, 98, 122, 170, 179, 186, 209]. However, these are beyond the scope of this book.

1.3 Connectivity in Vehicular Traffic

Wireless communication may be utilized to establish higher levels of cooperation between automated vehicles, as highlighted in Fig. 1.1f. This is often referred to as cooperative adaptive cruise control (CACC) and has gained popularity in the research community in the early 2000s due to its potential of improving performance (in terms of safety, energy efficiency and reduced following distance). This led to the development of a large number of distributed control algorithms for platoons of CAVs [41, 47, 131, 145, 167, 192, 193, 212, 232, 236], many of which have been demonstrated experimentally. While introducing connectivity brings many potential benefits, it also comes with limitations (e.g., packet drops and interruptions in communication), which can make the control design more challenging [214]. This gave rise to different compensation strategies [30, 256, 257]. In particular, it was shown that the time-varying time delays arising in the control loops may be compensated using predictors [16, 17, 39]. In addition, cybersecurity also needed to be addressed [23, 163, 225] to make such platooning feasible.

Forming platoons requires multiple CAVs to follow each other which rarely occurs in real traffic when the CAV penetration is low. This created a demand for flexible

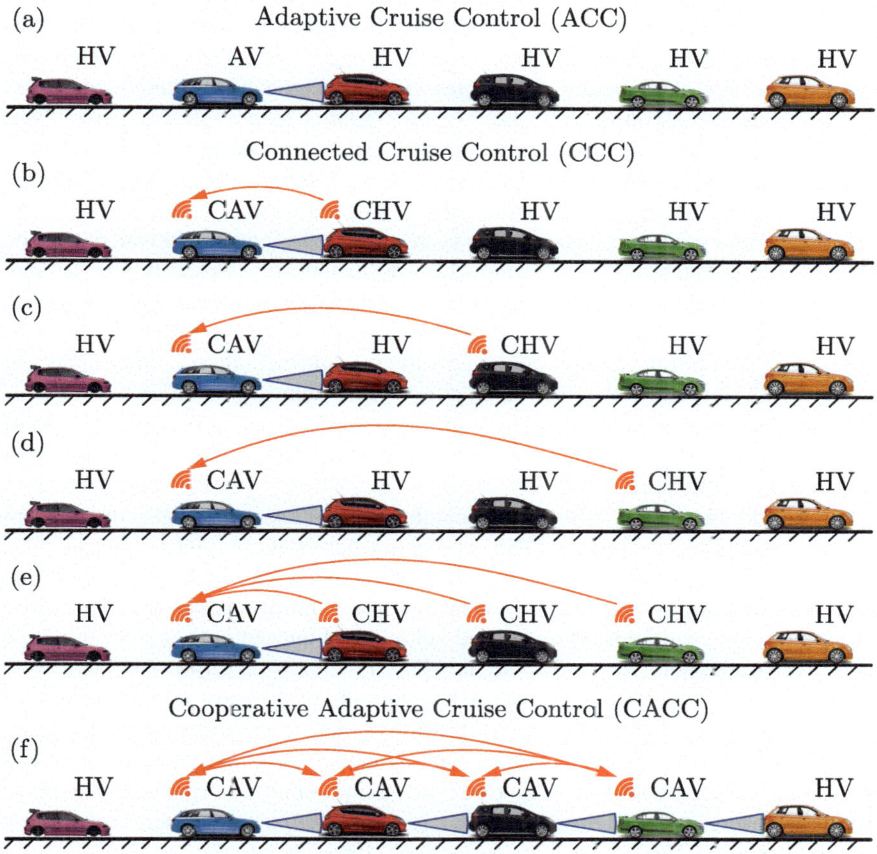

Fig. 1.1 Different strategies once an automated vehicle (AV) or a connected automated vehicle (CAV) is inserted into the flow of human-driven vehicles (HVs). Once a human-driven vehicle is capable of wireless communication, we refer to it as connected human-driven vehicle (CHV)

communications strategies which can provide benefits through connectivity for varying penetrations of automation and connectivity. Connected cruise control (CCC), highlighted in Fig. 1.1b–e, can be utilized to benefit CAVs in a large variety of mixed traffic scenarios [62, 91, 152, 178, 235, 251, 255]. The corresponding control algorithms can be optimized for control effort and energy efficiency [63, 64, 85, 190, 191, 233, 234], and they can be robustified against uncertainties in the behavior of the human-driven vehicles involved [32, 79] and against stochastic delays arising due to packet drops in communication [138, 172, 174, 204]. CCC can also be extended to incorporate nonlinearities [8, 9, 136, 253] as well as state and parameter estimation [100, 101, 134, 190, 252]. CCC strategies will be discussed in detail in Chap. 4 while experimental realizations on closed test tracks and public roads [2, 7, 20, 61, 153, 175] will be shown in Chap. 5.

We remark that there exist a large variety of different communication topologies in the literature; see, for example, [69, 78, 133, 214, 220, 221, 260, 262], which may give rise to different cooperation strategies in mixed traffic. The detailed discussion of these strategies is omitted for brevity. Our goal is to provide the reader with the fundamental tools needed for the analysis and control design of mixed traffic. Accordingly, the mathematical techniques presented in this book are applicable for any communication topology that may arise in the real world.

When analyzing connected vehicle systems and designing controllers for the embedded CAVs, the notion of string stability (instability) is used to characterize how velocity fluctuations are attenuated (amplified) as they propagate along the vehicle chain. There exist many different notions of string stability [22, 55, 151, 168, 169, 188, 202] depending on the how the size of fluctuations is measured and which signals are utilized in the calculation. Time delays [35, 42, 62, 163, 213, 251] and nonlinearities [142, 152] may further complicate the definitions. In case of mixed traffic systems, it is important to notice that one cannot control the motion of human-driven vehicles.

In order to ensure flexibility while maintaining mathematical rigor, the concept of head-to-tail string stability (HTSS) has been established [61–63, 152, 153, 175, 251]. This allows one to incorporate human-driven vehicles which may amplify the perturbations of the vehicles ahead, and also enables control design which can ensure attenuation between the head connected vehicle and the tail CAV. HTSS has been successfully adopted by many researchers [70, 117, 123–126, 140, 141, 177, 178, 185, 201, 226, 231, 249, 265] and has become one of the most important metrics when evaluating the behavior of mixed traffic systems. In the forthcoming chapters, we will use this concept to characterize the dynamics of human-driven, automated and mixed vehicle chains and relate the results to the stability of ring configurations. We emphasize that the analysis tools and design principles discussed in this book are applicable at different levels of automation [195].

1.4 Present and Future of Road Transportation

After the success of AHS experiments in the late 1990s with level 2 automated vehicles [180], the concept of highly-automated (often called self-driving) vehicles was put forward through three grand challenges in the mid 2000s [27]. These were enabled by the rise of affordable digital sensors (including digital cameras and lidars) as well as by the increased on-board computational capabilities of vehicles, and gave rise to a huge development in the car industry. Nowadays, multiple automakers are selling level 3 AVs: these can drive themselves under certain circumstances but may give back the control to the human driver when necessary. Other companies are deploying level 4 AVs, which are operating in geo-fenced areas without human drivers on board (though remote operators monitor the vehicle so they can provide instructions in challenging situations).

Cooperative driving of CAVs was also accelerated through two grand challenges in the 2010s [166] which demonstrated the benefits of adding wireless connectivity on the top of automation. Recent experimental efforts include the deployment of infrastructure based sensors [68, 129] as well as roadside communication units [102, 219]. On one hand, such installations may be used to support the operation of AVs and CAVs and improve their safety and energy efficiency. On the other hand, these vehicles may be utilized to actuate the traffic flow and benefit the whole system through the control of individual CAVs [61, 82, 133, 136, 198, 248] or platoons of CAVs [33, 103, 164, 251]. For recent reviews about connected automated vehicles and their potential impact on road transportation, we refer the reader to [40, 50, 74, 119].

Time and energy efficiency of CAVs may benefit from connectivity which can be realized by optimizing their longitudinal controllers based on the motion information of the vehicles ahead on the highway [1, 5, 85, 86, 120, 189–191, 208, 211, 222, 223]. We will discuss some aspects of this in Chaps. 4 and 5 but we remark that this area is much wider and may be a topic of a book on its own [187]. Similar strategies can be applied in urban environments when information about the traffic signal phase and timing is available [6, 12, 53, 200, 237] and also when CAVs cooperate at signal-free intersections [127, 128, 183]. These scenarios are, however, beyond the scope of this book.

Another benefit of connectivity is the increased traffic safety which was the original motivation of making vehicles connected to each other [28]. There are different levels of cooperation possible [196]: CAVs may share their status and intent with each other and may even negotiate for road resources to make decisions such that potential conflicts are avoided [215–218] and the rules of the road are maintained [26]. Design tools developed in the last decade also allow the synthesis of controllers that can formally guarantee the safety of CAVs even in the presence of nonlinearities and time delays [2–4, 31, 87, 135, 137, 148, 238, 258, 259]. We will discuss the safety implications of connectivity and automation in Chaps. 4 and 5 but the formal analyses of conflict and safety are kept for a future monograph.

Finally, we remark that the field of connected and automated vehicles, or in more general, connected and automated transportation, has many open challenges and is expected to be an active area of research during the forthcoming decades. For example, the deployment of highly-automated vehicles is hindered by limitations of perception systems (especially during inclement weather such as rain and snow). While connectivity may come to rescue, its low penetration rate is calling for new technologies that enable vehicles to communicate with each other and with the infrastructure. Probably most importantly, the business models behind connected and automated vehicle technologies have not yet been crystallized. Still, the efforts made by academia, industry, and policy makers point toward safer, more efficient, and more comfortable road transportation.

Chapter 2
Human Driving Behavior

In this chapter we focus on describing the traffic environment consisting of human-driven vehicles since such vehicles will likely dominate road transportation in the forthcoming decades and provide the "playground" for automated vehicles. Our goal here is not to provide the reader with a comprehensive review about vehicular traffic modeling [206], but rather to establish a baseline for the following technical sections in terms of modeling and analysis of connected vehicle systems. For this reason, we consider simplified scenarios where vehicles are following each other on a single lane and omit overtaking, lane changes, road junctions, and traffic lights.

2.1 Dynamics of Vehicle Chains

In this section we consider a chain of human drivers following each other on a single lane as depicted in Fig. 2.1. There exists a large variety of different models depending on whether one considers time, position, and velocity to be discrete or continuous and whether the models are deterministic or stochastic. Cellular automata models like [143] use discretized quantities and stochastic rules to prescribe driver behavior. These allow fast simulation but drop many details. Instead, in this book, we use so-called car-following models where all quantities are continuous. We omit all stochastic effects but one may easily extend the presented models by incorporating noise in the human parameters.

G. Orosz and T. G. Molnár, *Dynamics and Control of Connected Vehicles*,
Surveys and Tutorials in the Applied Mathematical Sciences 15,
https://doi.org/10.1007/978-3-031-94555-7_2

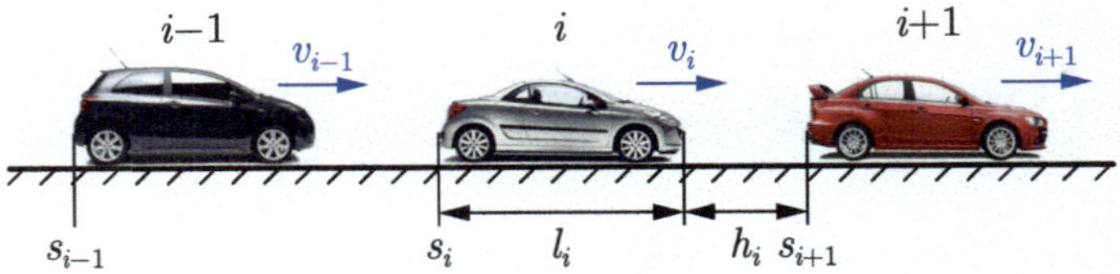

Fig. 2.1 A chain of vehicles following each other on a single lane

2.1.1 Car-Following Models

In Fig. 2.1 the position of the rear bumper of vehicle i is denoted by s_i while the corresponding longitudinal velocity is denoted by v_i. The bumper-to-bumper distance between the vehicles is called the headway and it can be calculated as

$$h_i = s_{i+1} - s_i - l_i \, , \tag{2.1}$$

where l_i denotes the length of the i-th vehicle.

Let as assume that each driver only responds to the motion of the vehicle immediately ahead (that often happens on real roads when following an SUV or a pickup truck). Then assume that each driver responds to the headway, the derivative of the headway and its own velocity, and they respond to these stimuli with a reaction time delay. Then the motion of the i-th vehicle is described by the delay differential equation

$$\begin{aligned} \dot{h}_i(t) &= v_{i+1}(t) - v_i(t) \, , \\ \dot{v}_i(t) &= f_i\Big(h_i(t - \tau_i), \dot{h}_i(t - \tau_i), v_i(t - \tau_i)\Big) \, , \end{aligned} \tag{2.2}$$

where the first equation expresses that the derivative of the headway is equal to the relative velocity, while the second equation encodes the human behavior. The time delay $\tau_i \approx 0.5 - 1.5$ [s] is a lumped quantity which includes the driver reaction time as well as the mechanical (actuation) delays in the system, but we will often refer to this as reaction time delay. The reaction time of different drivers may differ significantly from each other and even for a single driver it may change as time evolves. Similarly, there is a large variety of functions f_i that can be used to describe how a driver responds to stimuli, and individual drivers typically behave differently from each other. Finally, we remark that the acceleration $\dot{v}_{i+1}$ may also be considered in car-following models as human drivers have binary information about this quantity when monitoring the tail light of the vehicle ahead.

Early car-following models, developed in the 1950s and 1960s [60], were given in the form

$$f(h, \dot{h}, v) = \alpha \frac{\dot{h}^q v^r}{h^p} \, , \quad \alpha \in \mathbb{R}^+, \; p, q, r \in \mathbb{N} \, . \tag{2.3}$$

These are able to capture some important human response to different stimuli but have a significant shortcoming: for zero range rate $\dot{h}$ they command zero acceleration which allow vehicles to travel with very short constant headway h.

The next wave of car-following models came in the 1990s and 2000s when personal computers became available yielding a plethora of new models. One of these is called the optimal velocity model (OVM) [14, 15] which, as a matter of fact, is equivalent to the early model [29] and can be formulated as

$$f(h, \dot{h}, v) = \alpha\big(V(h) - v\big) + \beta\dot{h}\,, \quad \alpha, \beta \in \mathbb{R}^+\,. \tag{2.4}$$

The first term represents the driver's intention to keep a velocity-dependent distance from the vehicle ahead while the second term describes the intention to match the velocity to that of the preceding vehicle.

The desired headway-velocity relationship is given by the optimal velocity function $V(h)$ that is also called range policy. While there exist many range policies in the literature [8, 14, 15, 29, 159, 181] one may require the following properties: (i) continuous; (ii) monotonously increasing; (iii) saturates at zero velocity and at the speed limit $v_{\max}$. In particular, in this book, we use the form

$$V(h) = \begin{cases} 0 & \text{if} \quad h \le h_{\text{st}}\,, \\ F(h) & \text{if} \quad h_{\text{st}} < h < h_{\text{go}}\,, \\ v_{\max} & \text{if} \quad h \ge h_{\text{go}}\,, \end{cases} \tag{2.5}$$

where h_{st} and h_{go} may be referred to as the standstill headway and the free-flow headway, respectively.

The simplest choice for the mid-section is the affine function

$$F(h) = v_{\max}\frac{h - h_{\text{st}}}{h_{\text{go}} - h_{\text{st}}}\,, \tag{2.6}$$

depicted in Fig. 2.2a with the gradient

$$\kappa = \frac{v_{\max}}{h_{\text{go}} - h_{\text{st}}}\,, \tag{2.7}$$

highlighted. We remark that the inverse of the gradient $t_{\text{h}} = 1/\kappa$ is often referred as time headway though strictly speaking this only true for $h_{\text{st}} = 0$. Typical values of these parameters for highways are $h_{\text{st}} \approx 3 - 7\,[\text{m}]$, $\kappa \approx 0.5 - 1.5\,[1/\text{s}]$, and $v_{\max} \approx 25 - 40\,[\text{m/s}]$.

The function (2.5), (2.6) is non-smooth at the locations h_{st} and h_{go}, which may not be so typical human behavior. This motivates the use of the quadratic function

$$F(h) = v_{\max}\frac{(2h_{\text{go}} - h_{\text{st}} - h)(h - h_{\text{st}})}{(h_{\text{go}} - h_{\text{st}})^2}\,, \tag{2.8}$$

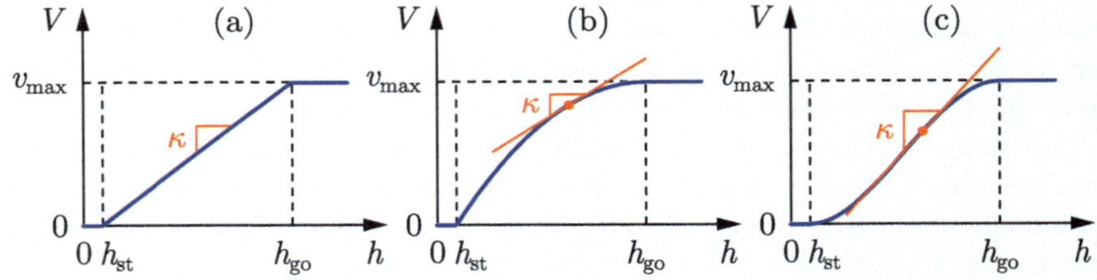

Fig. 2.2 Range policies of the form (2.5) with different mid-sections: **a** affine (2.6), **b** quadratic (2.8), and **c** cubic (2.9). The parameters used are $h_{st} = 5\,[\mathrm{m}]$, $h_{go} = 55\,[\mathrm{m}]$, $v_{max} = 30\,[\mathrm{m/s}]$ which results in the gradient $\kappa = 0.6\,[1/\mathrm{s}]$ in the affine case

shown in Fig. 2.2b. Notice that the gradient is zero at $h = h_{go}$ and it takes the maximum value $\kappa_{max} = 2v_{max}/(h_{go} - h_{st})$ at $h = h_{st}$. In order to provide further smoothness one may choose the cubic function

$$F(h) = v_{max} \frac{(3h_{go} - h_{st} - 2\,h)(h - h_{st})^2}{(h_{go} - h_{st})^3}\,, \tag{2.9}$$

displayed in Fig. 2.2c, whose derivative is zero at $h = h_{st}$ and at $h = h_{go}$. The gradient takes the maximum $\kappa_{max} = (3/2)\,v_{max}/(h_{go} - h_{st})$ at the mid-point $h = (h_{go} + h_{st})/2$.

While the OVM has been widely used for modeling human car-following behavior their exist many other models. One of these is called the intelligent driver model (IDM) [205] that can be formulated as

$$f(h, \dot{h}, v) = a\left(1 - \left(\frac{v}{v_{max}}\right)^4 - \left(\frac{h_{st} + v\,t_h - \dot{h}v/\sqrt{4ab}}{h}\right)^2\right)\,, \tag{2.10}$$

where the parameters a, b, t_h play similar roles to α, β, $1/\kappa$ in the OVM. Commercial traffic simulation softwares like SUMO and VISSIM often come with built-in car-following models but they also allow the users to specify their own models.

2.1.2 *Uniform Flow Equilibrium*

In order to understand the implications of car-following rules (2.2) on the traffic flow, let us search for the equilibrium of the form

$$\begin{aligned} h_i(t) &\equiv h_i^*\,, \\ v_i(t) &\equiv v_i^*\,. \end{aligned} \tag{2.11}$$

Substituting these into (2.2) we obtain

$$\begin{aligned} v_i^* &= v^* \,, \\ 0 &= f_i(h_i^*, 0, v^*) \,, \end{aligned} \tag{2.12}$$

where v^* is the equilibrium speed shared by all vehicles. Assuming that the implicit function theorem holds, the second equation yields

$$v^* = V_i(h_i^*) \,, \tag{2.13}$$

or

$$h_i^* = H_i(v^*) \,, \tag{2.14}$$

where $H_i = V_i^{-1}$ on the domain where the inverse exist.

Assuming identical drivers we have $f_i = f \Rightarrow V_i = V,\, H_i = H$. That is, we obtain the so-called uniform flow equilibrium

$$\begin{aligned} h_i(t) &\equiv h^* \,, \\ v_i(t) &\equiv v^* \,, \end{aligned} \tag{2.15}$$

where vehicles are traveling equidistantly with the same speed so that

$$v^* = V(h^*) \,, \tag{2.16}$$

or

$$h^* = H(v^*) \,, \tag{2.17}$$

where $H = V^{-1}$.

For example, in case of the OVM (2.4) the function V is explicitly given by the range policy, while for the IDM (2.10) we obtain

$$H(v^*) = \frac{h_{\mathrm{st}} + v^* t_{\mathrm{h}}}{\sqrt{1 - \left(\dfrac{v^*}{v_{\max}}\right)^4}} \,. \tag{2.18}$$

Its inverse $V = H^{-1}$ is similar to the range policy shown in Fig. 2.2b with the exception that $v_{\max}$ is only approached as $h \to \infty$ and that $V(h) < 0$ for $h < h_{\mathrm{st}}$; see Fig. 2.4a, b. Having negative equilibrium speed for small headway can be eliminated with a slight modification of IDM model (2.10) as shown in [108]. At $h = h_{\mathrm{st}}$ the gradient takes the maximum value $1/t_{\mathrm{h}}$.

The uniform flow may also be represented by a macroscopic view that describes traffic using density ρ (number of vehicles on a unit length of the road) and flux q (number of vehicles passing a cross section of the road in unit time). In the equilibrium situation these quantities can be defined as

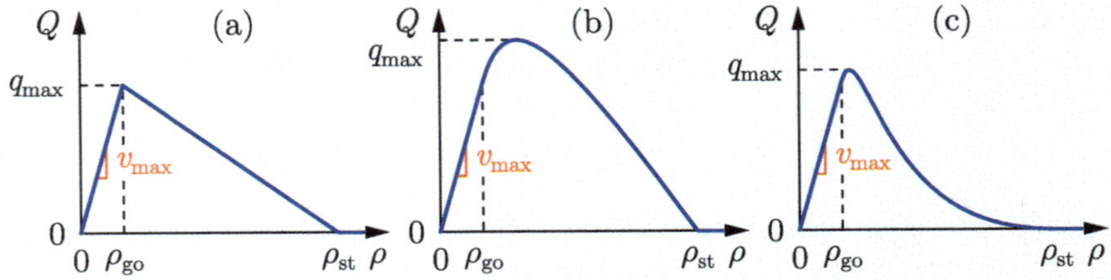

Fig. 2.3 Fundamental diagrams corresponding to the range policies in Fig. 2.2 obtained by using (2.19), (2.20) with $l = 5$ [m]. The parameters are $v_{max} = 30$ [m/s], $\rho_{go} = 16.7$ [veh/km], and $\rho_{st} = 100$ [veh/km]. The corresponding lane capacities are $q_{max} = 1800$ [veh/h] for panel (**a**), $q_{max} = 2318$ [veh/h] for panel (**b**), and $q_{max} = 1937$ [veh/h] for panel (**c**)

$$\rho^* = \frac{1}{h^* + l} , \tag{2.19}$$

and

$$q^* = \rho^* v^* = \frac{V(h^*)}{h^* + l} = \rho^* \mathcal{V}(\rho^*) = Q(\rho^*) , \tag{2.20}$$

where l denotes the vehicle length and $\mathcal{V}(\rho^*) = V(1/\rho^* - l)$.

The function Q is often referred to as the fundamental diagram [71] that has been widely used in traffic engineering practice. Figure 2.3a–c shows the fundamental diagrams corresponding to the range policies in Fig. 2.2a–c; see (2.5). Notice that the constant-speed section of the range policy at $h \geq h_{go}$ corresponds to the rising part of the fundamental diagram at $\rho \leq \rho_{go}$. This is often referred as free flow where adding more vehicles to the lane does not influence their speed but increases the flux. The increasing section of the range policy between $h_{st} < h < h_{go}$ corresponds to $\rho_{go} < \rho < \rho_{st}$ in the fundamental diagram. This is often referred as jammed state where adding more vehicles influences their speed and may decrease the flux. Finally, the zero section of the range policy at $h < h_{st}$ corresponds to $\rho \geq \rho_{st}$ and is often referred to as gridlock.

The maximum flux q_{max} is called the capacity of the lane. For the affine range policy the capacity is achieved at ρ_{go} while for the quadratic and cubic range policies they occur for slightly larger densities. The capacity given by the quadratic range policy (see Fig. 2.3b) corresponds to the experimental results in Chap. 5. Figure 2.4c, d shows the fundamental diagrams of the IDM associated with the range policies in Fig. 2.4a, b.

We remark that the fundamental diagrams can also be obtained as steady states of continuum models that describe the time evolution of the density and flux. These models are typically written as partial differential equations. The link between these equations and (2.2) is an area of active research [139].

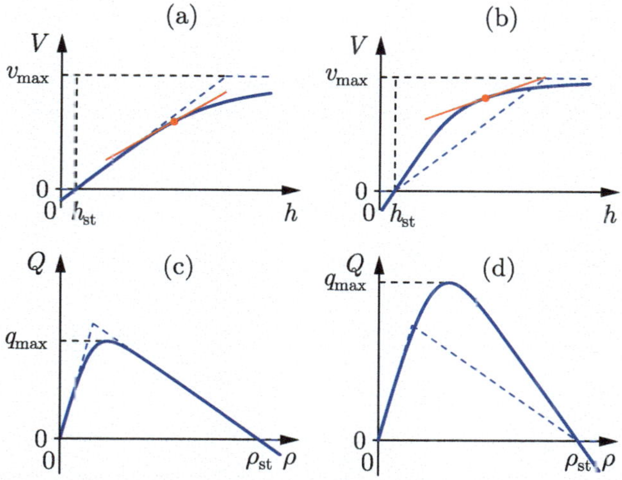

Fig. 2.4 Range policies (top) and fundamental diagrams (bottom) corresponding to the IDM (2.10). The parameters are $v_{max} = 30\,[m/s]$, $h_{st} = 5\,[m]$, $l = 5\,[m]$, $\rho_{st} = 100\,[veh/km]$. In panels **a**, **c** $t_h = 1.67\,[s]$ is assumed, which corresponds to the lane capacity $q_{max} = 1524\,[veh/h]$, while panels **b**, **d** are associated with $t_h = 0.83\,[s]$ and $q_{max} = 2468\,[veh/h]$. The piecewise linear functions in Figs. 2.2a and 2.3a are indicated by dashed blue lines for comparison

2.2 Human Driving Without Driver Reaction Time

In this section we consider identical vehicles and drop the reaction time delay of drivers. That is, we analyze the equations

$$\begin{aligned} \dot{h}_i &= v_{i+1} - v_i \,, \\ \dot{v}_i &= f(h_i, \dot{h}_i, v_i) \,, \end{aligned} \tag{2.21}$$

cf. (2.2), where the argument t is dropped for the sake simplicity. These simplifications allow us to present a set of definitions and mathematical methods which can be used to analyze the dynamics of vehicle chains. These methods will be extended to the case with time delay in the next section.

2.2.1 Plant Stability and String Stability for Open Chain of Vehicles

In order to characterize the behavior of vehicle chains typically two types of stability are introduced. Plant stability concerns about whether a vehicle is able to approach the equilibrium when having no excitations from the vehicles ahead. This corresponds to the traditional notion of stability in dynamical systems. String stability, on the other hand, concerns about whether excitations, introduced by some vehicles ahead, decay as they propagate along the vehicle chain [55].

In order to explain these concepts mathematically we define the following perturbations

$$\begin{aligned}\tilde{h}_i(t) &= h_i(t) - h^* \,, \\ \tilde{v}_i(t) &= v_i(t) - v^* \,.\end{aligned} \tag{2.22}$$

Substituting these into (2.21) and using Taylor expansion we obtain

$$\begin{aligned}\dot{\tilde{h}}_i &= \tilde{v}_{i+1} - \tilde{v}_i \,, \\ \dot{\tilde{v}}_i &= f_h^* \tilde{h}_i + f_{\dot{h}}^* \dot{\tilde{h}}_i + f_v^* \tilde{v}_i + \text{h.o.t.} \,,\end{aligned} \tag{2.23}$$

where we used the abbreviated notation $f_h^* = \partial_h f(h^*, 0, v^*)$, $f_{\dot{h}}^* = \partial_{\dot{h}} f(h^*, 0, v^*)$, $f_v^* = \partial_v f(h^*, 0, v^*)$, while h.o.t. stands for the higher order terms. By defining the state, the input and the output as

$$\mathbf{x}_i = \begin{bmatrix} \tilde{h}_i \\ \tilde{v}_i \end{bmatrix}, \quad u_i = \tilde{v}_{i+1} \,, \quad y_i = \tilde{v}_i \,, \tag{2.24}$$

we obtain the system

$$\begin{aligned}\dot{\mathbf{x}}_i &= \mathbf{a}\,\mathbf{x}_i + \mathbf{b}\,u_i + \text{h.o.t.} \,, \\ y_i &= \mathbf{c}\,\mathbf{x}_i \,,\end{aligned} \tag{2.25}$$

where the matrices are defined as

$$\mathbf{a} = \begin{bmatrix} 0 & -1 \\ f_h^* & f_v^* - f_{\dot{h}}^* \end{bmatrix}, \quad \mathbf{b} = \begin{bmatrix} 1 \\ f_{\dot{h}}^* \end{bmatrix}, \quad \mathbf{c} = \begin{bmatrix} 0 & 1 \end{bmatrix}. \tag{2.26}$$

Equation (2.25) with

$$u_i = y_{i+1} \,, \tag{2.27}$$

approximates the dynamics of the vehicle chain.

For example, in case of the OVM (2.4) we have

$$f_h^* = \alpha\kappa \,, \quad f_{\dot{h}}^* = \beta \,, \quad f_v^* = -\alpha \,, \tag{2.28}$$

where the gradient $\kappa = V'(h^*)$ is calculated at the equilibrium headway h^*. That is, the matrices in (2.26) become

$$\mathbf{a} = \begin{bmatrix} 0 & -1 \\ \alpha\kappa & -(\alpha + \beta) \end{bmatrix}, \quad \mathbf{b} = \begin{bmatrix} 1 \\ \beta \end{bmatrix}, \quad \mathbf{c} = \begin{bmatrix} 0 & 1 \end{bmatrix}. \tag{2.29}$$

Notice that the matrices in (2.26) and in (2.29) both contain three parameters. Since α, β, and κ have clear physical meaning we will use them when presenting the results and in the rest of the book we refer to this as the *OVM parametrization* of linearized car-following models. In other words, using (2.28) the linearization of any car-following model can be represented as (2.29). For example, for the IDM (2.10) we obtain

$$f_h^* = 2a\frac{(h_{\rm st}+v^*t_{\rm h})^2}{h^{*3}}, \quad f_{\dot h}^* = \sqrt{\frac{a}{b}}v^*\frac{h_{\rm st}+v^*t_{\rm h}}{h^{*2}}, \quad f_v^* = -a\left(\frac{4v^{*3}}{v_{\max}^4}+2t_{\rm h}\frac{h_{\rm st}+v^*t_{\rm h}}{h^{*2}}\right), \tag{2.30}$$

which enter the matrices in (2.26), while the corresponding α, β, κ can be obtained from (2.28).

Before providing formal definitions of plant and string stability, let us drop the higher order terms in (2.25) and solve the system for the output:

$$y_i(t) = \mathbf{c}\mathrm{e}^{\mathbf{a}t}\mathbf{x}_i(0) + \int_0^t \mathbf{c}\mathrm{e}^{\mathbf{a}(t-\tilde t)}\mathbf{b}u_i(\tilde t)\mathrm{d}\tilde t\ . \tag{2.31}$$

Here the matrix exponential can be calculated as

$$\mathrm{e}^{\mathbf{a}t} = \mathbf{P}\mathrm{e}^{\mathbf{J}t}\mathbf{P}^{-1}\ , \tag{2.32}$$

where

$$\mathbf{J} = \mathbf{P}^{-1}\mathbf{a}\mathbf{P}\ , \tag{2.33}$$

is the Jordan normal form of matrix **a** while matrix **P** is constructed using the eigenvectors of **a**.

Let us re-write matrix **a** as

$$\mathbf{a} = \begin{bmatrix} 0 & -1 \\ \omega_{\rm n}^2 & -2\zeta\omega_{\rm n} \end{bmatrix}, \tag{2.34}$$

where

$$\omega_{\rm n} = \sqrt{\alpha\kappa}\ , \quad \zeta = \frac{\alpha+\beta}{2\sqrt{\alpha\kappa}}\ , \tag{2.35}$$

cf. (2.29). Note that $\omega_{\rm n}$ and ζ are real when $\alpha\kappa > 0$ (while the analysis below is valid also for $\alpha\kappa < 0$). We also remark that (2.34) resembles the state matrix of a classical mass-spring-damper system—with the off-diagonal elements having the opposite signs compared to the classical example. The corresponding characteristic equation is

$$\det(s\mathbf{I}-\mathbf{a}) = s^2 + 2\zeta\omega_{\rm n}s + \omega_{\rm n}^2 = 0\ , \tag{2.36}$$

where **I** is the two-dimensional identity matrix. This yields the eigenvalues

$$s_{1,2} = -\zeta\omega_\mathrm{n} \pm \mathrm{j}\omega_\mathrm{d}\,, \quad \omega_\mathrm{d} = \omega_\mathrm{n}\sqrt{1-\zeta^2}\,, \tag{2.37}$$

with $\mathrm{j}^2 = -1$ that are complex conjugates for $\zeta < 1$ and real for $\zeta \geq 1$.

When $\zeta \neq 1$, we have $s_1 \neq s_2$ and the corresponding eigenvectors are given by

$$\mathbf{p}_{1,2} = \begin{bmatrix} 1 \\ -s_{1,2} \end{bmatrix}. \tag{2.38}$$

Then, one may obtain

$$\mathbf{P} = [\mathbf{p}_1\ \mathbf{p}_2] \quad \Rightarrow \quad \mathbf{J} = \mathbf{P}^{-1}\mathbf{a}\mathbf{P} = \begin{bmatrix} s_1 & 0 \\ 0 & s_2 \end{bmatrix} \quad \Rightarrow \quad \mathrm{e}^{\mathbf{J}t} = \begin{bmatrix} \mathrm{e}^{s_1 t} & 0 \\ 0 & \mathrm{e}^{s_2 t} \end{bmatrix}, \tag{2.39}$$

yielding

$$\mathrm{e}^{\mathbf{a}t} = \mathbf{P}\mathrm{e}^{\mathbf{J}t}\mathbf{P}^{-1} = \frac{1}{s_1 - s_2}\begin{bmatrix} -s_2\mathrm{e}^{s_1 t} + s_1\mathrm{e}^{s_2 t} & -\mathrm{e}^{s_1 t} + \mathrm{e}^{s_2 t} \\ s_1 s_2(\mathrm{e}^{s_1 t} - \mathrm{e}^{s_2 t}) & s_1\mathrm{e}^{s_1 t} - s_2\mathrm{e}^{s_2 t} \end{bmatrix}. \tag{2.40}$$

For $\zeta = 1$ the eigenvalues are given as $s_1 = s_2 = -\omega_\mathrm{n}$, that is, they have algebraic multiplicity two. However, there is only a single eigenvector corresponding to this case meaning that the geometric multiplicity is one. This yields a different Jordan form compared to (2.39) (with a 1 in the top right corner) and we omit the calculation of the matrix exponential in this special case.

In order to test how perturbations propagate along the chain of vehicles, let us consider zero initial condition and the Dirac-delta input $\delta(t)$:

$$\mathbf{x}_i(0) = \mathbf{0}\,, \tag{2.41}$$

$$u_i(t) = A\delta(t)\,. \tag{2.42}$$

Then, evaluating the convolution integral (2.31) we obtain the form

$$y_i(t) = \int_0^t \mathbf{c}\,\mathrm{e}^{\mathbf{a}(t-\tilde{t})}\mathbf{b}u_i(\tilde{t})\,\mathrm{d}\tilde{t} = A\,\mathbf{c}\,\mathrm{e}^{\mathbf{a}t}\mathbf{b} = A\left(k_1\mathrm{e}^{s_1 t} + k_2\mathrm{e}^{s_2 t}\right), \tag{2.43}$$

for the velocity of the i-th vehicle, where the constants k_1 and k_2 are functions of the system parameters:

$$\begin{aligned} k_1 &= \frac{s_1 s_2 + \beta s_1}{s_1 - s_2} = \frac{\beta}{2} + \frac{2\alpha\kappa - \beta(\alpha+\beta)}{2\sqrt{(\alpha+\beta)^2 - 4\alpha\kappa}}\,, \\ k_2 &= \frac{s_1 s_2 + \beta s_2}{s_2 - s_1} = \frac{\beta}{2} - \frac{2\alpha\kappa - \beta(\alpha+\beta)}{2\sqrt{(\alpha+\beta)^2 - 4\alpha\kappa}}\,. \end{aligned} \tag{2.44}$$

Then for the $(i-1)$-st vehicle we consider zero initial condition and the input

$$\mathbf{x}_{i-1}(0) = \mathbf{0}\,, \tag{2.45}$$

$$u_{i-1}(t) = y_i(t)\,, \tag{2.46}$$

and applying (2.31) we obtain

$$y_{i-1}(t) = \int_0^t \mathbf{c}\,\mathrm{e}^{\mathbf{a}(t-\tilde{t})}\mathbf{b}\,u_{i-1}(\tilde{t})\mathrm{d}\tilde{t} = A\left(\left(k_1^2 t + \frac{2k_1k_2}{s_1-s_2}\right)\mathrm{e}^{s_1 t} + \left(k_2^2 t + \frac{2k_1k_2}{s_2-s_1}\right)\mathrm{e}^{s_2 t}\right). \tag{2.47}$$

Following the above steps, the response of the $(i-2)$-nd vehicle is determined by

$$\mathbf{x}_{i-2}(0) = \mathbf{0}\,, \tag{2.48}$$

$$u_{i-2}(t) = y_{i-1}(t)\,, \tag{2.49}$$

which leads to the form

$$\begin{aligned} y_{i-2}(t) &= \int_0^t \mathbf{c}\,\mathrm{e}^{\mathbf{a}(t-\tilde{t})}\mathbf{b}\,u_{i-2}(\tilde{t})\mathrm{d}\tilde{t} \\ &= A\left(\left(\frac{k_1^3}{2}t^2 + \frac{3k_1^2k_2}{s_1-s_2}t + \frac{3k_1k_2}{(s_1-s_2)^2}(k_2-k_1)\right)\mathrm{e}^{s_1 t}\right. \\ &\quad \left. + \left(\frac{k_2^3}{2}t^2 + \frac{3k_1k_2^2}{s_2-s_1}t + \frac{3k_1k_2}{(s_2-s_1)^2}(k_1-k_2)\right)\mathrm{e}^{s_2 t}\right). \end{aligned} \tag{2.50}$$

As we continue along the chain for the $(i-k)$-th vehicle we have

$$\mathbf{x}_{i-k}(0) = \mathbf{0}\,, \tag{2.51}$$

$$u_{i-k}(t) = y_{i-k+1}(t)\,, \tag{2.52}$$

yielding

$$y_{i-k}(t) = \int_0^t \mathbf{c}\,\mathrm{e}^{\mathbf{a}(t-\tilde{t})}\mathbf{b}\,u_{i-k}(\tilde{t})\,\mathrm{d}\tilde{t} = A\left(p_1^{(k)}(t)\,\mathrm{e}^{s_1 t} + p_2^{(k)}(t)\,\mathrm{e}^{s_2 t}\right), \tag{2.53}$$

where $p_1^{(k)}(t)$ and $p_2^{(k)}(t)$ are k-th order polynomials in t; cf. (2.43), (2.47) and (2.50).

For $\zeta < 1$ (i.e., when the eigenvalues (2.37) are complex conjugates) one can re-write (2.40) as

$$\mathrm{e}^{\mathbf{a}t} = \mathrm{e}^{-\zeta\omega_{\mathrm{n}}t}\begin{bmatrix} \cos(\omega_{\mathrm{d}}t) + \frac{\zeta\omega_{\mathrm{n}}}{\omega_{\mathrm{d}}}\sin(\omega_{\mathrm{d}}t) & -\frac{1}{\omega_{\mathrm{d}}}\sin(\omega_{\mathrm{d}}t) \\ \frac{\omega_{\mathrm{n}}^2}{\omega_{\mathrm{d}}}\sin(\omega_{\mathrm{d}}t) & \cos(\omega_{\mathrm{d}}t) - \frac{\zeta\omega_{\mathrm{n}}}{\omega_{\mathrm{d}}}\sin(\omega_{\mathrm{d}}t) \end{bmatrix}. \tag{2.54}$$

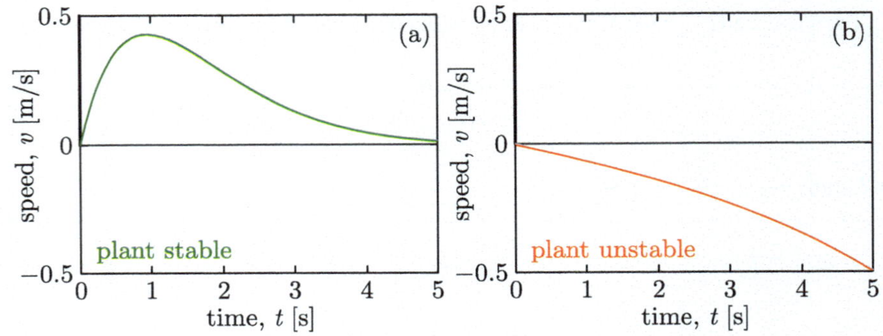

Fig. 2.5 Speed perturbations of the i-th vehicle responding to an impulse. The parameters $\kappa = 0.6\,[1/\mathrm{s}]$, $\beta = 0\,[1/\mathrm{s}]$, $\alpha = 2\,[1/\mathrm{s}]$ are used in panel (**a**) and while we set $\alpha = -0.1\,[1/\mathrm{s}]$ in panel (**b**)

Using this, the velocity perturbation (2.43) for the i-th vehicle is expressed as

$$y_i(t) = A\mathrm{e}^{-\zeta\omega_\mathrm{n}t}\big(k_1\cos(\omega_\mathrm{d}t) + k_2\sin(\omega_\mathrm{d}t)\big)\,, \tag{2.55}$$

where

$$k_1 = \beta\,, \qquad k_2 = (\omega_\mathrm{n} - \beta\zeta)\frac{\omega_\mathrm{n}}{\omega_\mathrm{d}} = \frac{2\alpha\kappa - \beta(\alpha+\beta)}{\sqrt{4\alpha\kappa - (\alpha+\beta)^2}}\,. \tag{2.56}$$

Figure 2.5 shows this solution for two different sets of parameters as indicated. For panel (a) the parameters are chosen such that the perturbation approaches zero as the time goes to infinity. We refer to this solution as plant stable. On the other hand, for panel (b) the parameter choice leads to diverging solution that we refer to as plant unstable.

One may also calculate the solutions for the following vehicles

$$y_{i-1}(t) = A\mathrm{e}^{-\zeta\omega_\mathrm{n}t}\left(\frac{k_1^2 - k_2^2}{2}t\cos(\omega_\mathrm{d}t) + \left(\frac{k_1^2 + k_2^2}{2\omega_\mathrm{d}} + k_1k_2t\right)\sin(\omega_\mathrm{d}t)\right), \tag{2.57}$$

and

$$\begin{aligned} y_{i-2}(t) = A\mathrm{e}^{-\zeta\omega_\mathrm{n}t}\Bigg(\bigg(&-\frac{3k_1^2k_2 + 3k_2^3}{8\omega_\mathrm{d}}t + \frac{k_1^3 - 3k_1k_2^2}{8}t^2\bigg)\cos(\omega_\mathrm{d}t) \\ &+\bigg(\frac{3k_1^2k_2 + 3k_2^3}{8\omega_\mathrm{d}^2} + \frac{3k_1^3 + 3k_1k_2^2}{8\omega_\mathrm{d}}t + \frac{3k_1^2k_2 - k_2^3}{8}t^2\bigg)\sin(\omega_\mathrm{d}t)\Bigg), \end{aligned} \tag{2.58}$$

etc., cf. (2.47), (2.50) and (2.53).

We display these solutions in Fig. 2.6 where for both panels the parameters are chosen to ensure plant stability, that is, the perturbations approach zero as time goes

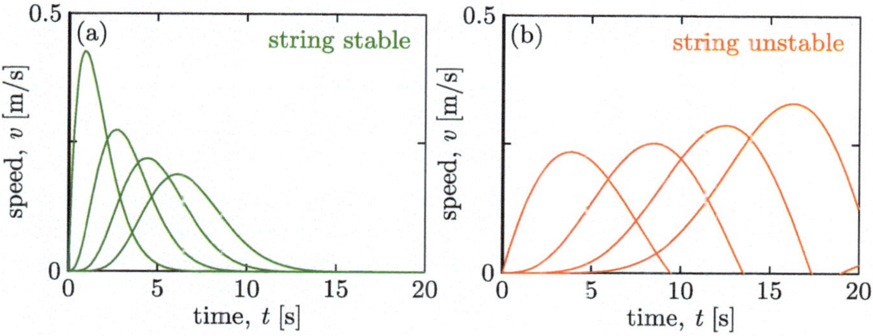

Fig. 2.6 Speed perturbations along a vehicle chain when the i-th vehicle is responding to an impulse. The parameters $\kappa = 0.6\,[1/\mathrm{s}]$, $\beta = 0\,[1/\mathrm{s}]$, $\alpha = 2\,[1/\mathrm{s}]$ are used in panel (**a**) while we set $\alpha = 0.2\,[1/\mathrm{s}]$ in panel (**b**)

to infinity. In panel (a) the parameters are chosen such that the perturbations are attenuated as they propagate along the vehicle chain. We refer to this as a string stable solution. In contrast, in panel (b) the perturbations are amplified along the chain and we refer to this as a string unstable solution.

The examples above were constructed to illustrate the notions of plant and string stability. Now we are ready to state some more general definitions and theorems for the system (2.25). Let us start with plant stability.

Definition 2.1 Vehicle i is called plant stable if the trivial solution $\mathbf{x}_i(t) \equiv \mathbf{0}$ is asymptotically stable (i.e., it is Lyapunov stable and $\lim_{t\to\infty} \mathbf{x}_i(t) = \mathbf{0}$) for all initial conditions $\mathbf{x}_i(0) \in \mathbb{R}^2$ when $u_i(t) \equiv 0$.

Moreover, at the linear level, i.e., when the higher order terms are neglected in (2.25), we can state the following theorem.

Theorem 2.1 *Vehicle i is plant stable if and only if the eigenvalues of matrix* **a** *have negative real parts, i.e.,* $\mathrm{Re}(s_1) < 0$ *and* $\mathrm{Re}(s_2) < 0$.

Indeed, the eigenvalues can be obtained by solving the characteristic equation

$$\det(s\mathbf{I} - \mathbf{a}) = 0\,. \tag{2.59}$$

We remark that a vehicle chain is plant stable if and only if all vehicles in the chain are plant stable. Since above we considered identical vehicles, Theorem 2.1 is adequate to ensure plant stability at the linear level for the whole chain. Also note that Definition 2.1 and Theorem 2.1 can be extended to higher dimensional state spaces, i.e., $\mathbf{x}_i \in \mathbb{R}^n$, and can also be restricted to a smaller domain in the state space, i.e., $\mathbf{x}_i \in D \subset \mathbb{R}^n$ (that contains the origin $\mathbf{0} \in D$).

While plant stability is equivalent to asymptotic stability when no input is presented to the vehicles, string stability is related to input–output stability given by the following definition.

Definition 2.2 Vehicle i is input–output stable if there exist class-$\mathcal{K}$ functions α and β such that

$$\|y_i\|_{\mathcal{L}_p} < \alpha\big(\|u_i\|_{\mathcal{L}_p}\big) + \beta\big(\|\mathbf{x}_i(0)\|\big)\,, \tag{2.60}$$

for all inputs $u_i \in \mathcal{L}_p$ and initial conditions $\mathbf{x}_i(0) \in \mathbb{R}^2$.

Here class-$\mathcal{K}$ function refers to $\alpha\colon [0, a) \to [0, \infty)$ that is continuous, strictly monotonously increasing and $\alpha(0) = 0$. Note that a may be finite or extended to infinity. Also, $\mathcal{L}_p$ denotes the function space whose elements have finite norm defined by

$$\|y\|_{\mathcal{L}_p} = \sqrt[p]{\int_0^\infty |y(t)|^p \mathrm{d}t}\,, \tag{2.61}$$

and for $p \to \infty$ this simplifies to

$$\|y\|_{\mathcal{L}_\infty} = \sup_{t\geq 0} |y(t)|\,, \tag{2.62}$$

while for the state $\|\mathbf{x}\|$ refers to a standard vector norm on $\mathbb{R}^2$. Definition 2.2 can be easily extended to vehicle chains as follows.

Definition 2.3 The vehicle chain consisting of vehicles $i, i-1, \ldots, i-k$ is input–output stable if there exist class-$\mathcal{K}$ functions α and β such that

$$\|y_{i-k}\|_{\mathcal{L}_p} < \alpha\big(\|u_i\|_{\mathcal{L}_p}\big) + \beta\left(\left\|\begin{bmatrix}\mathbf{x}_i(0)\\ \mathbf{x}_{i-1}(0)\\ \vdots \\ \mathbf{x}_{i-k}(0)\end{bmatrix}\right\|\right)\,, \tag{2.63}$$

for all inputs $u_i \in \mathcal{L}_p$ and initial conditions $\mathbf{x}_j(0) \in \mathbb{R}^2$ for $j = i, i-1, \ldots, i-k$.

The norm in the last term is a standard vector norm on $\mathbb{R}^{2k}$. We remark that if Definition 2.2 holds for vehicles $i, i-1, \ldots, i-k$, Definition 2.3 also holds but the converse is not necessarily true. Again, Definitions 2.2 and 2.3 can be extended to higher dimensional state spaces, i.e., $\mathbf{x}_i \in \mathbb{R}^n$, can be restricted to a smaller set in the state space, i.e., $\mathbf{x}_i \in D \subset \mathbb{R}^n$, and restricted to a smaller input set $u_i \in \mathcal{D} \subset \mathcal{L}_p$.

While Definitions 2.2 and 2.3 provide us with the general notion of disturbance attenuation along vehicle chains it is difficult to turn them into practically applicable conditions. Finding the class-$\mathcal{K}$ functions typically requires the use of a Lyapunov function that often results in conservative conditions. Since the plant stability definition 2.1 already deals with the initial conditions we assume that plant stability holds and define string stability while assuming zero initial conditions. In particular, for two consecutive vehicles we have the following definition.

Definition 2.4 The chain of two consecutive vehicles $i+1$ and i is called $(i+1)$-to-i string stable if

$$\|y_i\|_{\mathcal{L}_p} < \|y_{i+1}\|_{\mathcal{L}_p}\,, \tag{2.64}$$

for all inputs $y_{i+1} = u_i \in \mathcal{L}_p$ and initial condition $\mathbf{x}_i(0) = \mathbf{0}$.

We may extend this to a vehicle chain as follows.

Definition 2.5 The vehicle chain consisting of vehicles $i+1, i, \ldots, i-k$ is $(i+1)$-to-$(i-k)$ string stable if

$$\|y_{i-k}\|_{\mathcal{L}_p} < \|y_{i+1}\|_{\mathcal{L}_p}\,, \tag{2.65}$$

for all inputs $y_{i+1} = u_i \in \mathcal{L}_p$ and initial conditions $\mathbf{x}_j(0) = \mathbf{0}$ for $j = i, i-1, \ldots, i-k$.

Again, if Definition 2.4 holds for vehicles $i, i-1, \ldots, i-k$ then Definition 2.5 also holds but the converse is not necessarily true. Indeed, if all vehicles from i to $i-k$ attenuate the velocity fluctuations, then we observe attenuation when comparing the fluctuations of vehicle $i+1$ to those of vehicle $i-k$. However, one may consider cases when some vehicles in the chain amplify the fluctuations of the vehicle immediately ahead, but the effect of these amplifications is compensated by other vehicles who attenuate the fluctuations of the vehicle immediately ahead, leading to attenuation between vehicle $i+1$ and vehicle $i-k$. This property has important consequences when mixing automated and connected automated vehicles into the traffic flow.

While Definitions 2.4 and 2.5 are more practical than Definitions 2.2 and 2.3, they are still quite general as they require attenuation for any input signal $y_{i+1} = u_i \in \mathcal{L}_p$. Depending on the input signal as well as on which $\mathcal{L}_p$ norm is used, one may get different results for string stability. Since functions in $\mathcal{L}_p$ may be represented using their Fourier components, one may test string stability by using the harmonic input

$$u_i(t) = v_{\text{amp}} \cos(\omega t)\,, \tag{2.66}$$

with angular frequency $\omega > 0$. When dropping the nonlinearities in (2.25) the steady-state solution can be written as

$$y_i^{(\text{s})}(t) = M v_{\text{amp}} \cos(\omega t + \theta)\,, \tag{2.67}$$

where the transfer function

$$T_{i,i+1}(s) = \mathbf{c}\,(s\mathbf{I} - \mathbf{a})^{-1}\,\mathbf{b}\,, \tag{2.68}$$

can be used to calculate the amplification and the phase shift as

$$M = \left|T_{i,i+1}(\mathrm{j}\omega)\right|, \quad \theta = \angle T_{i,i+1}(\mathrm{j}\omega)\,, \tag{2.69}$$

with $j^2 = -1$. Assuming plant stability for the linear system also means that transients decay as time evolves, that is, $\lim_{t\to\infty}(y_i(t) - y_i^{(s)}(t)) = 0$. These arguments lead to the following theorems that hold at the linear level.

Theorem 2.2 *If and only if*

$$\left|T_{i,i+1}(j\omega)\right| < 1 , \quad \forall\omega > 0 , \tag{2.70}$$

then the pair of consecutive vehicles $i+1$ *and* i *is*

- $(i+1)$*-to-*i *string stable in the* $\mathcal{L}_2$ *norm for any* $y_{i+1} = u_i \in \mathcal{L}_2$*;*
- $(i+1)$*-to-*i *string stable in the* $\mathcal{L}_\infty$ *norm for the harmonic input (2.66) when* $t \to \infty$*, that is,*

$$\left\| y_i^{(s)} \right\|_{\mathcal{L}_\infty} < \left\| y_{i+1} \right\|_{\mathcal{L}_\infty} . \tag{2.71}$$

For the proof see [169]. We remark that while having string stability in $\mathcal{L}_\infty$ norm is more desired, the stability conditions are more restrictive in this case. As will be shown below, condition (2.70) provides a practical way to compute string stability and it is widely used in the literature. Theorem 2.2 can also be extended to vehicle chains as follows.

Theorem 2.3 *If and only if*

$$\left|G_{i-k,i+1}(j\omega)\right| < 1 , \quad \forall\omega > 0 , \tag{2.72}$$

then the chain of consecutive vehicles $i+1, i, \ldots i-k$ *is*

- $(i+1)$*-to-*$(i-k)$ *string stable in the* $\mathcal{L}_2$ *norm for any* $y_{i+1} = u_i \in \mathcal{L}_2$*;*
- $(i+1)$*-to-*$(i-k)$ *string stable in the* $\mathcal{L}_\infty$ *norm for the harmonic input (2.66) when* $t \to \infty$*, that is,*

$$\left\| y_{i-k}^{(s)} \right\|_{\mathcal{L}_\infty} < \left\| y_{i+1} \right\|_{\mathcal{L}_\infty} . \tag{2.73}$$

The transfer function in (2.72) can be calculated as

$$G_{i-k,i+1}(s) = T_{i-k,i-k+1}(s) \cdots T_{i-1,i}(s)\, T_{i,i+1}(s) , \tag{2.74}$$

and for identical vehicles this simplifies to

$$G_{i-k,i+1}(s) = \big(T_{i,i+1}(s)\big)^{k+1} . \tag{2.75}$$

Similarly to Definitions 2.4 and 2.5, if Theorem 2.2 holds for vehicles $i, i-1, \ldots, i-k$ then Theorem 2.3 holds but the converse is not true.

Finally, we remark that the $(i+1)$-to-$(i-k)$ transfer function (2.74) can also be used to reason about the plant stability of the vehicle chain, which can be formulated by the following theorem.

Theorem 2.4 *The chain of consecutive vehicles $i, i-1, \dots i-k$ is plant stable if and only if the roots of the characteristic equation*

$$\mathrm{Den}(G_{i-k,i+1}(s)) = 0\,, \tag{2.76}$$

are located in the left-half plane, that is, $\mathrm{Re}(s_j) < 0$, $j = 1, \dots, 2(k+1)$, *or equivalently, if the roots of the equations*

$$\begin{aligned} \mathrm{Den}(T_{i-k,i-k+1}(s)) &= 0\,, \\ &\vdots \\ \mathrm{Den}(T_{i-1,i}(s)) &= 0\,, \\ \mathrm{Den}(T_{i,i+1}(s)) &= 0\,, \end{aligned} \tag{2.77}$$

are located in the left-half plane.

Here we used the notation Den to denote the denominator of the transfer function. Notice that for $k = 0$ (2.74) yields the transfer function $G_{i,i+1}(s) = T_{i,i+1}(s)$ (see (2.68)) and Theorem 2.4 is equivalent to Theorem 2.1. These transfer function calculations will be generalized in Chap. 4 for connected vehicle systems that may contain vehicles with and without connectivity and automation.

Now we will turn the above conditions into stability charts where we visualize the domains of plant stability and string stability in parameter space. We will use the OVM parametrization for this purpose, and recall that at the linear level any car-following model can be represented by (2.25) with matrices (2.26) or with matrices (2.29). In order to simplify the notation, we drop the subscripts in the transfer function (2.68) and obtain

$$\begin{aligned} T(s) &= \mathbf{c}\,(s\mathbf{I} - \mathbf{a})^{-1}\,\mathbf{b} = \frac{\mathbf{c}\,\mathrm{adj}\,(s\mathbf{I} - \mathbf{a})\,\mathbf{b}}{\det\,(s\mathbf{I} - \mathbf{a})} \\ &= \frac{[0\ 1]\begin{bmatrix} s+\alpha+\beta & -1 \\ \alpha\kappa & s \end{bmatrix}\begin{bmatrix} 1 \\ \beta \end{bmatrix}}{\det\begin{bmatrix} s & 1 \\ -\alpha\kappa & s+\alpha+\beta \end{bmatrix}} = \frac{\beta s + \alpha\kappa}{s^2 + (\alpha+\beta)s + \alpha\kappa}\,, \end{aligned} \tag{2.78}$$

where adj stands for the adjugate of the matrix.

In order to evaluate plant stability we look at the denominator of the transfer function (2.78), that is, the characteristic equation

$$\det\,(s\mathbf{I} - \mathbf{a}) = s^2 + (\alpha+\beta)s + \alpha\kappa = 0\,, \tag{2.79}$$

cf. (2.59). Applying the Routh–Hurwitz criteria (see Appendix A and [52]) one may show that Theorem 2.1 holds if and only if

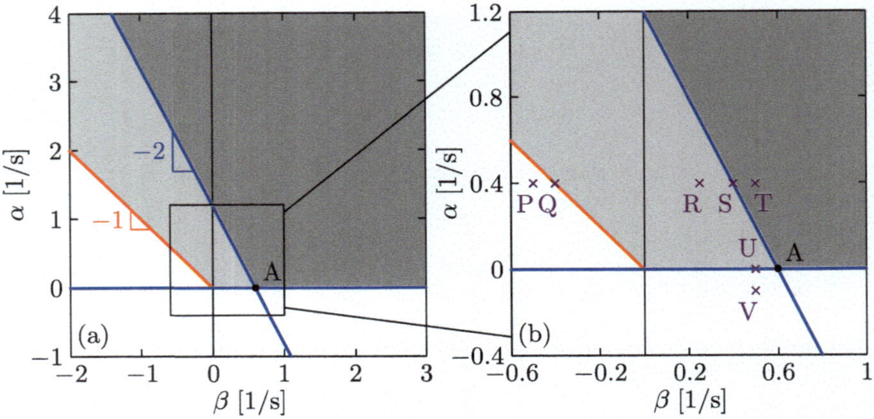

Fig. 2.7 Stability chart of system (2.25), (2.29) for $\kappa = 0.6\,[1/\text{s}]$. The right panel shows a zoom of the left panel. Red curves indicate plant stability boundaries while blue curves denote string stability boundaries. The plant and string stable regions are shaded by light and dark gray, respectively. Point A is located at $(\kappa, 0)$ while points P(−0.5,0.4), Q(−0.4,0.4), R(0.25,0.4), S(0.4,0.4), T(0.5,0.4), U(0.5,0) and V(0.5,−0.1) correspond to the spectra and frequency response plots in Figs. 2.8 and 2.10

$$\begin{aligned} \alpha + \beta &> 0\,, \\ \alpha\kappa > 0 \quad &\Rightarrow \quad \alpha > 0\,, \end{aligned} \tag{2.80}$$

where in the second line we assumed $\kappa > 0$.

The plant stable domain defined by (2.80) is shown in the (β, α) plane in Fig. 2.7; see the union of the light gray shaded and the dark gray shaded regions. The boundaries of plant stability are indicated by red lines. (Note that the line $\alpha = 0$ overlaps with a string stability boundary shown in blue as discussed below.)

Figure 2.8 shows the eigenvalues s_1 and s_2 in the complex plane, which are the solutions of the characteristic equation (2.79). The panels of Fig. 2.8 correspond to points P, Q, T, U and V in Fig. 2.7, where P and V are located in the plant unstable region, Q and U lie at the plant stability boundaries, and T is plant stable.

Figure 2.9 shows numerical simulations corresponding to points P and T in Fig. 2.7. The car-following model (2.21) was simulated for a pair of vehicles (vehicles 0 and 1) by utilizing the OVM (2.4) with the piecewise linear range policy (2.5), (2.6). The lead vehicle was assumed to be cruising at a constant speed $v_1(t) \equiv v^*$ while the initial state of the tail vehicle was slightly perturbed from the equilibrium as $h_0(0) = h^* + \delta h$ and $v_0(0) = v^*$. For point P the plant unstable system shows growing oscillations, while for point T the plant stable system approaches the trivial equilibrium as the perturbation decays.

For string stability we utilize the condition given in Theorem 2.2 that is equivalent to

$$|T(\mathrm{j}\omega)|^2 < 1\,, \quad \forall \omega > 0\,, \tag{2.81}$$

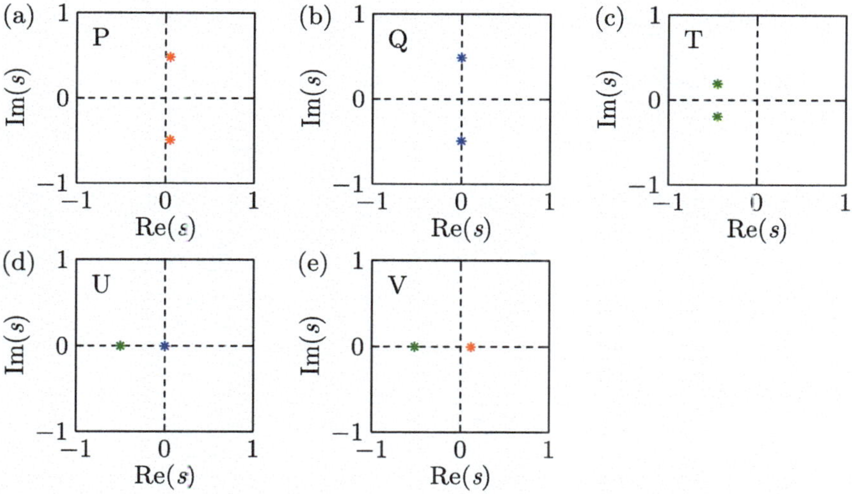

Fig. 2.8 Spectrum plots obtained from the characteristic equation (2.79) with $\kappa = 0.6\,[1/\mathrm{s}]$. Panels **a–e** correspond to points P(−0.5,0.4), Q(−0.4,0.4), T(0.5,0.4), U(0.5,0) and V(0.5,−0.1) in Fig. 2.7, respectively

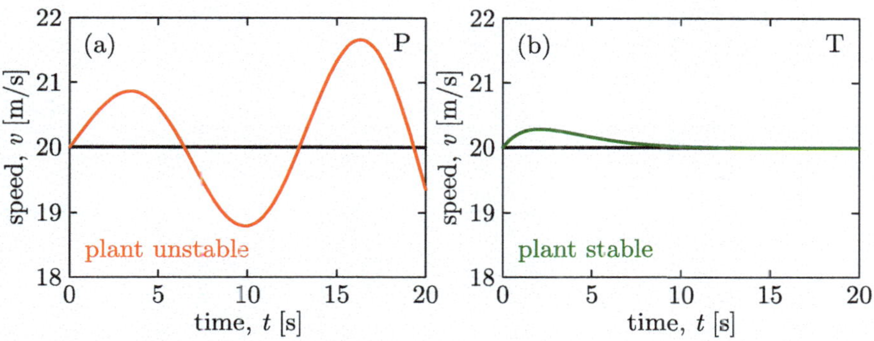

Fig. 2.9 Numerical simulation of the OVM (2.21), (2.4) with piecewise linear range policy (2.5), (2.6) for the case of two vehicles. The speed of the lead vehicle is $v_1(t) \equiv v^* = 20\,[\mathrm{m/s}]$, the initial conditions are $v_0(0) = v^*$ and $h_0(0) = h^* + \delta h = 38.33 + 1.5\,[\mathrm{m}]$ while the rest of the parameters are $h_{\mathrm{st}} = 5\,[\mathrm{m}]$, $h_{\mathrm{go}} = 55\,[\mathrm{m}]$ and $v_{\max} = 30\,[\mathrm{m/s}]$. Panel **a** corresponds to point P(−0.5,0.4) while panel **b** is associated with point T(0.5,0.4) in Fig. 2.7

and for the transfer function (2.78) we obtain

$$\frac{|\beta \mathrm{j}\omega + \alpha\kappa|^2}{\left|-\omega^2 + (\alpha+\beta)\mathrm{j}\omega + \alpha\kappa\right|^2} < 1\,. \tag{2.82}$$

This is equivalent to

$$\beta^2\omega^2 + \alpha^2\kappa^2 < \omega^4 + \alpha^2\kappa^2 - 2\alpha\kappa\omega^2 + \alpha^2\omega^2 + \beta^2\omega^2 + 2\alpha\beta\omega^2 \,, \tag{2.83}$$

which can be simplified to

$$\omega^2 - 2\alpha\kappa + \alpha^2 + 2\alpha\beta > 0 \,. \tag{2.84}$$

Since the left-hand side of this inequality is a strictly monotonously increasing function of the angular frequency ω, we can find the string stability condition by substituting $\omega = 0$, that is,

$$\alpha\,(\alpha + 2\beta - 2\kappa) > 0 \,. \tag{2.85}$$

This breaks down to

$$\begin{aligned} \alpha &> 0 \,, \\ \alpha &> 2(\kappa - \beta) \,, \end{aligned} \tag{2.86}$$

and

$$\begin{aligned} \alpha &< 0 \,, \\ \alpha &< 2(\kappa - \beta) \,, \end{aligned} \tag{2.87}$$

where only condition (2.86) guarantees string stability, since in case of (2.87) the system is plant unstable; cf. (2.80).

We remark that for zero frequency we have

$$|T(0)| = 1 \,, \tag{2.88}$$

and

$$\left. \frac{\mathrm{d}\,|T(\mathrm{j}\omega)|}{\mathrm{d}\omega} \right|_{\omega=0} = 0 \,. \tag{2.89}$$

That is, the string stability condition is in fact given by

$$\left. \frac{\mathrm{d}^2\,|T(\mathrm{j}\omega)|}{\mathrm{d}\omega^2} \right|_{\omega=0} < 0 \,. \tag{2.90}$$

The string stable domain defined by (2.86) is shown in Fig. 2.7; see the dark gray shaded region. The string stability boundaries are denoted by blue lines. The frequency responses corresponding to points R, S, and T are indicated in Fig. 2.10a–c, where R lies in the string unstable region, S is located at the string stability boundary, and T corresponds to string stable parameters.

Figure 2.11 shows numerical simulation results corresponding to points R and T in Fig. 2.7. The car-following model (2.21) was simulated considering an open chain of 24 follower vehicles driven by the OVM (2.4) with the piecewise linear range policy (2.5), (2.6). The speed of the head vehicle is assumed to oscillate harmonically around an equilibrium speed. The initial state of the follower vehicles was set at the equilibrium. For the string unstable point R the amplitude of the speed

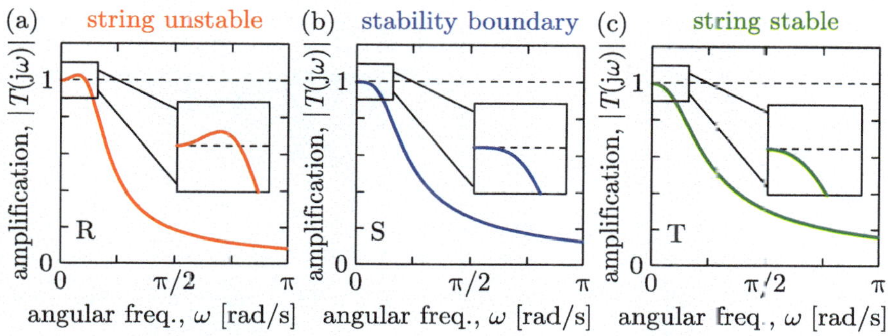

Fig. 2.10 Frequency response plots corresponding to the transfer function (2.78) with $\kappa = 0.6\,[1/\text{s}]$. Panels **a–c** are associated with points R(0.25,0.4), S(0.4,0.4) and T(0.5,0.4) in Fig. 2.7, respectively

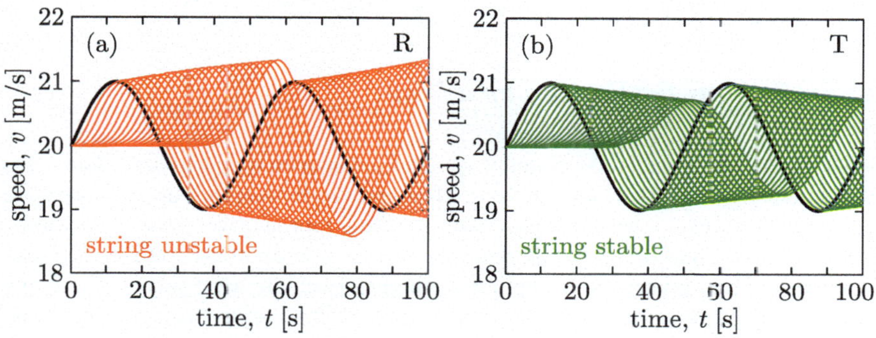

Fig. 2.11 Numerical simulation of the OVM (2.21), (2.4) with piecewise linear range policy (2.5), (2.6) for the case of 24 follower vehicles. The speed of the head vehicle is given by $v_{24}(t) = v^* + v_{\text{amp}} \sin(\omega t)$, $v^* = 20\,[\text{m/s}]$, $v_{\text{amp}} = 1\,[\text{m/s}]$, $\omega = 0.04\pi\,[\text{rad/s}]$, the initial conditions are $v_i(0) = v^*$ and $h_i(0) = h^* = 38.33\,[\text{m}]$, $i = 0, 1, \ldots, 23$, while the rest of the parameters are $h_{\text{st}} = 5\,[\text{m}]$, $h_{\text{go}} = 55\,[\text{m}]$ and $v_{\max} = 30\,[\text{m/s}]$. Panel **a** corresponds to point R(0.25,0.4) while panel **b** is associated with point T(0.5,0.4) in Fig. 2.7

fluctuations increases upstream along the chain of vehicles. Note that this behavior can be observed for certain perturbation frequencies only: for ω values which violate (2.81); cf. Fig. 2.10a. For the string stable point T the speed fluctuations are attenuated upstream (regardless of the angular frequency ω).

Figure 2.12 shows simulation results for the same 24 vehicle setup by considering a large perturbation in the speed of the head vehicle. The perturbation consists of a section of constant deceleration followed by a section of constant acceleration. Despite its non-harmonic nature, the perturbation is again amplified along the vehicle chain for the string unstable point R while it is attenuated by the string stable system in point T.

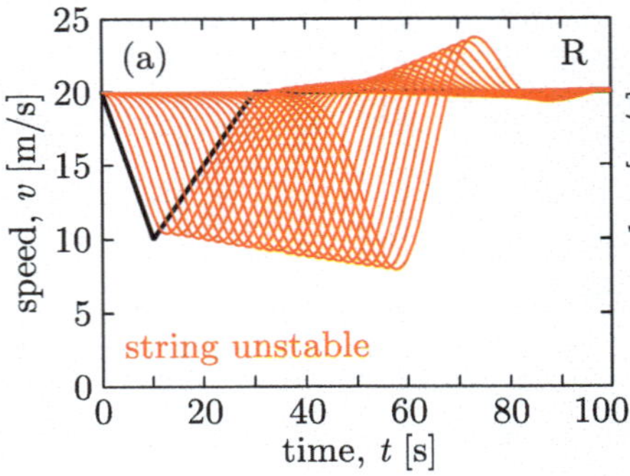

Fig. 2.12 Numerical simulation of the OVM (2.21), (2.4) with piecewise linear range policy (2.5), (2.6) for the case of 24 follower vehicles. The speed of the head vehicle is subject to a 10 [m/s] perturbation, whereas the initial conditions and parameters are the same as in Fig. 2.11. Panel **a** corresponds to point R(0.25,0.4) while panel **b** is associated with point T(0.5,0.4) in Fig. 2.7

2.2.2 *Stability for Vehicles on a Ring*

In Sect. 2.2.1 we considered a chain of vehicles on an open road and described some mathematical techniques that can be used to analyze how perturbations propagate along the chain of vehicles. Two shortcomings of this approach were that (i) the results depended on the norm used to measure the size of perturbations and (ii) they also depended on the perturbations applied by the head vehicle. One way to overcome these shortcomings is to apply periodic boundary conditions that force the head vehicle to follow the tail vehicle. This is often referred to as the ring configuration; see Fig. 2.13 for graphical representation. There are no external inputs given to the system, but the ring length is introduced as a new parameter. We remark that these

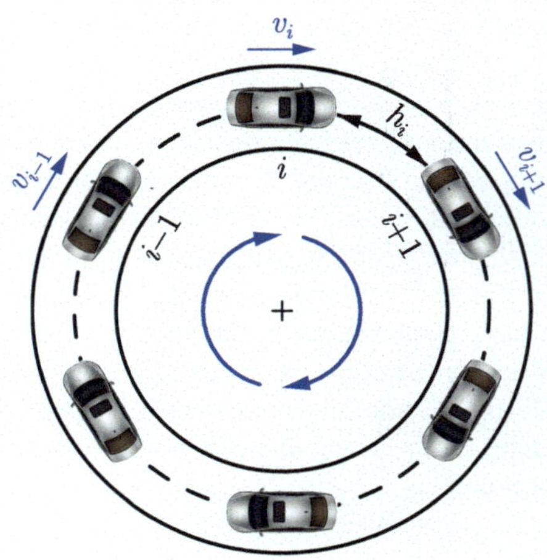

Fig. 2.13 Vehicles following each other on a ring

periodic boundary conditions are introduced for the sake of analysis, while driving vehicles on an actual ring road would require adding the lateral dynamics. Recent experiments demonstrated that periodic boundary conditions can actually be realized on open roads with the help of vehicle-to-everything (V2X) connectivity [7].

Consider N identical vehicles of length l placed on a ring as illustrated in Fig. 2.13. Let the circumference of the ring be $L + Nl$ where we call L as the net ring length. The dynamics of vehicles $i = 1, \ldots, N$ can be described by (2.21) that yields $2N$ scalar differential equations. These are supplemented by the periodic boundary condition

$$\begin{aligned} h_{N+1} &= h_1 \,, \\ v_{N+1} &= v_1 \,. \end{aligned} \tag{2.91}$$

Furthermore, the sum of the headways is equal to the net ring length

$$\sum_{i=1}^{N} h_i = L \,. \tag{2.92}$$

There are two options to satisfy this constraint: (i) express one of the headways from (2.92) and disregard the differential equation governing this headway; (ii) use all $2N$ differential equations and ensure that they are solved accurately with initial conditions satisfying (2.92). We will use the latter option to simplify our analysis.

Figure 2.14 shows numerical simulations for $N = 24$ identical vehicles using the ring configuration. The OVM (2.4) was used with the piecewise linear range policy (2.5), (2.6). The two panels are shown for two different sets of driver parameters

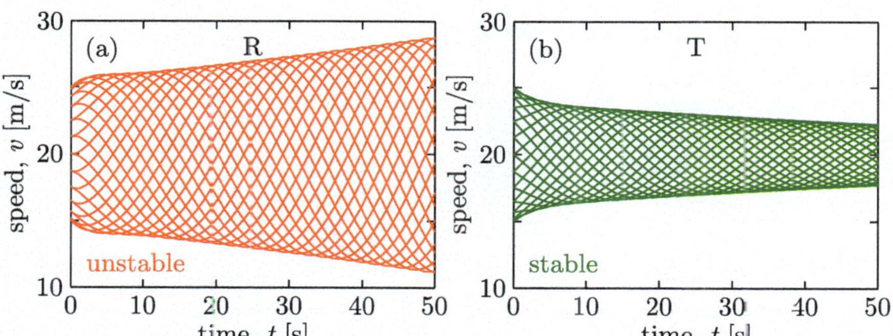

Fig. 2.14 Numerical simulation of the OVM (2.21), (2.4) using the ring configuration given by (2.91), (2.92) with piecewise linear range policy (2.5), (2.6) for $N = 24$ vehicles. The initial conditions are given by $v_i(0) = v^* + v_{\rm amp} \cos(2k\pi i/N)$ and $h_i(0) = h^* + v_{\rm amp}/\omega(\sin(2k\pi(i+1)/N) - \sin(2k\pi i/N))$, $i = 1, \ldots, N$, where the parameters are $v^* = 20\,[\rm m/s]$, $h^* = 38.33\,[\rm m]$, $v_{\rm amp} = 5\,[\rm m/s]$ and $\omega = (\alpha + 2\beta)\tan(k\pi/N)$ with $k = 1$. The rest of the parameters are $h_{\rm st} = 5\,[\rm m]$, $h_{\rm go} = 55\,[\rm m]$ and $v_{\rm max} = 30\,[\rm m/s]$. Panel **a** corresponds to point R(0.25,0.4) while panel **b** is associated with point T(0.5,0.4) in Fig. 2.15

(β and α), which result in an unstable and a stable scenario (corresponding to points R and T in Fig. 2.15 discussed later). The initial speed and headway were set non-uniformly among the vehicles as described in the caption (with more explanation further below in the text). For the unstable case this leads to growing oscillations in the speed of the vehicles, whereas in the stable scenario the system settles to an equilibrium of uniform speed and headway. In what follows, we derive the equilibrium of the ring configuration and analyze its stability by linearization to explain this phenomenon. We will also link the results to those obtained for the open chain configuration above.

The ring of identical vehicles has the uniform flow equilibrium given by solution (2.15). According to (2.16) and (2.92), the equilibrium headway and speed satisfy

$$h^* = L/N \ , \quad v^* = V(h^*) \ . \tag{2.93}$$

As opposed to the open chain of vehicles, the equilibrium of the ring is now determined by the net ring length L and the number N of vehicles rather than by the prescribed motion of the head vehicle.

Similarly to the open chain, the dynamics can be linearized around the uniform flow by introducing the perturbations (2.22) which leads to the linearized system (2.25). Since the input of each vehicle's dynamics is considered to be the output (the speed perturbation) of the preceding vehicle, i.e., $u_i = y_{i+1} = \mathbf{c}\,\mathbf{x}_{i+1}$, the linearized system becomes

$$\dot{\mathbf{x}}_i = \mathbf{a}\,\mathbf{x}_i + \mathbf{b}\mathbf{c}\,\mathbf{x}_{i+1} \ , \quad i = 1, \ldots, N \ . \tag{2.94}$$

For the OVM parametrization the coefficient matrices are given by (2.29) which yield

$$\mathbf{b}\mathbf{c} = \begin{bmatrix} 0 & 1 \\ 0 & \beta \end{bmatrix} . \tag{2.95}$$

Note that the solution of (2.94) must obey the constraint (2.92) on the sum of the headways which is equivalent to

$$\sum_{i=1}^{N} \tilde{h}_i = \sum_{i=1}^{N} x_{i,1} = 0 \ , \tag{2.96}$$

where $x_{i,1}$ is the first component of $\mathbf{x}_i$.

Furthermore, the periodic boundary condition (2.91) yields

$$\mathbf{x}_{N+1} = \mathbf{x}_1 \ , \tag{2.97}$$

for the perturbations. This allows us to formulate the linearized dynamics (2.94), (2.97) of the entire ring as a coupled set of differential equations in the matrix form

$$\begin{bmatrix} \dot{\mathbf{x}}_1 \\ \dot{\mathbf{x}}_2 \\ \vdots \\ \dot{\mathbf{x}}_N \end{bmatrix} = \begin{bmatrix} \mathbf{a} & \mathbf{bc} & & \\ & \ddots & \ddots & \\ & & \mathbf{a} & \mathbf{bc} \\ \mathbf{bc} & & & \mathbf{a} \end{bmatrix} \begin{bmatrix} \mathbf{x}_1 \\ \mathbf{x}_2 \\ \vdots \\ \mathbf{x}_N \end{bmatrix}, \tag{2.98}$$

where only the nonzero elements are indicated. This shortly reads as

$$\dot{\mathbf{X}} = \mathbf{A}\mathbf{X}\,. \tag{2.99}$$

The linear stability of the uniform flow is determined by the eigenvalues of the coefficient matrix **A** given by the characteristic equation

$$\det(\lambda\mathbf{I} - \mathbf{A}) = 0\,. \tag{2.100}$$

This can be summarized in the following theorem.

Theorem 2.5 *The uniform flow solution of the ring configuration is asymptotically stable at the linear level if and only if the eigenvalues λ_l have negative real parts, that is,* $\mathrm{Re}(\lambda_l) < 0$, $l = 1, \ldots, 2N$.

We remark that while this theorem is similar to the plant stability Theorem 2.1, it in fact provides conditions equivalent to the string stability Theorems 2.2 and 2.3 as will be shown below.

The blocks of matrix **A** in (2.98) are organized into a special structure which can be represented by the help of the N-dimensional identity matrix $\mathbf{I}_N$ and the N-dimensional circulant matrix

$$\mathbf{R}_N = \begin{bmatrix} 0 & 1 & & \\ & \ddots & \ddots & \\ & & 0 & 1 \\ 1 & & & 0 \end{bmatrix}, \tag{2.101}$$

where the zero elements are only spelled out in the diagonal. Using these matrices and the Kronecker-product $\otimes$ we can write

$$\mathbf{A} = \mathbf{I}_N \otimes \mathbf{a} + \mathbf{R}_N \otimes \mathbf{bc}\,. \tag{2.102}$$

Due to the special structure of **A**, its eigenvalues can be analyzed for any number of vehicles N. They are associated with the eigenvalues of $\mathbf{R}_N$, given as $\mathrm{e}^{\mathrm{j}\frac{2k\pi}{N}}$, $k = 0, 1, \ldots, N-1$. The corresponding eigenvectors of $\mathbf{R}_N$ are collected in the columns of the matrix

$$\mathbf{P}_N = \begin{bmatrix} 1 & 1 & & 1 & & 1 \\ 1 & e^{j\frac{2\pi}{N}} & & e^{j\frac{2\pi}{N}k} & & e^{j\frac{2\pi}{N}(N-1)} \\ 1 & e^{j\frac{2\pi}{N}2} & \dots & e^{j\frac{2\pi}{N}k2} & \dots & e^{j\frac{2\pi}{N}(N-1)2} \\ \vdots & \vdots & & \vdots & & \vdots \\ 1 & e^{j\frac{2\pi}{N}(N-1)} & & e^{j\frac{2\pi}{N}k(N-1)} & & e^{j\frac{2\pi}{N}(N-1)^2} \end{bmatrix}, \tag{2.103}$$

that is, the l-th element of the k-th eigenvector is $e^{j\frac{2\pi}{N}kl}$ with $k, l = 0, 1, \dots, N-1$; see [150] for more details on circulant matrices. One can utilize matrix $\mathbf{P}_N$ to introduce the coordinate transformation

$$\mathbf{X} = (\mathbf{P}_N \otimes \mathbf{I})\mathbf{Z} , \tag{2.104}$$

which transforms (2.99) to the decoupled set of differential equations

$$\dot{\mathbf{Z}} = \begin{bmatrix} \mathbf{a}+\mathbf{bc} & & & \\ & \mathbf{a}+\mathbf{bc}\, e^{j\frac{2\pi}{N}} & & \\ & & \ddots & \\ & & & \mathbf{a}+\mathbf{bc}\, e^{j\frac{2(N-1)\pi}{N}} \end{bmatrix} \mathbf{Z} . \tag{2.105}$$

Since the coordinate transformation (2.104) results in a similarity transformation of the system matrix, the eigenvalues of the matrix in (2.105) are the same as the eigenvalues of $\mathbf{A}$. These can be obtained from the characteristic equation

$$\prod_{k=0}^{N-1} \det\left(\lambda\mathbf{I} - \mathbf{a} - \mathbf{bc}\, e^{j\frac{2k\pi}{N}}\right) = 0 , \tag{2.106}$$

which is equivalent to (2.100).

This formulation decouples the problem into N modes which we distinguish with index k. Each mode is associated with two eigenvalues given by

$$\det\left(\lambda\mathbf{I} - \mathbf{a} - \mathbf{bc}\, e^{j\frac{2k\pi}{N}}\right) = 0 , \tag{2.107}$$

for $k = 0, 1, \dots, N-1$. Note that this equation has complex coefficients, thus two complex eigenvalues do not necessarily form a complex conjugate pair. Furthermore, stability criteria (such as the Routh–Hurwitz criteria) are typically formulated for equations with real coefficients.

When considering the mode for $k = 0$, one obtains

$$\det(\lambda\mathbf{I} - \mathbf{a} - \mathbf{bc}) = 0 , \tag{2.108}$$

which has real coefficients. The other modes for $k > 0$ can be combined with their conjugate modes for $N - k$ as

$$\det\left(\lambda\mathbf{I} - \mathbf{a} - \mathbf{bc}\,\mathrm{e}^{\mathrm{j}\frac{2k\pi}{N}}\right)\det\left(\lambda\mathbf{I} - \mathbf{a} - \mathbf{bc}\,\mathrm{e}^{\mathrm{j}\frac{2(N-k)\pi}{N}}\right) = 0\,. \tag{2.109}$$

We remark that for the special mode $k = N/2$, that only appears for even N, we still obtain an equation with real coefficients, but we omit the detailed discussion of this special case.

Now let us turn the characteristic equations (2.108), (2.109) of different modes into the expression of the stability boundary for the case of the OVM parametrization with coefficient matrices (2.29), (2.95). For $k = 0$, substitution of the coefficient matrices into (2.108) leads to

$$\det\begin{bmatrix} \lambda & 0 \\ -\alpha\kappa & \lambda+\alpha \end{bmatrix} = \lambda(\lambda+\alpha) = 0 \quad \Rightarrow \quad \lambda_1 = 0\,, \quad \lambda_2 = -\alpha\,. \tag{2.110}$$

That is, there is a zero eigenvalue associated with mode $k = 0$ regardless of the model parameters. This holds for any car-following model with the ring configuration and it originates from a continuous transnational symmetry along the ring: the system is invariant to shifting the position of all vehicles along the ring with an arbitrary distance. This eigenvalue could be removed by replacing the differential equation for one of the headways with (2.92), while more sophisticated techniques were used in [156, 157]. Therefore, we disregard this zero eigenvalue and focus on the rest of the eigenvalues only. The second eigenvalue in (2.110) leads to the stability condition $\alpha > 0$.

For $k > 0$, considering the OVM parametrization and using $\mathrm{e}^{\mathrm{j}\frac{2(N-k)\pi}{N}} = \mathrm{e}^{-\mathrm{j}\frac{2k\pi}{N}}$, (2.109) takes the form

$$\det\begin{bmatrix} \lambda & 1-\mathrm{e}^{\mathrm{j}\frac{2k\pi}{N}} \\ -\alpha\kappa & \lambda+\alpha+\beta-\beta\mathrm{e}^{\mathrm{j}\frac{2k\pi}{N}} \end{bmatrix}\det\begin{bmatrix} \lambda & 1-\mathrm{e}^{-\mathrm{j}\frac{2k\pi}{N}} \\ -\alpha\kappa & \lambda+\alpha+\beta-\beta\mathrm{e}^{-\mathrm{j}\frac{2k\pi}{N}} \end{bmatrix} = 0\,, \tag{2.111}$$

which simplifies to

$$\left(\lambda^2+\alpha\lambda+(\beta\lambda+\alpha\kappa)\left(1-\mathrm{e}^{\mathrm{j}\frac{2k\pi}{N}}\right)\right)\left(\lambda^2+\alpha\lambda+(\beta\lambda+\alpha\kappa)\left(1-\mathrm{e}^{-\mathrm{j}\frac{2k\pi}{N}}\right)\right) = 0\,. \tag{2.112}$$

This yields a fourth-order characteristic polynomial in the form

$$a_4\lambda^4 + a_3\lambda^3 + a_2\lambda^2 + a_1\lambda + a_0 = 0\,, \tag{2.113}$$

where the coefficients are the following real-valued expressions

$$
\begin{aligned}
a_4 &= 1\,,\\
a_3 &= 2\alpha + 2\beta\left(1-\cos\frac{2k\pi}{N}\right),\\
a_2 &= \alpha^2 + 2\left(\alpha\beta+\beta^2+\alpha\kappa\right)\left(1-\cos\frac{2k\pi}{N}\right),\\
a_1 &= 2\alpha\kappa(\alpha+2\beta)\left(1-\cos\frac{2k\pi}{N}\right),\\
a_0 &= 2\alpha^2\kappa^2\left(1-\cos\frac{2k\pi}{N}\right).
\end{aligned}
\tag{2.114}
$$

One may apply the Routh–Hurwitz criteria $a_4 > 0$, $a_3 > 0$, $a_2 > 0$, $a_1 > 0$, $a_0 > 0$, $a_3(a_2a_1 - a_3a_0) - a_4a_1^2 > 0$ (see (A.11) in Appendix A) and the trigonometric identity $\left(1-\cos\frac{2k\pi}{N}\right) = 2\sin^2\frac{k\pi}{N}$ to obtain the conditions

$$
\begin{aligned}
&2\alpha + 4\beta\sin^2\frac{k\pi}{N} > 0\,,\\
&\alpha^2 + 4\left(\alpha\beta+\beta^2+\alpha\kappa\right)\sin^2\frac{k\pi}{N} > 0\,,\\
&4\alpha\kappa(\alpha+2\beta)\sin^2\frac{k\pi}{N} > 0\,,\\
&8\alpha\kappa\sin^2\frac{k\pi}{N}\left(\alpha^2+4\left(\alpha\beta+\beta^2\right)\sin^2\frac{k\pi}{N}\right)\\
&\quad\times\left(\alpha^2+2\alpha\beta-2\alpha\kappa+2\left(\alpha\beta+2\beta^2+\alpha\kappa\right)\sin^2\frac{k\pi}{N}\right) > 0\,.
\end{aligned}
\tag{2.115}
$$

Assuming $\kappa > 0$ the first and the third condition in (2.115) yield

$$
\begin{aligned}
\alpha &> 0\,,\\
\alpha &> -2\beta\,.
\end{aligned}
\tag{2.116}
$$

Note, however, that $\alpha > 0$ was already required for the stability of the $k = 0$ mode.

One may also show that

$$
\alpha^2 + 4\left(\alpha\beta+\beta^2\right)\sin^2\frac{k\pi}{N} = \left(\alpha + 2\beta\sin^2\frac{k\pi}{N}\right)^2 + 4\beta^2\sin^2\frac{k\pi}{N}\cos^2\frac{k\pi}{N} > 0\,.
\tag{2.117}
$$

Thus, considering $\kappa > 0$ and $\alpha > 0$, the second condition in (2.115) holds while the last condition simplifies to

$$
\alpha^2 + 2\left(\beta\left(1+\sin^2\frac{k\pi}{N}\right) - \kappa\cos^2\frac{k\pi}{N}\right)\alpha + 4\beta^2\sin^2\frac{k\pi}{N} > 0\,.
\tag{2.118}
$$

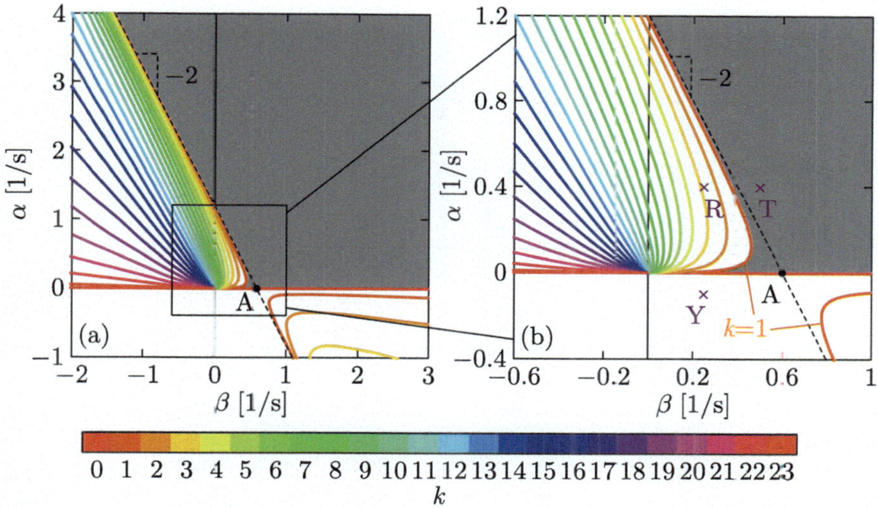

Fig. 2.15 Stability chart of system (2.98), (2.29) for $N = 24$ vehicles and $\kappa = 0.6$ [1/s]. The wave numbers k are indicated using color bar. Point A is located at $(\kappa, 0)$ while points R(0.25,0.4), T(0.5,0.4) and Y(0.25,−0.1) correspond to the spectra in Fig. 2.16

The corresponding stability curves are depicted in Fig. 2.15 for $N = 24$. These are ordered according to the modes k as indicated. Notice that the stability limits are given by $\alpha = 0$ (that is a stability boundary for any mode k) and by the left-hand side of (2.118) for the $k = 1$ mode, as the corresponding curves bound the shaded stable area. The asymptote of the $k > 0$ curves is indicated by the dashed line and it coincides with the string stability boundary given by (2.86) for the open chain configuration of vehicles. Note that as the number of vehicles increases ($N \to \infty$), the stable region of the ring configuration (shown in Fig. 2.15) converges to the string stability domain of the open chain of vehicles (shown in Fig. 2.7).

We remark that if one is only interested in finding the stability curves, they may be obtained from (2.107) without combining the conjugate modes. In particular, using the coefficient matrices (2.29), (2.95) of the OVM (2.4) yields

$$\lambda^2 + (\alpha + \beta)\lambda + \alpha\kappa - (\beta\lambda + \alpha\kappa)\, \mathrm{e}^{\mathrm{j}\frac{2k\pi}{N}} = 0\,. \tag{2.119}$$

For $k > 0$, substituting $\lambda = 0$ yields

$$\alpha = 0\,, \tag{2.120}$$

cf. (2.116), while substituting $\lambda = \mathrm{j}\omega$ with $\omega > 0$ and separating the real and imaginary parts one may obtain

$$
\begin{aligned}
-\omega^2 + \alpha\kappa - \alpha\kappa \cos\frac{2k\pi}{N} + \beta\omega \sin\frac{2k\pi}{N} &= 0\,, \\
(\alpha+\beta)\omega - \beta\omega \cos\frac{2k\pi}{N} - \alpha\kappa \sin\frac{2k\pi}{N} &= 0\,.
\end{aligned}
\tag{2.121}
$$

This can be solved for α and β in the form

$$
\begin{aligned}
\alpha &= \frac{\omega^2\big(1-\cos\frac{2k\pi}{N}\big)}{-\omega\sin\frac{2k\pi}{N} + 2\kappa\big(1-\cos\frac{2k\pi}{N}\big)}\,, \\
\beta &= \frac{-\omega^2 + \omega\kappa\sin\frac{2k\pi}{N}}{-\omega\sin\frac{2k\pi}{N} + 2\kappa\big(1-\cos\frac{2k\pi}{N}\big)}\,,
\end{aligned}
\tag{2.122}
$$

for $k = 1, \ldots, N-1$, that provide the stability boundaries in parametric form (parameterized by the angular frequency ω).

Note that the stability boundary of mode $N-k$ differs only in a minus sign compared to that of mode k (ω is replaced by $-\omega$). Thus, if one plots the stability curves for both $\omega > 0$ and $\omega < 0$, the stability boundary of mode k overlaps with that of mode $N-k$. In Fig. 2.15 stability curves are plotted for $\omega > 0$ only.

Alternatively, using the trigonometric identities $\big(1-\cos\frac{2k\pi}{N}\big) = 2\sin^2\frac{k\pi}{N}$ and $\sin\frac{2k\pi}{N} = 2\sin\frac{k\pi}{N}\cos\frac{k\pi}{N}$ one may re-write (2.121) as

$$
\begin{aligned}
(\alpha+2\beta)\sin\frac{k\pi}{N} - \omega\cos\frac{k\pi}{N} &= 0\,, \\
(2\alpha\kappa-\omega^2)\sin\frac{k\pi}{N} - \alpha\omega\cos\frac{k\pi}{N} &= 0\,.
\end{aligned}
\tag{2.123}
$$

This can be solved for α and β as

$$
\begin{aligned}
\alpha &= \frac{\omega^2\tan\frac{k\pi}{N}}{2\kappa\tan\frac{k\pi}{N} - \omega}\,, \\
\beta &= \frac{\omega}{2\tan\frac{k\pi}{N}} - \frac{\alpha}{2}\,,
\end{aligned}
\tag{2.124}
$$

which is equivalent to (2.122).

Furthermore, from the first line of (2.123) one can also express the angular frequency ω for a given (β, α) pair along the stability boundary:

$$
\omega = (\alpha+2\beta)\tan\frac{k\pi}{N}\,, \tag{2.125}
$$

which is the angular frequency value that was used in Fig. 2.14. This formula highlights that for $k \ll N$ the angular frequency ω is proportional to the wave number k. Substituting this into the second line of (2.123) yields

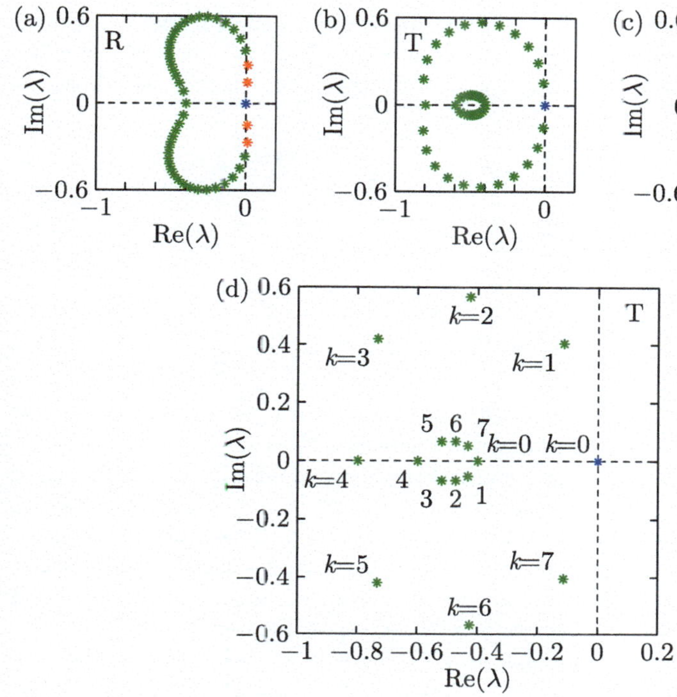

Fig. 2.16 Spectrum plots obtained from the characteristic equation (2.106) for $N = 24$ (top) and $N = 8$ (bottom) for $\kappa = 0.6\,[1/\mathrm{s}]$. The different panels correspond to the points R(0.25, 0.4), T(0.5, 0.4) and Y(0.25, −0.1) marked in Fig. 2.15

$$\left(1+\tan^2\frac{k\pi}{N}\right)\alpha^2+2\left(\beta\left(1+2\tan^2\frac{k\pi}{N}\right)-\kappa\right)\alpha+4\beta^2\tan^2\frac{k\pi}{N}=0\,, \tag{2.126}$$

cf. (2.118).

Figure 2.16 shows the eigenvalues of matrix $\mathbf{A}$ in the complex plane. Recall that there are $2N$ eigenvalues where one of them is zero. Panels (a)–(c) show the eigenvalues corresponding to points R, T, and Y in Fig. 2.15, respectively. Since point R is beyond the $k = 1$ and $k = 2$ stability boundaries in Fig. 2.15, there are two pairs of eigenvalues on the right of the imaginary axis in Fig. 2.16a. Point T is located in the stable region in Fig. 2.15, thus all eigenvalues lie in the left-half complex plane in Fig. 2.16b (except for the single zero eigenvalue). Point Y is below the $\alpha = 0$ stability boundary in Fig. 2.15, this gives rise to a set of N unstable eigenvalues in Fig. 2.16c.

The eigenvalues are the solutions of (2.107) for different modes k as depicted in Fig. 2.16d for $N = 8$. Parameter k is also called wave number and its physical meaning can be understood as follows. First, notice that the eigenvalues corresponding to mode k and mode $N - k$ are complex conjugates. Consider a complex conjugate pair of eigenvalues $\nu \pm \mathrm{j}\omega$ with $\omega > 0$ for a particular value of k and $N-k$ such that $0 < k < N/2$. Then in the modal coordinates in (2.105) we have the solutions $\mathbf{z}_k(t) = \mathbf{z}_k^0 \mathrm{e}^{(\nu+\mathrm{j}\omega)t}$ and $\mathbf{z}_{N-k}(t) = \mathbf{z}_{N-k}^0 \mathrm{e}^{(\nu-\mathrm{j}\omega)t}$. Using (2.103) and (2.104) yields

$$\mathbf{x}_i(t) = \mathrm{e}^{\mathrm{j}\frac{2k\pi}{N}i}\mathbf{z}_k(t) + \mathrm{e}^{\mathrm{j}\frac{2(N-k)\pi}{N}i}\mathbf{z}_{N-k}(t) = \mathrm{e}^{\nu t}\left(\mathbf{z}_k^0 \mathrm{e}^{\mathrm{j}\left(\frac{2k\pi}{N}i+\omega t\right)} + \bar{\mathbf{z}}_k^0 \mathrm{e}^{-\mathrm{j}\left(\frac{2k\pi}{N}i+\omega t\right)}\right), \tag{2.127}$$

where we exploited $\mathrm{e}^{\mathrm{j}\frac{2(N-k)\pi}{N}i} = \mathrm{e}^{-\mathrm{j}\frac{2k\pi}{N}i}$ and $\mathbf{z}_{N-k}^0 = \bar{\mathbf{z}}_k^0$. Extracting the velocity $\tilde{v}_i = x_{i,2}$ results in

$$\tilde{v}_i(t) = \mathrm{e}^{\nu t} v_{\mathrm{amp}} \cos\left(\frac{2k\pi}{N}i + \omega t + \theta\right), \tag{2.128}$$

where the amplitude v_{amp} and the phase θ are given by the initial conditions. Similar expression can be derived for the headway fluctuation $\tilde{h}_i = x_{i,1}$.

Formula (2.128) represents a backward traveling congestion wave. The wave grows as time evolves if $\nu > 0$ and decays if $\nu < 0$, while $\nu = 0$ indicates the stability boundary. The corresponding wavelength is $(L + Nl)/k$ if measured in distance (meters) or N/k if measured in number of vehicles. Loosely speaking, this implies that there are k traffic jams along the ring. For example, if $k = 2$ then the speed fluctuations are repeated along the ring twice corresponding to two traffic jams.

We remark that for an eigenvalue pair $\nu \pm \mathrm{j}\omega$ with $\omega > 0$ corresponding to k and $N-k$ such that $N/2 < k < N$, we obtain $-\frac{2k\pi}{N}i$ instead of $\frac{2k\pi}{N}i$ in (2.128). This corresponds to a forward traveling wave with wavelength $(L + Nl)/(N-k)$ if measured in distance (meters) or $N/(N-k)$ if measured in number of vehicles. Finally, notice that (2.128) motivated the choice of initial conditions in Fig. 2.14.

2.3 Human Driving with Driver Reaction Time

In this section we investigate the effect of driver reaction time on the dynamics of car following. We demonstrate that the corresponding time delay has significant effects on the system dynamics. In particular, this can lead to string instability and result in the formation of traveling congestion waves. These results will also motivate the rest of the book where automation and connectivity will be used to compensate for detrimental human behavior in the traffic flow.

For the sake of simplicity, we assume identical drivers with identical reaction time delay τ, that is, we consider

$$\begin{aligned}
\dot{h}_i(t) &= v_{i+1}(t) - v_i(t), \\
\dot{v}_i(t) &= f\left(h_i(t-\tau), \dot{h}_i(t-\tau), v_i(t-\tau)\right),
\end{aligned} \tag{2.129}$$

cf. (2.2) and (2.21). Similarly to the previous section, we analyze the stability of the uniform flow at the linear level, plot the corresponding spectra, and show simulation results for the open chain of vehicles as well as for the ring configuration. We also establish a rigorous relationship between the open chain and ring configurations, and discuss the dynamics when time delays are approximated by first-order lags.

2.3.1 Open Chain of Vehicles

Recall that the uniform flow equilibrium is given by (2.15). One may show that (2.129) may possess such equilibrium if the implicit function theorem is satisfied. In order to analyze stability, one can consider the perturbation (2.22) around the uniform flow and write (2.129) into the form

$$
\begin{aligned}
\dot{\tilde{h}}_i(t) &= \tilde{v}_{i+1}(t) - \tilde{v}_i(t)\,, \\
\dot{\tilde{v}}_i(t) &= f_h^* \tilde{h}_i(t-\tau) + f_{\dot{h}}^* \dot{\tilde{h}}_i(t-\tau) + f_v^* \tilde{v}_i(t-\tau) + \text{h.o.t.}\,,
\end{aligned}
\tag{2.130}
$$

cf. (2.23). By using the state, input, and output introduced in (2.24), we can represent the system equivalently as

$$
\begin{aligned}
\dot{\mathbf{x}}_i(t) &= \mathbf{a}_0\,\mathbf{x}_i(t) + \mathbf{a}_\tau\,\mathbf{x}_i(t-\tau) + \mathbf{b}_0\,u_i(t) + \mathbf{b}_\tau\,u_i(t-\tau) + \text{h.o.t.}\,, \\
y_i(t) &= \mathbf{c}\,\mathbf{x}_i(t)\,,
\end{aligned}
\tag{2.131}
$$

where the matrices read

$$
\mathbf{a}_0 = \begin{bmatrix} 0 & -1 \\ 0 & 0 \end{bmatrix}, \quad \mathbf{a}_\tau = \begin{bmatrix} 0 & 0 \\ f_h^* & f_v^* - f_{\dot{h}}^* \end{bmatrix}, \quad \mathbf{b}_0 = \begin{bmatrix} 1 \\ 0 \end{bmatrix}, \quad \mathbf{b}_\tau = \begin{bmatrix} 0 \\ f_{\dot{h}}^* \end{bmatrix}, \quad \mathbf{c} = \begin{bmatrix} 0 & 1 \end{bmatrix},
\tag{2.132}
$$

cf. (2.25), (2.26). Note that $\tau = 0$ recovers the delay-free case with $\mathbf{a} = \mathbf{a}_0 + \mathbf{a}_\tau$ and $\mathbf{b} = \mathbf{b}_0 + \mathbf{b}_\tau$.

For the sake of linear stability analysis, we drop the higher order terms and formulate the transfer function corresponding to (2.131):

$$
T(s) = \mathbf{c}\big(s\mathbf{I} - \mathbf{a}_0 - \mathbf{a}_\tau\,\mathrm{e}^{-s\tau}\big)^{-1}\big(\mathbf{b}_0 + \mathbf{b}_\tau\,\mathrm{e}^{-s\tau}\big)\,,
\tag{2.133}
$$

cf. (2.68), which will be used throughout this section. Note that for simplicity we have already dropped the subscript i, $i+1$.

The plant stability analysis requires the characteristic equation of the system which is defined by the denominator of the transfer function:

$$
\det\big(s\mathbf{I} - \mathbf{a}_0 - \mathbf{a}_\tau\,\mathrm{e}^{-s\tau}\big) = 0\,.
\tag{2.134}
$$

Note that this equation is now a transcendental one with infinitely many complex solutions for s due to the exponential term containing the delay; see Appendix B for more details about the infinite-dimensional nature of time delay systems. Still, the condition for plant stability is analogous to the one in Theorem 2.1 and can be stated as follows.

Theorem 2.6 *Vehicle i is plant stable if and only if the solutions of the characteristic equation (2.134) have negative real parts, that is,* $\mathrm{Re}(s_k) < 0$, $k = 1, \ldots, \infty$.

Moreover, for string stability analysis, we can still utilize Theorem 2.2.

In what follows, we present the results of plant and string stability analysis while using the OVM parametrization (2.28). That is, rather than using the coefficients f_h^*, $f_{\dot{h}}^*$, f_v^* we use the analogous parameters α, β, κ as the latter have clear physical interpretations. Thus, the expressions of the coefficient matrices become

$$\mathbf{a}_0 = \begin{bmatrix} 0 & -1 \\ 0 & 0 \end{bmatrix}, \quad \mathbf{a}_\tau = \begin{bmatrix} 0 & 0 \\ \alpha\kappa & -(\alpha+\beta) \end{bmatrix}, \quad \mathbf{b}_0 = \begin{bmatrix} 1 \\ 0 \end{bmatrix}, \quad \mathbf{b}_\tau = \begin{bmatrix} 0 \\ \beta \end{bmatrix}, \quad \mathbf{c} = \begin{bmatrix} 0 & 1 \end{bmatrix}, \tag{2.135}$$

cf. (2.29). Substituting these into (2.133) leads to the transfer function

$$T(s) = \frac{(\beta s + \alpha\kappa)\mathrm{e}^{-s\tau}}{s^2 + ((\alpha+\beta)s + \alpha\kappa)\,\mathrm{e}^{-s\tau}} = \frac{\beta s + \alpha\kappa}{s^2\mathrm{e}^{s\tau} + (\alpha+\beta)s + \alpha\kappa}, \tag{2.136}$$

while (2.134) yields the characteristic equation

$$s^2\mathrm{e}^{s\tau} + (\alpha+\beta)s + \alpha\kappa = 0\,, \tag{2.137}$$

cf. (2.78) and (2.79).

First, let us discuss the plant stability calculation. Since the left-hand side of the characteristic equation is not a polynomial of s, the Routh–Hurwitz criteria are no longer applicable for finding the plant stable domain [197]. Instead, one can derive the expression of the plant stability boundaries by analyzing the cases where (2.137) has solutions with zero real parts, i.e., either $s = 0$ or $s = \pm\mathrm{j}\Omega$, $\Omega > 0$. Substituting $s = 0$ into (2.137) gives

$$\alpha = 0\,, \tag{2.138}$$

whereas substituting $s = \mathrm{j}\Omega$ and decomposing the resulting equation into real and imaginary parts lead to

$$\begin{aligned} \alpha &= \frac{\Omega^2\cos(\Omega\tau)}{\kappa}\,, \\ \beta &= \Omega\sin(\Omega\tau) - \alpha\,. \end{aligned} \tag{2.139}$$

Notice that the $s = 0$ plant stability boundary (2.138) is identical to that of the delay-free system (with $\tau = 0$), cf. the second row of (2.80). Therefore, the analysis of the delay-free scenario also allows us to determine which side of this boundary

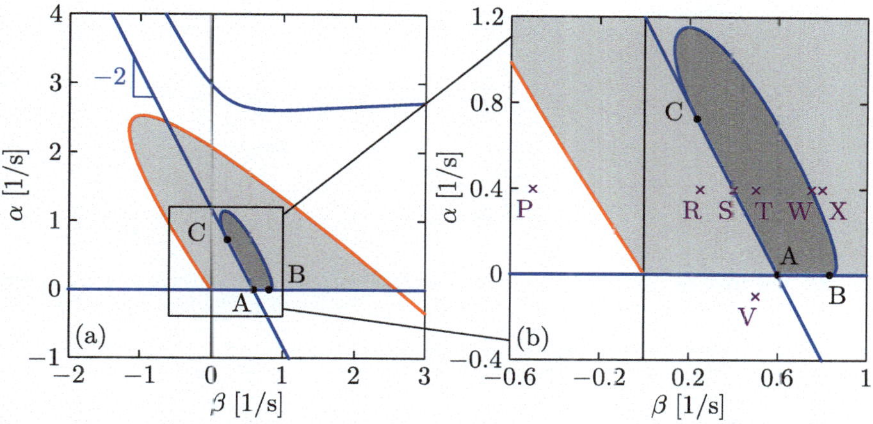

Fig. 2.17 Stability chart of system (2.131), (2.135) for $\kappa = 0.6\,[1/\mathrm{s}]$ and $\tau = 0.6\,[\mathrm{s}]$. The plant and string stable regions are shaded by light and dark gray, respectively. Point A is located at the intersection $(\kappa, 0)$, while points B and C lie at the starting points of the string stability boundaries. Points P(−0.5,0.4), R(0.25,0.4), S(0.4,0.4), T(0.5,0.4), V(0.5,−0.1), W(0.751,0.4) and X(0.8,0.4) correspond to the spectra and frequency response plots in Figs. 2.19, 2.20 and 2.21

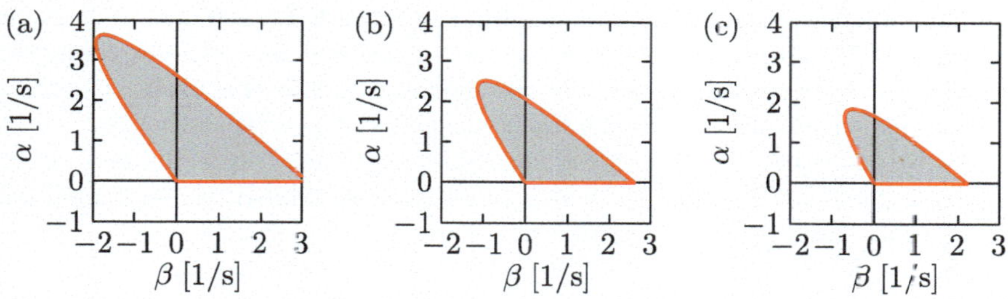

Fig. 2.18 Plant stability charts of system (2.131), (2.135) for $\kappa = 0.6\,[1/\mathrm{s}]$. Panels **a**–**c** correspond to time delays $\tau = 0.5\,[\mathrm{s}]$, $\tau = 0.6\,[\mathrm{s}]$ and $\tau = 0.7\,[\mathrm{s}]$, respectively. The light gray shaded plant stable region shrinks significantly as the delay is increased

the stable region lies on $(\alpha > 0)$. We remark that this property will hold for all models presented in this book, even for those describing the motion of automated and connected automated vehicles. This is due to a general property of time delay systems, namely, that the $s = 0$ stability boundaries are not influenced by the delays; see more details in Appendix B. One may also notice that the $s = \mathrm{j}\Omega$ stability boundary (2.139) simplifies to $\alpha = \Omega^2/\kappa$, $\beta = -\alpha$ for $\tau = 0$, cf. the first row of (2.80).

Equations (2.138) and (2.139) define the plant stability boundaries in the (β, α) plane where $\Omega > 0$ is used to parameterize the curves. The plant stability boundaries are depicted in Fig. 2.17 by red color. They enclose the light gray shaded plant stable region. Note that this region is bounded as opposed to the delay-free case in Fig. 2.7. The time delay therefore negatively impacts plant stability, and the plant stable domain shrinks significantly as the delay is increased; see Fig. 2.18.

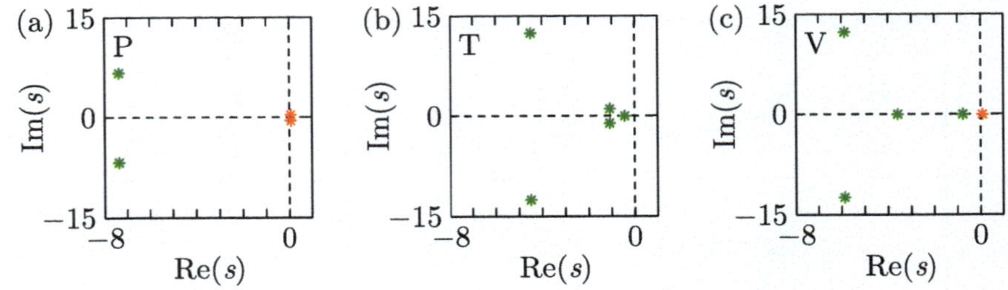

Fig. 2.19 Spectrum plots obtained from the characteristic equation (2.137) with $\kappa = 0.6\,[1/\mathrm{s}]$ and $\tau = 0.6\,[\mathrm{s}]$. Panels **a**–**c** correspond to points P(−0.5,0.4), T(0.5,0.4) and V(0.5,−0.1) in Fig. 2.17, respectively

Although the boundaries of plant stability separate the parameter space into stable and unstable domains, they do not directly tell us which region is the stable one. One can find the stable region by calculating the roots of the characteristic equation (2.137) numerically for specific points in the (β, α) plane. This is illustrated in Fig. 2.19 where the eigenvalues are depicted corresponding to points marked P, T, and V in Fig. 2.17. Note that there are infinitely many eigenvalues and the figure shows the rightmost ones only, which were calculated numerically using DDE-BIFTOOL [49, 194]. Point T is plant stable, while points P and V are plant unstable with a complex conjugate pair of unstable eigenvalues and a real unstable eigenvalue, respectively, since they are beyond the plant stability boundaries associated with $s = \mathrm{j}\Omega$ and $s = 0$.

Now we discuss the details of the string stability analysis. The condition of string stability is given by Theorem 2.2, which can be evaluated based on the magnitude $|T(\mathrm{j}\omega)|$ of the transfer function (2.136) at $s = \mathrm{j}\omega$:

$$T(\mathrm{j}\omega) = \frac{\mathrm{j}\beta\omega + \alpha\kappa}{-\omega^2\left(\cos(\omega\tau) + \mathrm{j}\sin(\omega\tau)\right) + \mathrm{j}(\alpha+\beta)\omega + \alpha\kappa} \,. \tag{2.140}$$

Accordingly, $|T(\mathrm{j}\omega)| < 1$ must hold for all $\omega > 0$ to ensure string stability. This is illustrated in Fig. 2.20c, which is the delayed counterpart of Fig. 2.10c. String stability can be lost in two qualitatively different ways: when $|T(\mathrm{j}\omega)|$ goes above one at $\omega = 0$, as in Fig. 2.20a, b, or at $\omega > 0$, as illustrated in Fig. 2.20d, e. We shortly refer to these cases as $\omega = 0$ and $\omega > 0$ string instability, respectively. Note that for the delay-free case in Fig. 2.10 only the $\omega = 0$ string instability is possible, while the $\omega > 0$ string instability is caused by the delay. In what follows, we give the detailed expressions of the string stability conditions, including the $\omega = 0$ and $\omega > 0$ string stability limits corresponding to Fig. 2.20b, d.

To simplify the subsequent formulae, we re-state the string stability condition in the following equivalent forms:

$$|T(\mathrm{j}\omega)| < 1 \iff |T(\mathrm{j}\omega)|^2 < 1 \iff P(\omega) > 0\,, \quad \forall\omega > 0\,, \tag{2.141}$$

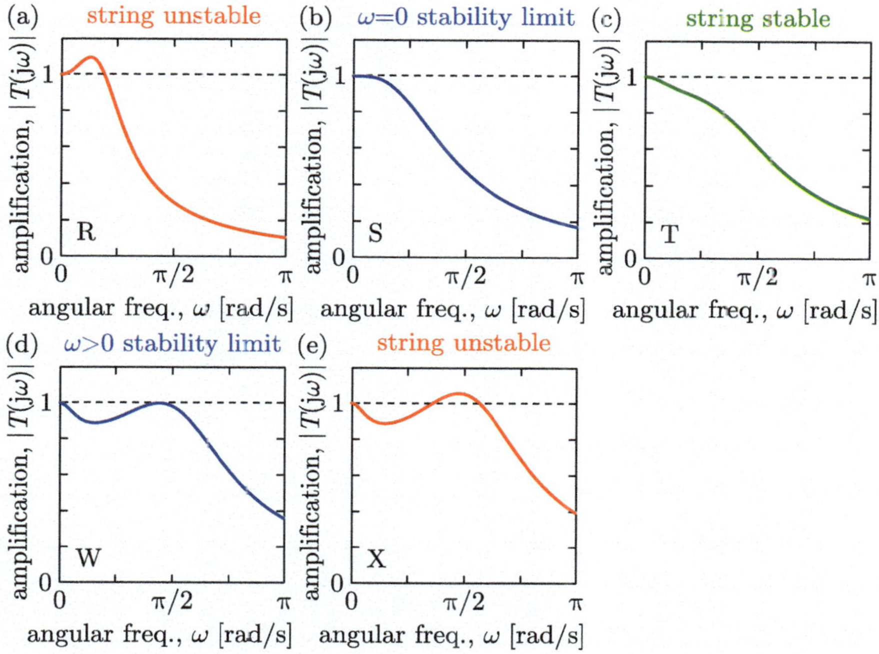

Fig. 2.20 Frequency response plots corresponding to the transfer function (2.136) with $\kappa = 0.6\,[1/\mathrm{s}]$ and $\tau = 0.6\,[\mathrm{s}]$. Panels **a–e** are associated with points R(0.25,0.4), S(0.4,0.4), T(0.5,0.4), W(0.751,0.4) and X(0.8,0.4) in Fig. 2.17, respectively

where a properly defined P can simplify the analysis significantly. As we will see throughout this book, the following construction will lead to simple results:

$$P(\omega) = \frac{1}{\omega^2}\left(\mathrm{Den}\left(|T(\mathrm{j}\omega)|^2\right) - \mathrm{Num}\left(|T(\mathrm{j}\omega)|^2\right)\right), \tag{2.142}$$

where Den and Num stand for denominator and numerator, respectively. The corresponding expression of P reads

$$P(\omega) = \omega^2 + \alpha^2 + 2\alpha\beta - 2(\alpha+\beta)\omega\sin(\omega\tau) - 2\alpha\kappa\cos(\omega\tau)\,. \tag{2.143}$$

Note that this gives back the left-hand side of (2.84) for $\tau = 0$, i.e., the construction of P is analogous to the derivation of (2.84) from (2.81) in the delay-free case. Figure 2.21 illustrates P as a function of ω where the panels are counterparts of those in Fig. 2.20.

Now we can state the $\omega = 0$ string stability conditions. Figure 2.20 shows that the graph of $|T(\mathrm{j}\omega)|$ starts from one with zero slope, i.e., properties (2.88) and (2.89) hold, even in the presence of delay. Thus, the $\omega = 0$ string stability condition becomes (2.90), which can be stated equivalently as

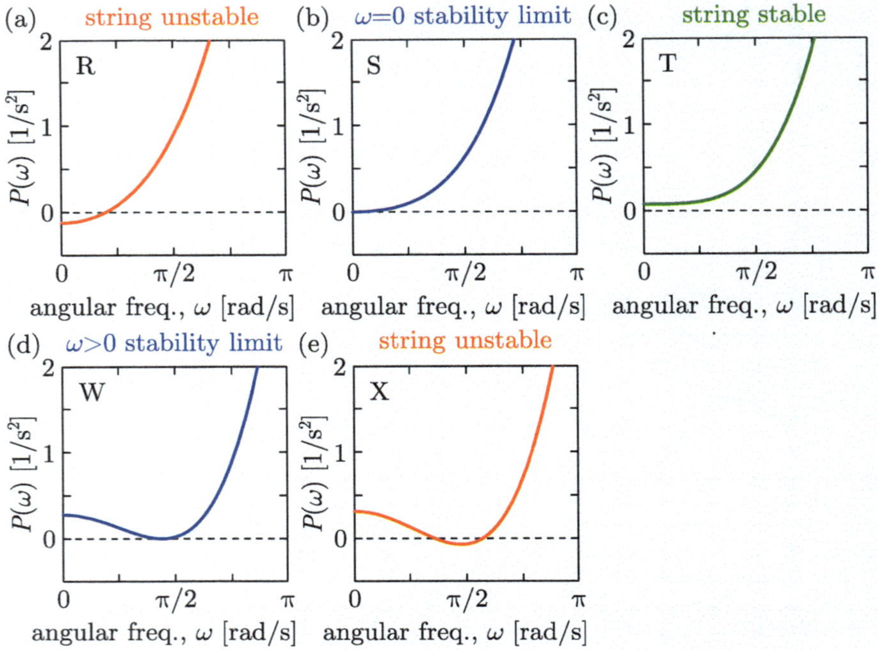

Fig. 2.21 Frequency response plots corresponding to $P(\omega)$ in (2.143) with $\kappa = 0.6\,[1/\mathrm{s}]$ and $\tau = 0.6\,[\mathrm{s}]$. Panels **a–e** are associated with points R(0.25,0.4), S(0.4,0.4), T(0.5,0.4), W(0.751,0.4) and X(0.8,0.4) in Fig. 2.17, respectively

$$P(0) > 0\,. \tag{2.144}$$

Substituting the expression (2.143) of P leads to the stability condition

$$\begin{aligned} &\alpha > 0\,,\\ &\alpha > 2(\kappa - \beta)\,, \end{aligned} \tag{2.145}$$

which is the same as in the delay-free case, cf. (2.86). These boundaries are depicted in Fig. 2.17 by blue straight lines; they are the exact same curves as the ones in Fig. 2.7 obtained for the delay-free case. We remark that (2.144) also yields (2.87), but this is still omitted as it does not satisfy plant stability.

Although the $\omega = 0$ string stability boundaries are the same as in the delay-free case, the time delay may give rise to $\omega > 0$ string instability. To analyze this we will look for the stability boundaries where the following conditions hold:

$$\begin{aligned} &\big|T(\mathrm{j}\hat{\omega})\big| = 1\,,\\ &\frac{\mathrm{d}\,\big|T(\mathrm{j}\hat{\omega})\big|}{\mathrm{d}\omega} = 0\,, \end{aligned} \tag{2.146}$$

for some $\omega = \hat{\omega} > 0$, which is equivalent to

$$\begin{aligned} P(\hat{\omega}) &= 0\,, \\ \frac{\mathrm{d}P}{\mathrm{d}\omega}(\hat{\omega}) &= 0\,, \end{aligned} \tag{2.147}$$

cf. Figs. 2.20d and 2.21d. Substituting (2.143) into (2.147) and expressing α and β give the string stability boundaries in the form

$$\begin{aligned} \alpha &= a \pm \sqrt{a^2 + b}\,, & a &= \frac{\hat{\omega}(\kappa\tau - 1) + \kappa\sin(\hat{\omega}\tau)\cos(\hat{\omega}\tau)}{(2\kappa\tau - 1)\sin(\hat{\omega}\tau) - \hat{\omega}\tau\cos(\hat{\omega}\tau)}\,, \\ & & b &= \frac{\hat{\omega}^2\left(\sin(\hat{\omega}\tau) - \hat{\omega}\tau\cos(\hat{\omega}\tau)\right)}{(2\kappa\tau - 1)\sin(\hat{\omega}\tau) - \hat{\omega}\tau\cos(\hat{\omega}\tau)}\,, \\ \beta &= \frac{\hat{\omega} + \alpha\kappa\tau\sin(\hat{\omega}\tau)}{\sin(\hat{\omega}\tau) + \hat{\omega}\tau\cos(\hat{\omega}\tau)} - \alpha\,, \end{aligned} \tag{2.148}$$

which are curves parameterized by $\hat{\omega} > 0$.

The string stability boundaries are depicted in Fig. 2.17 as blue curves while the string stable region is shaded dark gray. The string stable domain was identified by evaluating the frequency response for various points in the parameters space; see Figs. 2.20 and 2.21 for the frequency response of points R, S, T, W and X in Fig. 2.17. As opposed to the delay-free case in Fig. 2.7, now the string stable region is bounded due to the $\omega > 0$ string stability boundary induced by the delay.

Similarly to plant stability, the time delay negatively affects string stability and an increase in the delay shrinks the string stable domain as illustrated in Fig. 2.22. In fact, if a critical value τ_{cr} is exceeded by the time delay τ the string stable domain disappears. At the critical delay $\tau = \tau_{\mathrm{cr}}$ the string stable region becomes a single point, where points A, B, and C marked in Fig. 2.17 coincide.

The critical delay can be found by locating points A, B, and C. Point A is the intersection of the $\omega = 0$ string stability boundaries given by (2.145) while points B and C are the starting points of the $\omega > 0$ string stability boundaries. Thus, points

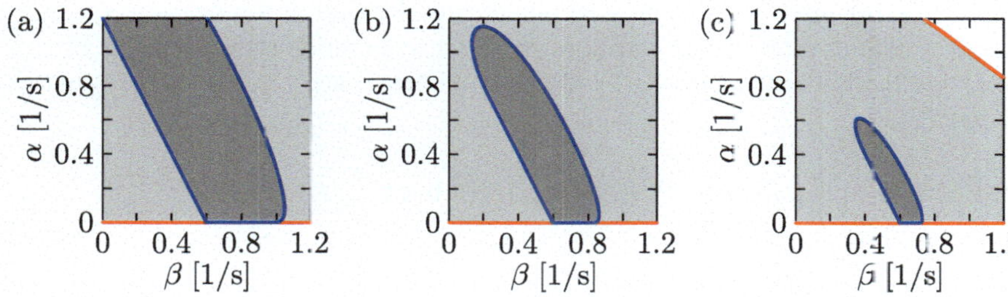

Fig. 2.22 String stability charts of system (2.131), (2.135) for $\kappa = 0.6$ [1/s]. Panels **a**–**c** correspond to time delays $\tau = 0.5$ [s], $\tau = 0.6$ [s] and $\tau = 0.7$ [s], respectively. The dark gray shaded string stable region shrinks significantly as the delay is increased

B and C can be found by taking the limit $\hat{\omega} \to 0$ in (2.148). Alternatively, one may notice that the derivative of P given in (2.143) can be written as $\frac{\mathrm{d}P}{\mathrm{d}\omega}(\omega) = 2\omega Q(\omega)$ where

$$Q(\omega) = 1 - (\alpha + \beta)\tau \frac{\sin(\omega\tau)}{\omega\tau} - (\alpha + \beta)\tau \cos(\omega\tau) + \alpha\kappa\tau^2 \frac{\sin(\omega\tau)}{\omega\tau} . \tag{2.149}$$

This form makes it convenient to take the limit $\hat{\omega} \to 0$ in (2.147) after dividing the second equation with $\hat{\omega}$. Solving for α and β leads to the points

$$\mathrm{A} : (\kappa, 0) , \quad \mathrm{B} : \left(\frac{1}{2\tau}, 0\right) , \quad \mathrm{C} : \left(\frac{2\kappa^2\tau^2 - 4\kappa\tau + 1}{2\tau(\kappa\tau - 1)}, \frac{2\kappa\tau - 1}{\tau(\kappa\tau - 1)}\right) , \tag{2.150}$$

which coincide at the critical delay

$$\tau_{\mathrm{cr}} = \frac{1}{2\kappa} . \tag{2.151}$$

For $\tau > \tau_{\mathrm{cr}}$ there exist no pair of α and β that could guarantee string stability. In other words, if the reaction time of the drivers is above half of the time headway $t_{\mathrm{h}} = 1/\kappa$, string stability cannot be guaranteed. For $\kappa = 0.6\,[1/\mathrm{s}]$ (the value used to generate the stability charts) the critical delay is $\tau_{\mathrm{cr}} = 0.83\,[\mathrm{s}]$, while the reaction time of human drivers is typically in the range $\tau \in [0.5, 1.5]\,[\mathrm{s}]$. This shows that driver reaction time is a critical factor that affects the stability of traffic flows consisting of human-driven vehicles, and human drivers may often behave in a string unstable manner that can lead to the formation of traffic jams on highways.

2.3.2 *Vehicles on a Ring*

Now we analyze how the dynamics of the ring configuration illustrated in Fig. 2.13 are affected by the time delay. We consider a ring of net length L with N vehicles whose driving behavior is governed by (2.129). Even when the delay τ is incorporated, the periodic boundary condition (2.91) holds and the sum of the headways is constant according to (2.92). The equilibrium of the ring configuration satisfies (2.15), (2.93) and the linear dynamics around this equilibrium are given by (2.131) which is equivalent to

$$\dot{\mathbf{x}}_i(t) = \mathbf{a}_0\, \mathbf{x}_i(t) + \mathbf{a}_\tau\, \mathbf{x}_i(t - \tau) + \mathbf{b}_0 \mathbf{c}\, \mathbf{x}_{i+1}(t) + \mathbf{b}_\tau \mathbf{c}\, \mathbf{x}_{i+1}(t - \tau) , \quad i = 1, \ldots, N , \tag{2.152}$$

since $u_i(t) = y_{i+1}(t) = \mathbf{c}\, \mathbf{x}_{i+1}(t)$; cf. (2.94).

The equations of motion for the N vehicles can be combined into

$$\begin{bmatrix} \dot{\mathbf{x}}_1(t) \\ \dot{\mathbf{x}}_2(t) \\ \vdots \\ \dot{\mathbf{x}}_N(t) \end{bmatrix} = \begin{bmatrix} \mathbf{a}_0 & \mathbf{b}_0\mathbf{c} & & \\ & \ddots & \ddots & \\ & & \mathbf{a}_0 & \mathbf{b}_0\mathbf{c} \\ \mathbf{b}_0\mathbf{c} & & & \mathbf{a}_0 \end{bmatrix} \begin{bmatrix} \mathbf{x}_1(t) \\ \mathbf{x}_2(t) \\ \vdots \\ \mathbf{x}_N(t) \end{bmatrix} + \begin{bmatrix} \mathbf{a}_\tau & \mathbf{b}_\tau\mathbf{c} & & \\ & \ddots & \ddots & \\ & & \mathbf{a}_\tau & \mathbf{b}_\tau\mathbf{c} \\ \mathbf{b}_\tau\mathbf{c} & & & \mathbf{a}_\tau \end{bmatrix} \begin{bmatrix} \mathbf{x}_1(t-\tau) \\ \mathbf{x}_2(t-\tau) \\ \vdots \\ \mathbf{x}_N(t-\tau) \end{bmatrix}, \tag{2.153}$$

or shortly

$$\dot{\mathbf{X}}(t) = \mathbf{A}_0\mathbf{X}(t) + \mathbf{A}_\tau\mathbf{X}(t-\tau)\,, \tag{2.154}$$

with

$$\begin{aligned} \mathbf{A}_0 &= \mathbf{I}_N \otimes \mathbf{a}_0 + \mathbf{R}_N \otimes \mathbf{b}_0\mathbf{c}\,, \\ \mathbf{A}_\tau &= \mathbf{I}_N \otimes \mathbf{a}_\tau + \mathbf{R}_N \otimes \mathbf{b}_\tau\mathbf{c}\,, \end{aligned} \tag{2.155}$$

cf. (2.98), (2.99) and (2.102). The corresponding characteristic equation, whose roots determine stability, reads

$$\det\left(\lambda\mathbf{I} - \mathbf{A}_0 - \mathbf{A}_\tau e^{-\lambda\tau}\right) = 0\,. \tag{2.156}$$

This is the delayed counterpart of (2.100) where $\mathbf{A} = \mathbf{A}_0 + \mathbf{A}_\tau$. Similarly, the stability condition of the ring configuration is given by the delayed counterpart of Theorem 2.5, which can be formulated based on the roots λ_l of (2.156) as follows.

Theorem 2.7 *The uniform flow solution of the ring configuration is asymptotically stable if and only if the characteristic roots λ_l have negative real parts, that is,* $\mathrm{Re}(\lambda_l) < 0$, $l = 1, \ldots, \infty$.

Notice that there are infinitely many eigenvalues due to the delay, which makes stability analysis more challenging.

We again exploit that the coefficient matrices $\mathbf{A}_0$ and $\mathbf{A}_\tau$ have special structures determined by the circulant matrix $\mathbf{R}_N$ defined in (2.101). With the help of the eigenvectors of $\mathbf{R}_N$ in (2.103) we apply the linear change of variables (2.104) to the system (2.154) to obtain

$$\begin{aligned} \dot{\mathbf{Z}}(t) = &\begin{bmatrix} \mathbf{a}_0 + \mathbf{b}_0\mathbf{c}\cdot 1 & & & \\ & \mathbf{a}_0 + \mathbf{b}_0\mathbf{c}\, e^{j\frac{2\pi}{N}} & & \\ & & \ddots & \\ & & & \mathbf{a}_0 + \mathbf{b}_0\mathbf{c}\, e^{j\frac{2(N-1)\pi}{N}} \end{bmatrix} \mathbf{Z}(t) \\ + &\begin{bmatrix} \mathbf{a}_\tau + \mathbf{b}_\tau\mathbf{c}\cdot 1 & & & \\ & \mathbf{a}_\tau + \mathbf{b}_\tau\mathbf{c}\, e^{j\frac{2\pi}{N}} & & \\ & & \ddots & \\ & & & \mathbf{a}_\tau + \mathbf{b}_\tau\mathbf{c}\, e^{j\frac{2(N-1)\pi}{N}} \end{bmatrix} \mathbf{Z}(t-\tau)\,. \end{aligned} \tag{2.157}$$

The characteristic equation thereof becomes

$$\prod_{k=0}^{N-1} \det\Big(\lambda \mathbf{I} - \big(\mathbf{a}_0 + \mathbf{b}_0 \mathbf{c}\, e^{j\frac{2k\pi}{N}}\big) - \big(\mathbf{a}_\tau + \mathbf{b}_\tau \mathbf{c}\, e^{j\frac{2k\pi}{N}}\big) e^{-\lambda\tau}\Big) = 0\,, \tag{2.158}$$

which decouples into N equations that are easier to analyze:

$$\det\Big(\lambda \mathbf{I} - \big(\mathbf{a}_0 + \mathbf{b}_0 \mathbf{c}\, e^{j\frac{2k\pi}{N}}\big) - \big(\mathbf{a}_\tau + \mathbf{b}_\tau \mathbf{c}\, e^{j\frac{2k\pi}{N}}\big) e^{-\lambda\tau}\Big) = 0\,, \tag{2.159}$$

for $k = 0, 1, \ldots, N-1$; cf. (2.105)–(2.107).

Now we discuss the stability of the ring configuration with time delay using the OVM parametrization with coefficient matrices (2.135). In this case (2.159) yields

$$\lambda^2 e^{\lambda\tau} + (\alpha+\beta)\lambda + \alpha\kappa - (\beta\lambda + \alpha\kappa)\, e^{j\frac{2k\pi}{N}} = 0\,, \tag{2.160}$$

cf. (2.119). To ensure stability, all the infinitely many roots of this equation have to be in the left half of the complex plane. Due to the delay it is no longer possible to use the Routh–Hurwitz criteria, however, one can analyze the stability limits where either a real root $\lambda = 0$ or a pair of complex conjugate roots $\lambda = \pm j\omega$, $\omega > 0$ is located on the imaginary axis.

Again $k = 0$ yields a root $\lambda = 0$ irrespective of the choice of the OVM parameters (and as a matter of fact irrespective of the car-following model itself). This root is associated with the translational symmetry along the ring, and it is therefore disregarded. The remaining part of the characteristic equation $\lambda e^{\lambda\tau} + \alpha = 0$ yields the stability condition

$$0 < \alpha < \frac{\pi}{2\tau}\,. \tag{2.161}$$

In case of $k > 0$ the substitution of $\lambda = 0$ into the characteristic equation (2.160) leads to the stability boundary

$$\alpha = 0\,, \tag{2.162}$$

which is the same as (2.120) obtained for the $\tau = 0$ case. On the other hand, the substitution of $\lambda = j\omega$ with $\omega > 0$ and the decomposition into real and imaginary parts give

$$\begin{aligned}
\alpha &= \frac{\omega^2\big(\cos(\omega\tau) - \cos\big(\omega\tau - \frac{2k\pi}{N}\big)\big)}{-\omega\sin\frac{2k\pi}{N} + 2\kappa\big(1-\cos\frac{2k\pi}{N}\big)}\,,\\
\beta &= \frac{-\omega^2\cos(\omega\tau) + \kappa\omega\big(\sin(\omega\tau) - \sin\big(\omega\tau - \frac{2k\pi}{N}\big)\big)}{-\omega\sin\frac{2k\pi}{N} + 2\kappa\big(1-\cos\frac{2k\pi}{N}\big)}\,,
\end{aligned} \tag{2.163}$$

$k = 1, \ldots, N-1$. Note that these boundaries reduce to the ones given in (2.122) for the delay-free case $\tau = 0$.

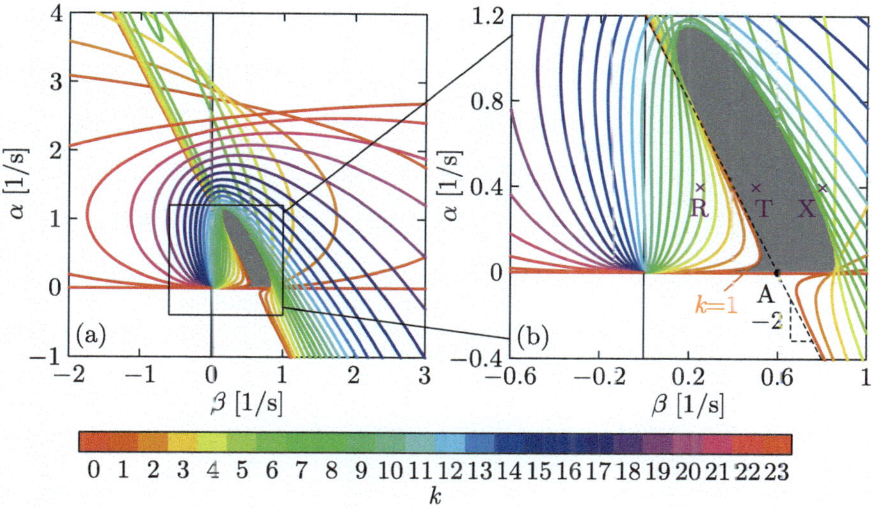

Fig. 2.23 Stability chart of system (2.153), (2.135) for $N = 24$ vehicles, $\kappa = 0.6\,[1/\mathrm{s}]$ and $\tau = 0.6\,[\mathrm{s}]$. The wave numbers k are indicated using a color bar. Point A is located at $(\kappa, 0)$ while points R(0.25,0.4), T(0.5,0.4) and X(0.8,0.4) correspond to the spectra in Fig. 2.25

Alternatively, using the trigonometric identities $\left(1 - \cos\frac{2k\pi}{N}\right) = 2\sin^2\frac{k\pi}{N}$ and $\sin\frac{2k\pi}{N} = 2\sin\frac{k\pi}{N}\cos\frac{k\pi}{N}$ one may obtain

$$
\begin{aligned}
\alpha &= \frac{\omega^2\left(\cos(\omega\tau)\tan\frac{k\pi}{N} - \sin(\omega\tau)\right)}{2\kappa\tan\frac{k\pi}{N} - \omega}\,, \\
\beta &= \frac{\omega\left(\cos(\omega\tau) + \sin(\omega\tau)\tan\frac{k\pi}{N}\right)}{2\tan\frac{k\pi}{N}} - \frac{\alpha}{2}\,,
\end{aligned}
\tag{2.164}
$$

cf. (2.124) for the delay-free case. Indeed, (2.163) and (2.164) are equivalent. Finally, the angular frequency ω can be expressed along the stability boundaries resulting in a similar (though more complicated) formula as (2.125).

The stability boundaries (2.162) and (2.163) are depicted in Fig. 2.23 where the curves are colored according to the wave number k as indicated. Note that the $\alpha = 0$ boundary is associated with all modes $k = 0, 1, \ldots, N-1$. The effect of the number N of vehicles on the stability boundaries is illustrated in Fig. 2.24; larger N gives rise to more densely packed stability boundaries. In the stability charts the dark gray shaded stable region was found by plotting the spectrum at given points of the parameter space, i.e., evaluating the rightmost roots of the characteristic equation (2.160); see Fig. 2.25 associated with points R, T and X marked in Fig. 2.23. Comparing these spectra to the non-delayed ones in Fig. 2.16 illustrates that including the delay results in (infinitely) many more eigenvalues, which spread toward higher imaginary values and thus bring in higher frequencies to the system dynamics.

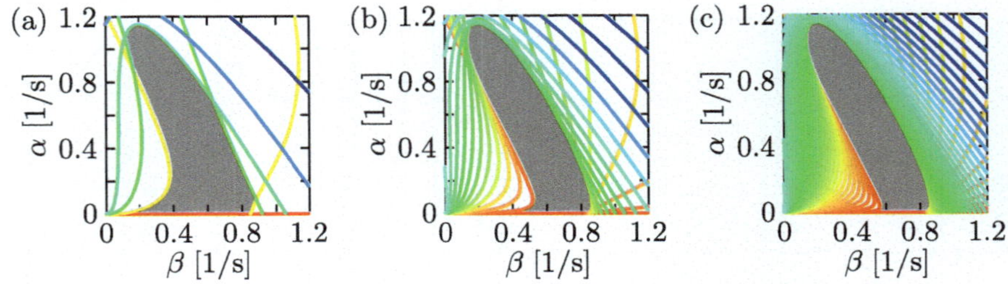

Fig. 2.24 Stability charts of system (2.153), (2.135) for $\kappa = 0.6\,[1/\mathrm{s}]$ and $\tau = 0.6\,[\mathrm{s}]$. Panels **a–c** correspond to $N = 8$, $N = 24$ and $N = 60$ vehicles on the ring, respectively. As the number of vehicles increases, more stability boundaries emerge and the dark gray shaded stable region tends to the string stable domain of the open chain configuration in Fig. 2.17

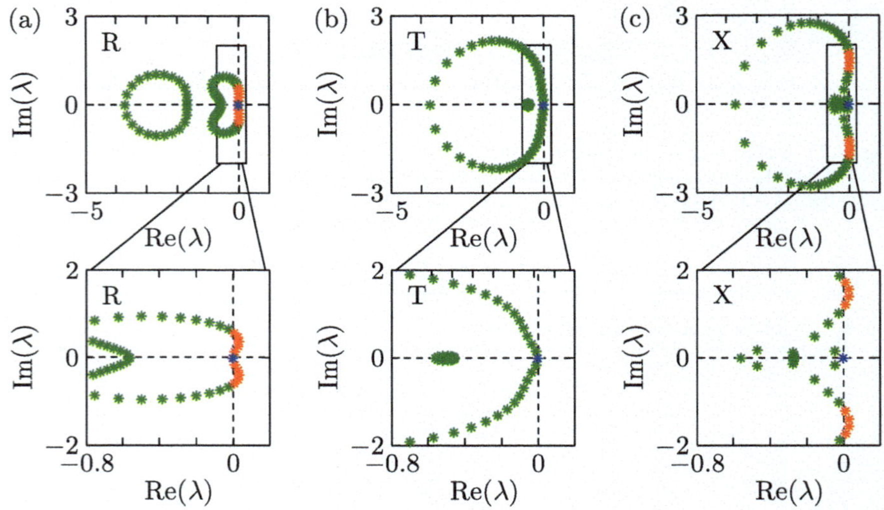

Fig. 2.25 Spectrum plots obtained from the characteristic equation (2.158) for $N = 24$, $\kappa = 0.6\,[1/\mathrm{s}]$ and $\tau = 0.6\,[\mathrm{s}]$. The different panels correspond to the points R(0.25, 0.4), T(0.5, 0.4) and X(0.8, 0.4) marked in Fig. 2.23

Comparing Fig. 2.23 to Fig. 2.17 reveals that the stable region of the ring configuration approximates the string stable domain of the open chain configuration for sufficiently large N, similarly to the delay-free cases in Figs. 2.15 and 2.7. The relationship of the open chain and ring configurations will be further discussed in the next subsection using rigorous analysis.

2.3.3 Connection Between the Open Chain and Ring Configurations

In what follows, we establish a rigorous connection between the stability of the ring configuration and the string stability of the open chain of vehicles based on [133]. This equivalency will be extended for more complex connected vehicle systems in later chapters where the calculations of stability boundaries are typically more complicated. The equivalency allows one to choose the method for analysis while aiming for lower complexity.

We begin by presenting an alternative way of calculating the $\omega > 0$ string stability boundaries of the open chain configuration, which will help relating the stability properties. We write the first line of (2.146) equivalently as

$$|T(\mathrm{j}\omega)| = 1 \iff T(\mathrm{j}\omega) = \mathrm{e}^{-\mathrm{j}K}\,, \tag{2.165}$$

where we dropped the hat from ω for simplicity. This equation can be interpreted as follows. At the boundary of string stability vehicles behave in a way that they copy the speed fluctuations of their predecessors with a phase lag $K \in [0, 2\pi)$, which corresponds to the traveling wave

$$\tilde{v}_i(t) = v_{\mathrm{amp}} \cos\left(Ki + \omega t + \theta\right). \tag{2.166}$$

cf. (2.128). Moreover, K can also be interpreted as a continuous (non-integer) wave number as $2\pi/K \in (1, \infty)$ measures the wavelength of congestion waves in number of vehicles. These congestion waves, associated with angular frequency $\omega > 0$, travel backward (upstream) along the chain of vehicles. As will be shown below, parameter K is related to parameter k defined in Sect. 2.3.2 for the ring configuration by $K = 2k\pi/N$ if $N \to \infty$.

To obtain the string stability boundaries in closed form for the OVM parametrization, we substitute (2.136) into (2.165) which gives

$$\frac{\mathrm{j}\beta\omega + \alpha\kappa}{-\omega^2\mathrm{e}^{\mathrm{j}\omega\tau} + \mathrm{j}(\alpha+\beta)\omega + \alpha\kappa} = \mathrm{e}^{-\mathrm{j}K}\,. \tag{2.167}$$

Rearrangement and decomposition into real and imaginary parts lead to

$$\begin{aligned} &\text{Re:} \; -\omega^2\cos(\omega\tau) + \beta\omega\sin K + \alpha\kappa(1-\cos K) = 0\,,\\ &\text{Im:} \; -\omega^2\sin(\omega\tau) + \alpha\omega + \beta\omega(1-\cos K) - \alpha\kappa\sin K = 0\,, \end{aligned} \tag{2.168}$$

which is a linear set of algebraic equations for α and β. The equations can be solved as

$$\alpha = \frac{\omega^2 \left(\cos(\omega\tau) - \cos(\omega\tau - K)\right)}{-\omega \sin K + 2\kappa(1 - \cos K)} ,$$
$$\beta = \frac{-\omega^2 \cos(\omega\tau) + \kappa\omega \left(\sin(\omega\tau) - \sin(\omega\tau - K)\right)}{-\omega \sin K + 2\kappa(1 - \cos K)} , \tag{2.169}$$

which specifies a surface parameterized by $\omega > 0$ and $K \in [0, 2\pi)$.

Notice that by substituting $K = 2k\pi/N$ into (2.169) and considering integer values of k, i.e., $k = 0, 1, \ldots, N-1$, one obtains the stability boundaries (2.163) of the ring configuration. That is, the stability boundaries of the ring configuration are in fact selected samples of the string stability boundaries of the open chain, where the associated wave numbers are the integer multiples of $2\pi/N$. Furthermore, note that the stability boundaries (2.169) were derived from the first line of (2.146) only. It can be shown that the second line of (2.146) gives the envelope of these boundaries that are shown by blue curves in Fig. 2.17. Finally, we remark that eliminating K from (2.168) would results in $P(\omega) = 0$ where P is defined by (2.143).

Alternatively, the link between the open chain and ring configurations can also be established by relating the characteristic equation (2.160) of the ring configuration to the transfer function (2.136) corresponding to the open chain. In fact, the characteristic equation is equivalent to

$$T(\lambda) = \frac{\beta\lambda + \alpha\kappa}{\lambda^2 e^{\lambda\tau} + (\alpha + \beta)\lambda + \alpha\kappa} = e^{-j\frac{2k\pi}{N}} . \tag{2.170}$$

This shows that at the stability boundary, where $\lambda = j\omega$, the characteristic equation of the ring exactly gives the string stability limit (2.165) of the open chain with $K = 2k\pi/N$. This highlights that the open chain and ring configurations indeed have closely related stability properties. A key difference is that the wave number is restricted to integer multiples of $2\pi/N$ in the case of the ring.

Furthermore, (2.170) also implies

$$T^N(\lambda) = 1 . \tag{2.171}$$

This is the general form of the characteristic equation for a ring of identical vehicles, and it follows from the (linearized) periodic boundary condition (2.97). For a nonhomogeneous ring it analogously reads $\prod_{i=1}^{N} T_{i,i+1}(\lambda) = 1$. Physically, this means that at the stability boundary associated with $\lambda = j\omega$ velocity fluctuations of frequency ω propagate around the ring and arrive back without a change in magnitude (if the dynamics are considered at the linear level). For $\lambda = j\omega$, (2.171) is equivalent to

$$|T(j\omega)| = 1 , \quad \angle T(j\omega) = \frac{2k\pi}{N} , \tag{2.172}$$

$k = 0, 1, \ldots, N-1$, which has the following limit as $N \to \infty$:

$$|T(j\omega)| = 1 , \quad \angle T(j\omega) \in [0, 2\pi) . \tag{2.173}$$

This supports that the stability of the ring configuration approximates the string stability of the open chain if the number of vehicles is large enough.

Finally, we remark that the string stability analysis of the open chain configuration essentially studies the growth of perturbations in space (along the vehicles, i.e., with respect to the vehicle index) such that perturbations are considered periodic in time. Conversely, the stability analysis of the ring configuration assumes periodicity in space (with respect to the vehicle index) and describes the growth of perturbations in time. The two approaches meet at the stability boundary where periodicity in both vehicle index and time is observed.

2.3.4 *First Order Lag Approximation*

As explained above, time delays increase the fidelity of car-following models and provide an explanation of why human-driven traffic is fundamentally string unstable. However, one may argue that the delay can be approximated by a first-order lag [35, 213], leading to ordinary differential equations rather than delay differential equations. Below we demonstrate the consequences of such approximation on the fidelity of the stability results as well as on the complexity of the calculation for the open chain and ring configurations. We will also demonstrate how the stability starts to change when one includes both time delay and first-order lag in the car-following models and draw some conclusions regarding how one may balance fidelity and complexity.

For simplicity, we still assume identical drivers and we re-write (2.129) in the form

$$
\begin{aligned}
\dot{h}_i(t) &= v_{i+1}(t) - v_i(t)\,, \\
\dot{v}_i(t) &= a_i(t)\,, \\
a_i(t) &= f\big(h_i(t-\tau), \dot{h}_i(t-\tau), v_i(t-\tau)\big)\,,
\end{aligned}
\tag{2.174}
$$

where we separated the terms involving the time delay τ. Lag approximations provide an approach to approximate such delay differential equation models with higher order ordinary differential equation models. The approximation is based on shifting the argument of the equations with the delay and then performing a Taylor series expansion with respect to time:

$$
\begin{aligned}
a_i(t+\tau) &= f\big(h_i(t), \dot{h}_i(t), v_i(t)\big)\,, \\
a_i(t+\tau) &\approx a_i(t) + \dot{a}_i(t)\tau + \frac{1}{2}\ddot{a}_i(t)\tau^2 + \dots\,.
\end{aligned}
\tag{2.175}
$$

While there exist no general proof of convergence when delayed arguments are expanded into Taylor series [93], in many practical applications a sufficiently accurate low-order approximation may be adequate.

Here we present a first-order lag approximation where the Taylor series is expanded up to first order in (2.175) while the higher order terms are neglected. Thus, the first-order lag approximation of (2.174) becomes

$$\begin{aligned}\dot{h}_i(t) &= v_{i+1}(t) - v_i(t)\,,\\ \dot{v}_i(t) &= a_i(t)\,,\\ \dot{a}_i(t) &= \frac{1}{\tau}\left(f\big(h_i(t), \dot{h}_i(t), v_i(t)\big) - a_i(t)\right),\end{aligned} \tag{2.176}$$

which is a third-order ordinary differential equation intended to approximate the dynamics of the second-order delay differential equation (2.174). In what follows, we analyze how this approximation affects the stability charts of the open chain and ring configurations.

First we look at the open chain of vehicles shown in Fig. 2.1. Equation (2.176) possess the uniform flow equilibrium (2.15) with $a_i(t) \equiv 0$. Linearization around the uniform flow leads to

$$\begin{aligned}\dot{\tilde{h}}_i(t) &= \tilde{v}_{i+1}(t) - \tilde{v}_i(t)\,,\\ \dot{\tilde{v}}_i(t) &= \tilde{a}_i(t)\,,\\ \dot{\tilde{a}}_i(t) &= \frac{1}{\tau}\left(f_h^* \tilde{h}_i(t) + f_{\dot{h}}^* \dot{\tilde{h}}_i(t) + f_v^* \tilde{v}_i(t) + \text{h.o.t.} - \tilde{a}_i(t)\right).\end{aligned} \tag{2.177}$$

This is of form

$$\begin{aligned}\dot{\mathbf{x}}_i &= \mathbf{a}\,\mathbf{x}_i + \mathbf{b}\,u_i + \text{h.o.t.}\,,\\ y_i &= \mathbf{c}\,\mathbf{x}_i\,,\end{aligned} \tag{2.178}$$

where the state, input and output are now defined as

$$\mathbf{x}_i = \begin{bmatrix}\tilde{h}_i\\ \tilde{v}_i\\ \tilde{a}_i\end{bmatrix}, \quad u_i = \tilde{v}_{i+1}\,, \quad y_i = \tilde{v}_i\,, \tag{2.179}$$

while the coefficient matrices read

$$\mathbf{a} = \begin{bmatrix}0 & -1 & 0\\ 0 & 0 & 1\\ \dfrac{f_h^*}{\tau} & \dfrac{f_v^* - f_{\dot{h}}^*}{\tau} & -\dfrac{1}{\tau}\end{bmatrix}, \quad \mathbf{b} = \begin{bmatrix}1\\ 0\\ \dfrac{f_{\dot{h}}^*}{\tau}\end{bmatrix}, \quad \mathbf{c} = \begin{bmatrix}0 & 1 & 0\end{bmatrix}. \tag{2.180}$$

Using the OVM parametrization (2.28), the coefficient matrices take the form

$$\mathbf{a} = \begin{bmatrix}0 & -1 & 0\\ 0 & 0 & 1\\ \dfrac{\alpha\kappa}{\tau} & -\dfrac{\alpha+\beta}{\tau} & -\dfrac{1}{\tau}\end{bmatrix}, \quad \mathbf{b} = \begin{bmatrix}1\\ 0\\ \dfrac{\beta}{\tau}\end{bmatrix}, \quad \mathbf{c} = \begin{bmatrix}0 & 1 & 0\end{bmatrix}. \tag{2.181}$$

Notice that the first-order lag does not modify the form (2.25) of the system but increases the dimension of the state by one compared to the lag-free case. As opposed, the time delay leaves the number of state variables unchanged (compared to the delay-free case), but it modifies the system equations to (2.131).

From this point on, the steps of stability analysis are the same as in Sect. 2.2.1, only the coefficient matrices are different. The transfer function of the system can be calculated from (2.68) in the form

$$T(s) = \mathbf{c}\,(s\mathbf{I} - \mathbf{a})^{-1}\,\mathbf{b} = \frac{\beta s + \alpha\kappa}{\tau s^3 + s^2 + (\alpha + \beta)s + \alpha\kappa}\,, \tag{2.182}$$

yielding the characteristic equation

$$\tau s^3 + s^2 + (\alpha + \beta)s + \alpha\kappa = 0\,. \tag{2.183}$$

By comparing these to the transfer function (2.136) and the characteristic equation (2.137) of the case with time delay, one can conclude that the first-order lag approximation is equivalent to approximating the $e^{s\tau}$ term with its first-order Taylor polynomial: $e^{s\tau} \approx 1 + s\tau$.

The plant stability boundaries associated with (2.183) are

$$\alpha = 0\,, \tag{2.184}$$

for $s = 0$ and

$$\begin{aligned} \alpha &= \frac{\Omega^2}{\kappa}\,, \\ \beta &= \Omega^2\tau - \alpha\,, \end{aligned} \tag{2.185}$$

for $s = \mathrm{j}\Omega$; cf. (2.138) and (2.139). The plant stable region

$$\begin{aligned} \alpha &> 0\,, \\ (1 - \kappa\tau)\alpha + \beta &> 0\,, \end{aligned} \tag{2.186}$$

can be verified by applying the Routh–Hurwitz criteria to (2.183), and it simplifies to (2.80) for $\tau = 0$.

The string stable domain can be found by evaluating the string stability condition (2.141) where the expression of P reads

$$P(\omega) = \tau^2\omega^4 + (1 - 2(\alpha + \beta)\tau)\omega^2 + \alpha^2 + 2\alpha\beta - 2\alpha\kappa\,. \tag{2.187}$$

If the coefficient of ω^2 is positive, $P(\omega)$ is a monotonously increasing function, thus $P(\omega) > 0$ holds for all $\omega > 0$ if $P(0) > 0$. If the coefficient of ω^2 is negative, then $P(\omega)$ has a minimum at $\omega = \hat{\omega} > 0$, thus $P(\omega) > 0$ holds for all $\omega > 0$ if $P(\hat{\omega}) > 0$ where $\frac{\mathrm{d}P}{\mathrm{d}\omega}(\hat{\omega}) = 0$. This train of thought leads to the following expressions of the string stable domain:

$$\begin{aligned} &\alpha > 0\,, \quad \alpha > 2(\kappa - \beta)\,, \quad \text{if} \;\; \alpha < \frac{1}{2\tau} - \beta\,, \\ &\alpha > \frac{\frac{1}{4\tau} + \beta(\beta\tau - 1)}{1 - 2\kappa\tau}\,, \quad \text{if} \;\; \alpha > \frac{1}{2\tau} - \beta\,, \end{aligned} \tag{2.188}$$

where we assumed $2\kappa\tau < 1$ in the second line.

The boundaries of string stability can also be found in parametric form by substituting (2.187) into (2.144) and (2.147) and solving for α and β. The $\omega = 0$ string stability conditions are again given by

$$\begin{aligned} &\alpha > 0\,, \\ &\alpha > 2(\kappa - \beta)\,, \end{aligned} \tag{2.189}$$

cf. (2.86), while the $\omega > 0$ string stability boundaries read

$$\begin{aligned} 3\alpha &= a \pm \sqrt{a^2 + b}\,, \quad a = \frac{1 + 2\tau^2\hat{\omega}^2}{2\tau} - \kappa\,, \\ &\qquad\qquad\qquad\quad\; b = -\tau^2\hat{\omega}^4\,, \\ \beta &= \frac{1 + 2\tau^2\hat{\omega}^2}{2\tau} - \alpha\,, \end{aligned} \tag{2.190}$$

which are curves parameterized by $\hat{\omega} > 0$; cf. (2.148).

The stability charts are depicted in Fig. 2.26, where the same color coding is used as before: red curves and light gray shading stand for plant stability whereas blue curves and dark gray shading correspond to string stability. Similarly to the case without delay in Fig. 2.7, the stable regions are large unbounded domains. The first-order lag, however, gives rise to $\omega > 0$ string stability boundaries which are similar to those obtained for time delay in Fig. 2.17 in the vicinity of points B and C.

Increasing the first-order lag makes the stable domains shrink as illustrated in Fig. 2.27. At a critical lag $\tau = \tau_{\rm cr}$ the string stable domain becomes a single point where the points A, B, and C in Fig. 2.26 coincide. The location of point A is the intersection of the $\omega = 0$ string stability boundaries in (2.189), whereas points B and C are the $\hat{\omega} \to 0$ starting points of the $\omega > 0$ string stability boundaries in (2.190). Their coordinates are

$$\text{A} : (\kappa, 0)\,, \quad \text{B} : \left(\frac{1}{2\tau}, 0\right), \quad \text{C} : \left(2\kappa - \frac{1}{2\tau}, -2\kappa + \frac{1}{\tau}\right), \tag{2.191}$$

and they coincide for the critical lag

$$\tau_{\rm cr} = \frac{1}{2\kappa}\,, \tag{2.192}$$

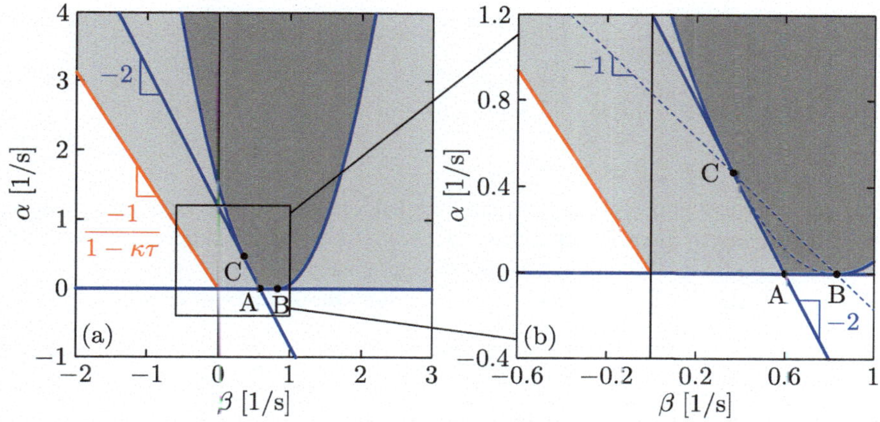

Fig. 2.26 Stability chart of system (2.178), (2.181) for $\kappa = 0.6\,[1/\mathrm{s}]$ and $\tau = 0.6\,[\mathrm{s}]$. The plant and string stable regions are shaded by light and dark gray, respectively. Point A is located at the intersection $(\kappa, 0)$ of the straight lines, while points B and C lie at the starting points of the string stability curves

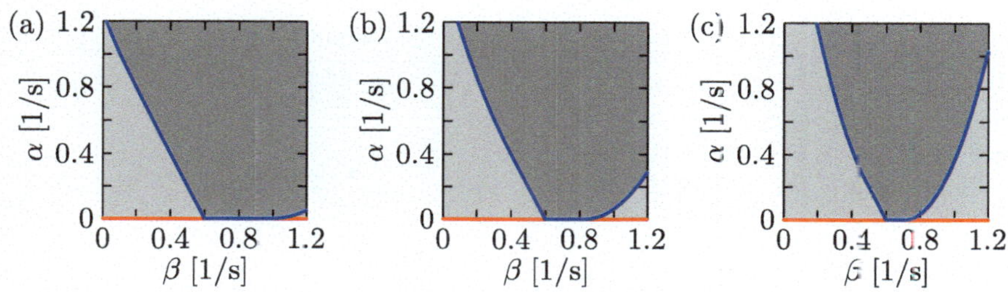

Fig. 2.27 String stability charts of system (2.178), (2.181) for $\kappa = 0.6\,[1/\mathrm{s}]$. Panels **a–c** correspond to first-order lags $\tau = 0.5\,[\mathrm{s}]$, $\tau = 0.6\,[\mathrm{s}]$ and $\tau = 0.7\,[\mathrm{s}]$, respectively. The dark gray shaded string stable region shrinks as the lag is increased

that is equal to the half of the time headway $t_\mathrm{h} = 1/\kappa$. If $\tau > \tau_\mathrm{cr}$, string stability cannot be guaranteed for any combination of α and β. Recall that this is the case for the time delay as well, and even the value of the critical delay in (2.151) matches the value of the critical lag in (2.192). That is, the analysis based on first-order lag approximation is able to capture how large delay is tolerable from point of view string stability but for $\tau < \tau_\mathrm{cr}$ it overestimates the string stable area significantly.

Now we analyze the ring configuration shown in Fig. 2.13 assuming a ring of net length L with N vehicles whose motion is governed by (2.176) involving a first-order lag. The periodic boundary condition (2.91) and the headway constraint (2.92) still apply, and the uniform flow equilibrium is given by (2.15), (2.93) with $a_i(t) \equiv 0$. Similarly to the delay-free case, the corresponding linearized system is of form (2.94) with periodic boundary condition (2.97) and characteristic equation (2.107); the only difference is the dimension and form of the coefficient matrices. For the OVM parametrization associated with matrices (2.181) the characteristic equation (2.107) takes the form

$$\det\left(\lambda \mathbf{I} - \mathbf{a} - \mathbf{bc}\,\mathrm{e}^{\mathrm{j}\frac{2k\pi}{N}}\right)\tau = \tau\lambda^3 + \lambda^2 + (\alpha+\beta)\lambda + \alpha\kappa - (\beta\lambda + \alpha\kappa)\,\mathrm{e}^{\mathrm{j}\frac{2k\pi}{N}} = 0\,, \tag{2.193}$$

for $k = 0, 1, \ldots, N-1$.

For $k = 0$ we obtain a zero eigenvalue $\lambda = 0$ independent of the model parameters while the other two eigenvalues given by $\tau\lambda^2 + \lambda + \alpha = 0$ can be kept on the left-half complex plane for $\alpha > 0$ independent of the value of τ.

For $k > 0$ the stability boundary associated with the $\lambda = 0$ still becomes

$$\alpha = 0\,, \tag{2.194}$$

while the stability boundaries corresponding to $\lambda = \mathrm{j}\omega$, $\omega > 0$ can be obtained from the real and imaginary parts of the characteristic equation:

$$\begin{aligned} \alpha &= \frac{\omega^2\left(1-\cos\frac{2k\pi}{N} - \omega\tau\sin\frac{2k\pi}{N}\right)}{-\omega\sin\frac{2k\pi}{N} + 2\kappa\left(1-\cos\frac{2k\pi}{N}\right)}\,,\\ \beta &= \frac{-\omega^2 + \kappa\omega\left(\omega\tau\left(1-\cos\frac{2k\pi}{N}\right) + \sin\frac{2k\pi}{N}\right)}{-\omega\sin\frac{2k\pi}{N} + 2\kappa\left(1-\cos\frac{2k\pi}{N}\right)}\,, \end{aligned} \tag{2.195}$$

for $k = 1, \ldots, N-1$, cf. (2.162) and (2.163).

Again, using the identities $\left(1-\cos\frac{2k\pi}{N}\right) = 2\sin^2\frac{k\pi}{N}$ and $\sin\frac{2k\pi}{N} = 2\sin\frac{k\pi}{N}\cos\frac{k\pi}{N}$ result in

$$\begin{aligned} \alpha &= \frac{\omega^2\left(\tan\frac{k\pi}{N} - \omega\tau\right)}{2\kappa\tan\frac{k\pi}{N} - \omega}\,,\\ \beta &= \frac{\omega\left(1+\omega\tau\tan\frac{k\pi}{N}\right)}{2\tan\frac{k\pi}{N}} - \frac{\alpha}{2}\,, \end{aligned} \tag{2.196}$$

cf. (2.164) for the delayed case. Indeed, (2.195) and (2.196) are equivalent.

The stability boundaries of the ring configuration are shown in Fig. 2.28. Similarly to the delay-free and delayed cases in Figs. 2.15 and 2.23, the stable region of the ring configuration approximates the string stable domain of the open chain if the number N of vehicles is large enough; cf. Figs. 2.26 and 2.28. The convergence of the stable domain with increasing N is illustrated in Fig. 2.29.

Finally, we remark that it is also possible to use both delays and lags to model human-driven vehicles in car following [65]. The lag is often used to model the response time of the vehicles (i.e., the time it takes for a vehicle to reach a desired acceleration), while the time delay captures the reaction time of human drivers. When incorporating a first-order lag ξ and a time delay τ, the governing equations with identical vehicles and drivers become

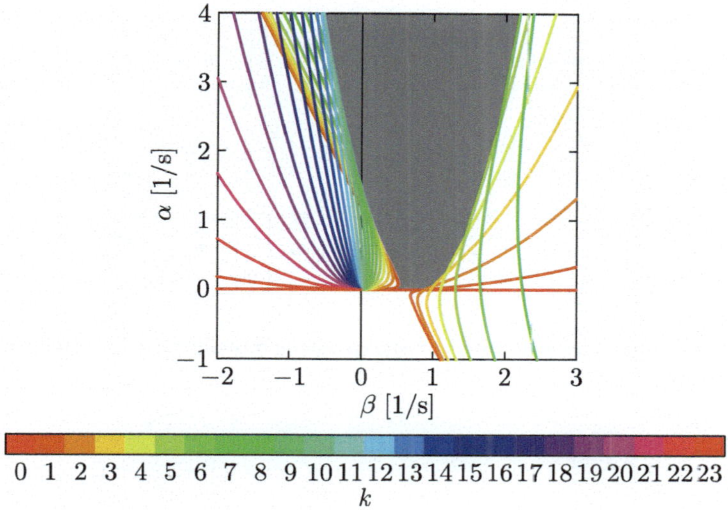

Fig. 2.28 Stability chart of system (2.98), (2.181) for $N = 24$ vehicles, $\kappa = 0.6\,[1/\mathrm{s}]$ and $\tau = 0.6\,[\mathrm{s}]$. The wave numbers k are indicated using color bar

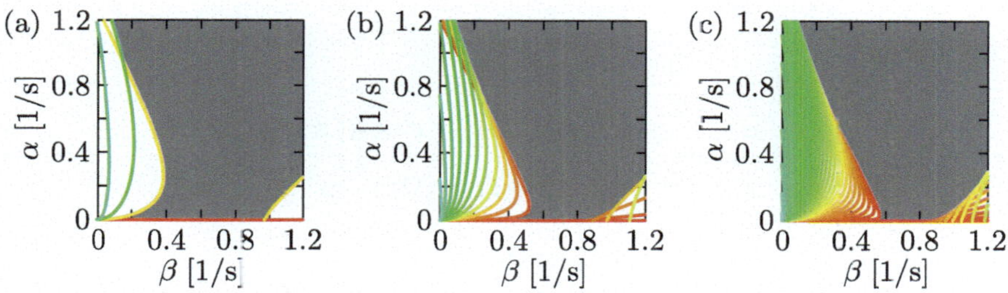

Fig. 2.29 Stability charts of system (2.98), (2.181) for $\kappa = 0.6\,[1/\mathrm{s}]$ and $\tau = 0.6\,[\mathrm{s}]$. Panels **a–c** correspond to $N = 8$, $N = 24$ and $N = 60$ vehicles in the ring configuration. As the number of vehicles increases, more stability boundaries emerge and the dark gray shaded stable region tends to the string stable domain of the open chain configuration in Fig. 2.26

$$
\begin{aligned}
\dot{h}_i(t) &= v_{i+1}(t) - v_i(t)\,, \\
\dot{v}_i(t) &= a_i(t)\,, \\
\dot{a}_i(t) &= \frac{1}{\xi}\Big(f\big(h_i(t-\tau), \dot{h}_i(t-\tau), v_i(t-\tau)\big) - a_i(t)\Big)\,.
\end{aligned}
\tag{2.197}
$$

Here we briefly illustrate the stability properties with both lag and delay for the open chain of vehicles only. The uniform flow is (2.15) with $a_i(t) \equiv 0$, while the linearized dynamics are given by (2.131) with coefficient matrices

$$\mathbf{a}_0 = \begin{bmatrix} 0 & -1 & 0 \\ 0 & 0 & 1 \\ 0 & 0 & -\dfrac{1}{\xi} \end{bmatrix}, \quad \mathbf{a}_\tau = \begin{bmatrix} 0 & 0 & 0 \\ 0 & 0 & 0 \\ \dfrac{f_h^*}{\xi} & \dfrac{f_v^* - f_h^*}{\xi} & 0 \end{bmatrix}, \quad \mathbf{b}_0 = \begin{bmatrix} 1 \\ 0 \\ 0 \end{bmatrix}, \quad \mathbf{b}_\tau = \begin{bmatrix} 0 \\ 0 \\ \dfrac{f_{\dot{h}}^*}{\xi} \end{bmatrix}, \quad \mathbf{c} = [0\ 1\ 0], \tag{2.198}$$

which become

$$\mathbf{a}_0 = \begin{bmatrix} 0 & -1 & 0 \\ 0 & 0 & 1 \\ 0 & 0 & -\dfrac{1}{\xi} \end{bmatrix}, \quad \mathbf{a}_\tau = \begin{bmatrix} 0 & 0 & 0 \\ 0 & 0 & 0 \\ \dfrac{\alpha\kappa}{\xi} & -\dfrac{\alpha+\beta}{\xi} & 0 \end{bmatrix}, \quad \mathbf{b}_0 = \begin{bmatrix} 1 \\ 0 \\ 0 \end{bmatrix}, \quad \mathbf{b}_\tau = \begin{bmatrix} 0 \\ 0 \\ \dfrac{\beta}{\xi} \end{bmatrix}, \quad \mathbf{c} = [0\ 1\ 0] \tag{2.199}$$

under the OVM parametrization. The corresponding transfer function can be obtained using (2.133) as

$$T(s) = \mathbf{c}\big(s\mathbf{I} - \mathbf{a}_0 - \mathbf{a}_\tau\, \mathrm{e}^{-s\tau}\big)^{-1}\big(\mathbf{b}_0 + \mathbf{b}_\tau\, \mathrm{e}^{-s\tau}\big) = \frac{\beta s + \alpha\kappa}{s^2 \mathrm{e}^{s\tau}(1+s\xi) + (\alpha+\beta)s + \alpha\kappa}\,. \tag{2.200}$$

The plant stability boundaries are

$$\alpha = 0\,, \tag{2.201}$$

for $s = 0$ and

$$\begin{aligned} \alpha &= \frac{1}{\kappa}\left(-\Omega^3\xi\sin(\Omega\tau) + \Omega^2\cos(\Omega\tau)\right), \\ \beta &= \Omega^2\xi\cos(\Omega\tau) + \Omega\sin(\Omega\tau) - \alpha\,, \end{aligned} \tag{2.202}$$

for $s = \mathrm{j}\Omega$. The string stability boundaries are

$$\begin{aligned} \alpha &= 0\,, \\ \alpha &= 2(\kappa - \beta)\,, \end{aligned} \tag{2.203}$$

for $\omega = 0$ and

$$\begin{aligned} \alpha = a \pm \sqrt{a^2 + b}\,, \quad a &= \frac{(d+1-\kappa\xi)\omega\sin(\omega\tau) + ((d+1)\omega^2\xi + \kappa)\cos(\omega\tau) - c}{2d+1}, \\ b &= \frac{2c\omega(\sin(\omega\tau) + \omega\xi\cos(\omega\tau)) - \omega^2 - \omega^4\xi^2}{2d+1}, \\ \beta = c + d\alpha\,, \quad c &= \frac{\omega(1+2\omega^2\xi^2)}{(1-\omega^2\xi\tau)\sin(\omega\tau) + \omega(\tau+2\xi)\cos(\omega\tau)}, \\ d &= \frac{\kappa\left((\tau+\xi)\sin(\omega\tau) + \omega\xi\tau\cos(\omega\tau)\right)}{(1-\omega^2\xi\tau)\sin(\omega\tau) + \omega(\tau+2\xi)\cos(\omega\tau)} - 1\,, \end{aligned} \tag{2.204}$$

for $\omega > 0$ that were obtained from

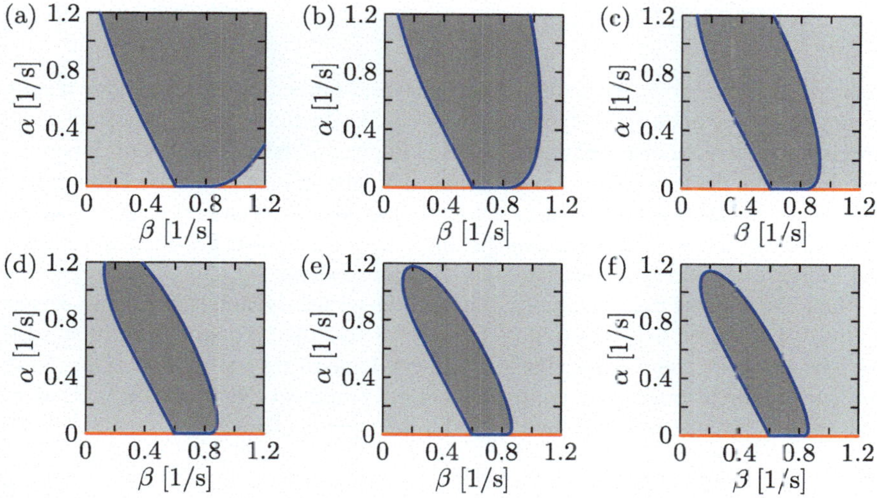

Fig. 2.30 Stability charts of system (2.131), (2.199) for $\kappa = 0.6\,[1/\text{s}]$. Panels **a–f** correspond to different combinations of time delay and first-order lag such that $\tau + \xi = 0.6\,[\text{s}]$: $\tau = 0, 0.12, 0.24, 0.36, 0.48, 0.6\,[\text{s}]$ and $\xi = 0.6, 0.48, 0.36, 0.24, 0.12, 0\,[\text{s}]$, respectively. Note that the lag-only case in panel **a** coincides with Fig. 2.26, while the delay-only case in panel **f** coincides with Fig. 2.17

$$
\begin{aligned}
P(\omega) = \omega^2 + \alpha^2 + 2\alpha\beta - 2\alpha\kappa\,(\cos(\omega\tau) - \omega\xi\,\sin(\omega\tau)) \\
- 2(\alpha + \beta)\omega\,(\sin(\omega\tau) + \omega\xi\,\cos(\omega\tau)) + \omega^4\xi^2\,.
\end{aligned}
\tag{2.205}
$$

The stability charts are illustrated in Fig. 2.30 for different combinations of lag and delay, which illustrates the transition between the lag-only case in Fig. 2.26 and the delay only case in Fig. 2.17. Although the sum of lag and delay is the same in each panel in Fig. 2.30, the size of the stable region is significantly smaller for larger delay. Still the lag and delay together determine whether a string stable domain exists: the string stable region disappears if their sum exceeds the critical value

$$
(\tau + \xi)_{\text{cr}} = \frac{1}{2\kappa}\,. \tag{2.206}
$$

Finally, we remark that while adding time lag ξ to the system significantly increases the complexity of the formulae, cf. (2.148) and (2.204), it does not necessarily increase the fidelity of the model. In particular, comparing Fig. 2.22 with 2.30c, d one may notice similar trends. This suggests that one may capture the string stability by simply varying the delay without including lag. On the other hand, considering lag only does not allow one to capture the qualitative shape of the string stable domain. Thus in the rest of this book we will omit the lag and lump all the dynamics into the delay.

2.4 Nonlinear Dynamics

In the previous sections, we analyzed the behavior of human-driven traffic by means of linearizing the underlying dynamical models. While this revealed important characteristics of driver behavior, traffic systems in general exhibit nonlinear dynamics that determine the overall motion of vehicles. Nonlinearities may arise from multiple sources, including the nonlinear driver characteristics captured by car-following models like the OVM or the IDM, and the limited acceleration capabilities of vehicles that imply saturation-type nonlinearities discussed below.

Next, we analyze the impact of nonlinearities on the dynamics of human-driven traffic. We focus on explaining the formation of so-called *phantom jams* that arise due to a certain mechanism specific to nonlinear dynamical systems: *bistability*. Note that this section is more involved in the theory of nonlinear dynamics, hence readers who are less interested in this field may skip to the next chapter.

2.4.1 *Bistability Phenomenon and Phantom Jams*

Nonlinearities in dynamical systems may cause periodic oscillations, so-called *limit cycles*, that do not exist in linear systems. In the context of vehicular traffic, these periodic motions manifest themselves as repeated, large-amplitude accelerations and decelerations of vehicles, ultimately leading to a traffic congestion also called as *phantom jam*. Phantom jams have been studied extensively by [7, 110, 157, 159, 160]. By following these works, we present results about the analysis of nonlinear traffic dynamics and explain the formation of phantom jams as published in [136].

To motivate the upcoming analysis, we first show simulation results about the nonlinear behavior of vehicular traffic. Consider human-driven traffic in an open chain, as shown in Fig. 2.1. Let the human driver behavior be described by

$$\begin{aligned} \dot{h}_i(t) &= v_{i+1}(t) - v_i(t)\,, \\ \dot{v}_i(t) &= \mathrm{sat}\left(f\big(h_i(t-\tau), \dot{h}_i(t-\tau), v_i(t-\tau)\big)\right). \end{aligned} \tag{2.207}$$

This model contains the delayed dynamics (2.129), while it also takes into account that the accelerations of vehicles saturate outside the physical braking limit $-a_{\min}$ and acceleration limit $a_{\max}$:

$$\mathrm{sat}(a) = \begin{cases} -a_{\min} & \text{if } \ a \le -a_{\min}\,, \\ a & \text{if } \ -a_{\min} < a < a_{\max}\,, \\ a_{\max} & \text{if } \ a \ge a_{\max}\,. \end{cases} \tag{2.208}$$

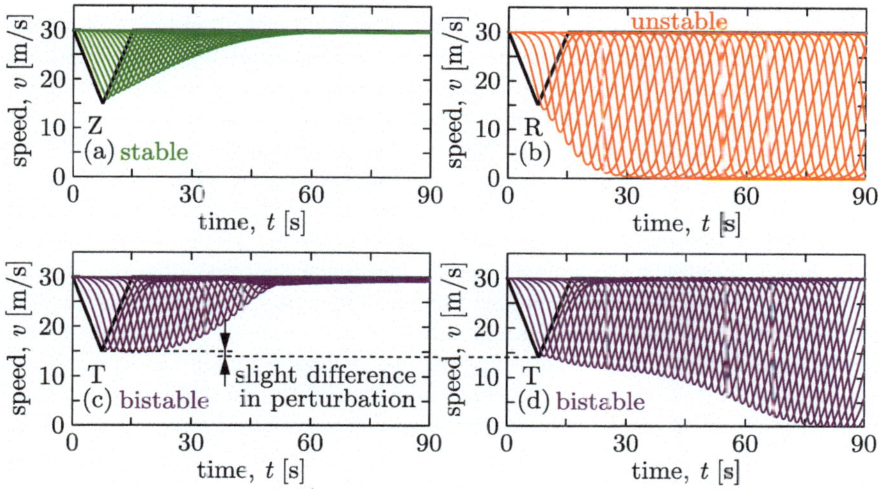

Fig. 2.31 Numerical simulation of the OVM (2.207), (2.4) with piecewise cubic range policy (2.5), (2.9) for the case of 48 follower vehicles. The speed of the lead vehicle $v_{48}(t)$ is subject to a 15 [m/s] perturbation in panels **a**–**c** and a 16 [m/s] perturbation in panel **d**. The initial conditions over $t \le 0$ are $v_i(t) = v_{\max}$ and $h_i(t) = h_{\text{go}}, i = 0, 1, \ldots, 47$, while the parameters are $a_{\min} = 7\,[\text{m/s}^2]$, $a_{\max} = 3\,[\text{m/s}^2]$, $h_{\text{st}} = 5\,[\text{m}]$, $h_{\text{go}} = 55\,[\text{m}]$, $v_{\max} = 30\,[\text{m/s}]$ and $\tau = 0.6\,[\text{s}]$. Panel **a** corresponds to the linearly string stable point Z(0.8,0.1), panel **b** is associated with the linearly string unstable point R(0.25,0.4), while panels **c**, **d** correspond to the bistable point T(0.5,0.4) in Fig. 2.33

To describe human driver behavior, we consider the OVM (2.4) with the nonlinear, piecewise cubic range policy in (2.5), (2.9). The saturation function sat and the range policy V give rise to rich nonlinear dynamic behavior as illustrated below.

Figure 2.31 shows simulation results for a scenario where 48 vehicles follow a lead vehicle. The simulations imitate a braking event, in which the leading vehicle slows down, then accelerates to recover its original speed and continues to cruise at that speed, while the subsequent vehicles respond to this perturbation (like in Fig. 2.12). Three different sets of human driver parameters are considered (corresponding to points Z, R, and T in Fig. 2.33 described later). Panel (a) shows a string stable scenario, in which each driver reduces its speed less than the vehicle ahead of it, and hence the perturbation ultimately dies out and smooth traffic flow is attained. Panel (b) demonstrates a string unstable scenario, where drivers overreact to the perturbation and reduce their speeds more than the vehicle ahead of them, resulting in large-amplitude speed oscillations, and eventually, a traffic congestion with stop-and-go motion. In Sect. 2.3.1, we analyzed string stability at the linear level and we predicted these phenomena.

For nonlinear dynamics, however, the onset of a traffic congestion may be hard to predict. While the two cases in Fig. 2.31a, b—smooth traffic and congestion—correspond to two different sets of human driver parameters, for nonlinear dynamics there exist scenarios in which even the *same drivers* can produce these two qualitatively different behaviors. This is illustrated in panels (c) and (d) of Fig. 2.31, where the same human driver parameters are simulated but the perturbation is different: the lead vehicle reduces its speed by 15 and 16 [m/s], respectively. The simulation highlights that a large enough perturbation triggers a traffic congestion, while the traffic flow smoothes out for a small perturbation. This phenomenon is called *bistability*, and the corresponding hard-to-predict congestion is a *phantom jam*.

Clearly, bistability is undesired as it can lead to unexpected phantom jams due to events like the braking of an individual driver. Importantly, bistability phenomena are unique to nonlinear dynamical systems, and linear analysis—that is restricted to small perturbations—cannot predict this phenomenon. Thus, nonlinearities like the nonlinear range policies or the saturation of accelerations have a crucial role in shaping the overall traffic behavior. Next, we analyze the nonlinear dynamics of human-driven traffic, explain the mechanism of bistability, and study the occurrence of large-amplitude periodic motions corresponding to phantom jams.

2.4.2 Vehicles on a Ring

To facilitate the analysis, we first consider the ring road setting in Fig. 2.13 with N vehicles. Consequently, the dynamics (2.207) are coupled with the periodic boundary condition (2.91) and the constraint (2.92) that headways sum up to the net ring length.

We study the equilibrium and periodic solutions (limit cycles) of this model. Recall that the equilibrium is given by the uniform flow (2.15), which corresponds to the ideal, smooth flow of traffic. If the equilibrium is stable, small perturbations like a braking of an individual driver are attenuated as in Fig. 2.31a, and smooth traffic is attained. As opposed, unstable equilibrium results in amplifying perturbations like in Fig. 2.31b. The linear analysis presented previously in Sect. 2.3.2 describes the local stability of the equilibrium.

Periodic solutions, on the other hand, may represent congested traffic with large-amplitude repeated (periodic) fluctuations in the speeds and headways of vehicles. Such periodic motions are called *limit cycles* in the nonlinear dynamical system. If a limit cycle is stable, then the system tends to this motion over time, and large-amplitude velocity fluctuations are showcased in traffic. If the limit cycle is unstable, then the traffic behavior diverges from the corresponding periodic motion (and instead, typically behaviors related to a stable equilibrium or another stable limit cycle are observed).

Moreover, a stable equilibrium and a stable limit cycle may coexist in certain scenarios (for certain driver parameters). This is called *bistability*. In such a case, the system converges to the equilibrium and showcases smooth traffic flow for small perturbations, while it approaches the limit cycle and exhibits phantom jam for large perturbations; cf. Fig. 2.31c, d. That is, both the equilibrium and the limit cycle are locally stable, with a domain of attraction determined by the nonlinear dynamics.

Now we investigate the nonlinear dynamics and analyze bistability. This study is conducted numerically, using a Matlab tool called DDE-BIFTOOL [49, 194].

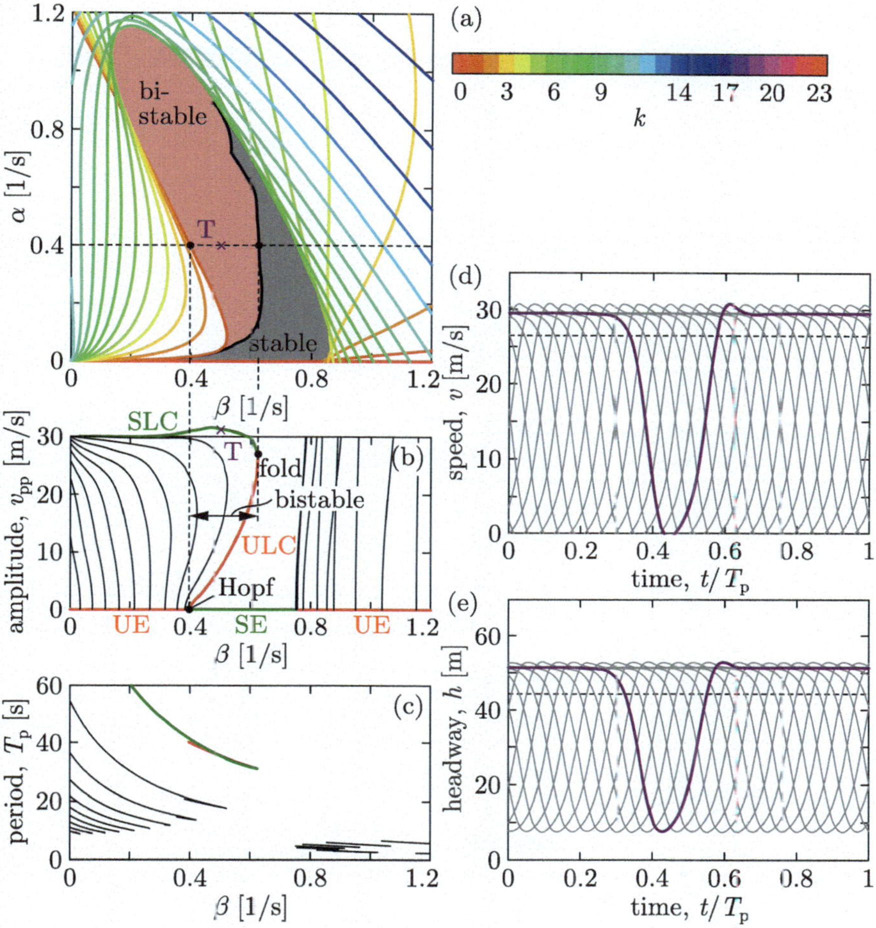

Fig. 2.32 Nonlinear dynamics of the OVM (2.129), (2.4) using the ring configuration given by (2.91), (2.92) with piecewise cubic range policy (2.5), (2.9) for $N = 24$ vehicles. The parameters are $h_{st} = 5\,[m]$, $h_{go} = 55\,[m]$, $v_{max} = 30\,[m/s]$ and $\tau = 0.6\,[s]$. The net ring length L is chosen such that $\kappa = V'(h^*) = 0.6\,[1/s]$, i.e., $h^* = 44.43\,[m]$ and $L = Nh^*$. Panels **b**, **c** are shown for $\alpha = 0.4\,[1/s]$, while panels **d**, **e** correspond to point T(0.5,0.4) marked in panels (**a**), (**b**)

This tool is able to analyze nonlinear dynamical systems given by delay differential equations, including the calculation of equilibria and their stability, as well as periodic solutions (limit cycles) and their stability. To this end, we implemented the model (2.207) with the OVM (2.4), range policy (2.5), (2.9), and ring conditions (2.91), (2.92). First, we neglect the saturation of accelerations (i.e., we simplify (2.207) to the expression (2.129) by taking $a_{min}, a_{max} \to \infty$), and then we highlight the effects of saturation.

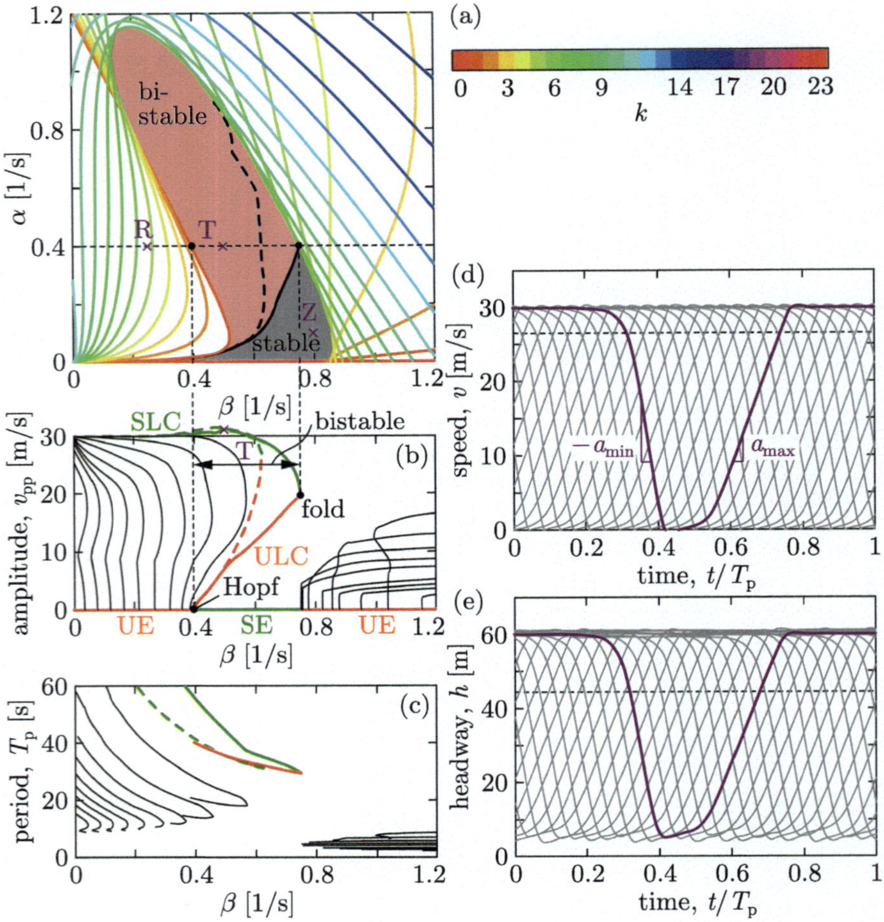

Fig. 2.33 Nonlinear dynamics of the OVM (2.207), (2.4) using the ring configuration given by (2.91), (2.92) with piecewise cubic range policy (2.5), (2.9) for $N = 24$ vehicles. Solid lines show the case where the accelerations of vehicles are saturated at $a_{min} = 7\,[\mathrm{m/s^2}]$ and $a_{max} = 3\,[\mathrm{m/s^2}]$, while dashed lines indicate the results of Fig. 2.32 where the saturation was neglected. The rest of the parameters are the same as in Fig. 2.32

Figure 2.32 shows the results without saturation, where the stability chart in panel (a) is repeated from Fig. 2.24b. In the linearly stable region (union of red and gray domains), the linearized dynamics suggest that the equilibrium traffic flow is stable, hence traffic congestion does not occur. However, analyzing the nonlinear dynamics reveals that large-amplitude oscillatory motions (limit cycles) exist for certain parameters. For example, point T in panel (a) yields such parameters, and panels (d) and (e) show one period of the corresponding oscillating motion. The motion of one of the vehicles is highlighted in purple. Clearly, the velocity and headway undergo large fluctuations that repeat over time as the vehicle drives around the ring. Remarkably, this stop-and-go phantom jam is revealed through the analysis of the nonlinear dynamics—the linearized dynamics could not predict it.

The phantom jam was found systematically through the following analysis. First, we calculated the equilibrium and analyzed its stability numerically for various (β, α) parameters. This way we reproduced the stability chart in Fig. 2.24b. We found that along the colorful curves in Fig. 2.32a a so-called Hopf bifurcation [75] happens that is associated with a complex pair of characteristic roots crossing the imaginary axis. The Hopf bifurcation gives birth to limit cycles that are responsible for the phantom jam.

Second, we computed limit cycles for various parameter values numerically. Specifically, we fixed parameter α and varied parameter β, see the dashed line in Fig. 2.32a. At each Hopf bifurcation point (located at the intersections of the dashed line and the colorful curves) a branch of limit cycles is born, which we continued by varying β. The amplitude and stability of the resulting limit cycles (i.e., the peak-to-peak amplitude v_{pp} of the corresponding velocity fluctuations) are indicated in Fig. 2.32b. The most relevant branch of limit cycles is highlighted by the thick colored curve. Along this branch, the limit cycles are first unstable (red) and their amplitude increases as β is increased. Then, the branch folds back, the limit cycles become stable (green), and the amplitude saturates around the speed limit $v_{\max}$ as β is decreased. These large-amplitude stable limit cycles are associated with oscillatory, stop-and-go motion—a phantom jam—where the speed repeatedly changes between zero and $v_{\max}$. The time period T_{p} of oscillations is plotted in panel (c), whereas one period of this motion is highlighted in panels (d) and (e) for point T.

Panel (b) of Fig. 2.32 reveals that for certain β parameters, i.e., between the Hopf bifurcation point and the point where the primary branch of limit cycles folds, a large-amplitude stable limit cycle (SLC) coexists with the stable equilibrium (SE) and also with an unstable limit cycle (ULC). The coexistence of the stable equilibrium and the stable limit cycle is called bistability—a phenomenon appearing in nonlinear systems only. It means that in practice two qualitatively different behaviors can be observed for the same human drivers: the traffic flow can either be stable and smooth (associated with convergence to the equilibrium) or a phantom jam can appear (associated with convergence to the large-amplitude stable limit cycle). Whether smooth flow or congestion is observed for a bistable traffic system depends on whether perturbations push the system to the domain of attraction of the equilibrium or to that of the stable limit cycle. Note that the unstable limit cycle plays a role in determining the domains of attraction and the topology of the system.

Importantly, bistability occurs in human-driven traffic for a relatively wide range of driver behaviors. By repeating the limit cycle computation of Fig. 2.32b for various values of α, we identified a large bistable region in the (β, α) parameter space, as indicated by red shading in Fig. 2.32a.

In Fig. 2.32 the only source of nonlinearity is the range policy of human drivers in the OVM. That is, nonlinear driver models with time delay are able to capture the formation of phantom jams without considering any other nonlinearity. Yet, other nonlinearities may quantitatively affect the occurrence of phantom jams. We

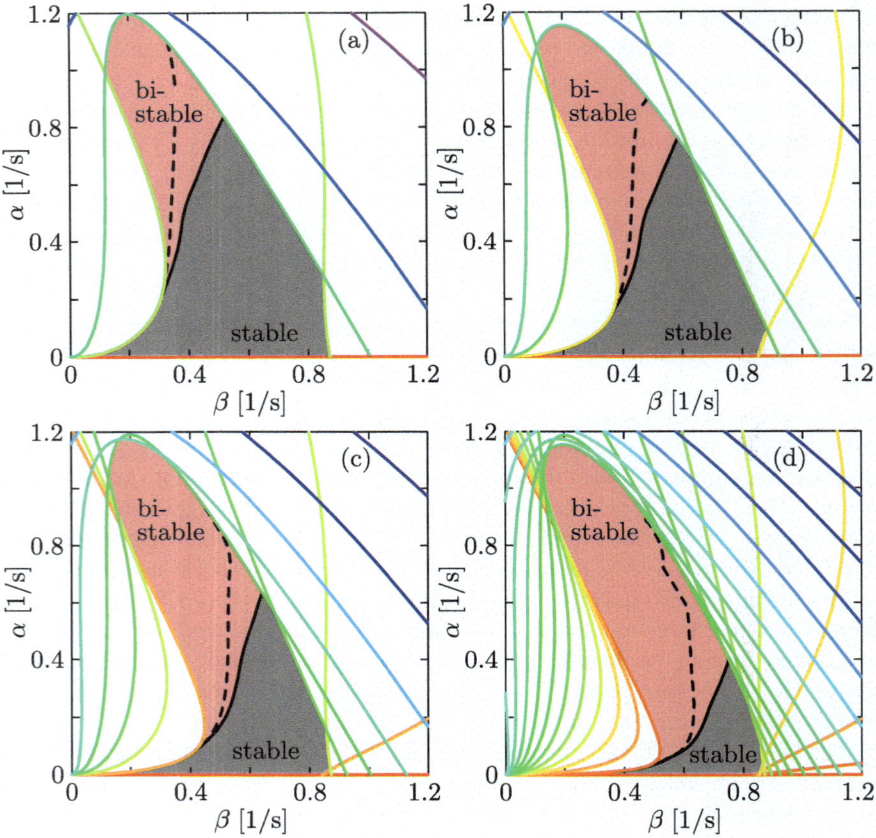

Fig. 2.34 Bistable domains of the OVM (2.207), (2.4) using the ring configuration given by (2.91), (2.92) with piecewise cubic range policy (2.5), (2.9). Panels **a–d** correspond to $N = 6$, $N = 8$, $N = 12$, and $N = 24$ vehicles on the ring, respectively, while the rest of the parameters are the same as in Fig. 2.33. As the number of vehicles increases, the red shaded bistable region grows

demonstrate this by studying the effect of the saturation-type nonlinearity in (2.207) that arises from the physical acceleration limits of vehicles. To this end, we repeat the above analysis by considering saturation at finite acceleration limits $a_{\min}$ and $a_{\max}$.

Figure 2.33 compares the nonlinear behavior with saturation (solid lines) and without saturation (dashed lines repeated from Fig. 2.32). While the qualitative behavior of traffic is not affected by the saturation, the quantitative results are significantly impacted. Panel (d) shows that the saturation limits the slope of the underlying speed profiles. Consequently, the amplitude of speed fluctuations decreases for certain β values in panel (b) and their period increases in panel (c) compared to the case without saturation. Most importantly, the size of the bistable domain in panel (a) significantly increases due to the saturation.

Figure 2.34 demonstrates how the nonlinear behavior is impacted by increasing the number of vehicles on the ring. Clearly, the bistable domain grows with the number of vehicles; hence, larger rings are more susceptible to phantom jams.

The above results of the ring configuration give intuition about the behavior of open chains of vehicles as well. In particular, we showed in Sect. 2.3.3 that the linear stability properties of large rings with many vehicles approximate the string stability properties of open chains. Similarly, one may consider large rings to capture the nonlinear behavior of open chains.

Chapter 3
Longitudinal Dynamics and Control of Automated Vehicles

In the previous chapter we discussed the dynamics of vehicle chains consisting of human-driven vehicles. Here, we focus on the control design when the longitudinal dynamics of vehicles are automated. The controllers designed in this section can be applied to vehicles of different levels of automation, including but not restricted to highly automated or self-driving vehicles. Nevertheless, cruise control and adaptive cruise control are currently available on many production vehicles and core parts of those algorithms are similar to those discussed in this chapter.

We use the concepts and mathematical techniques developed in the previous chapter including plant stability and string stability, and tie the analysis of human car-following models to the design of adaptive cruise control. While in the former case the vehicle and its driver were modeled together, in the latter case we separate the model of the longitudinal dynamics (the plant) from the algorithm that regulates the speed and the distance headway (the controller). We also demonstrate that using a unified framework for human-driven an automated vehicles enables us to study mixed traffic chains containing both types of vehicles.

3.1 Longitudinal Vehicle Dynamics

In this section we provide a model for the longitudinal dynamics of vehicles. Then, we use this model to lay out the fundamental principles of longitudinal control design for automated vehicles, including *cruise control* and *adaptive cruise control*.

In the previous chapter the car-following models presented for human-driven vehicles incorporated both the longitudinal vehicle dynamics and the human driver behavior. In particular, let us revisit the car-following model (2.21) without time delay. For $i = 0$ we may drop the index of the follower and re-write the equations as

G. Orosz and T. G. Molnár, *Dynamics and Control of Connected Vehicles*, Surveys and Tutorials in the Applied Mathematical Sciences 15,
https://doi.org/10.1007/978-3-031-94555-7_3

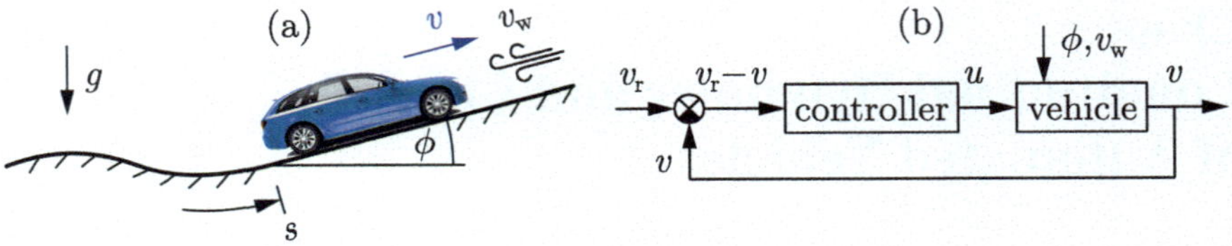

Fig. 3.1 **a** Longitudinal motion of an automated vehicle that is influenced by road elevation and resistance forces. **b** The control loop of this automated vehicle executing cruise control. The commanded acceleration u allows the vehicle to match its velocity v to the reference velocity v_r

$$
\begin{aligned}
\dot{h} &= v_1 - v \, , \\
\dot{v} &= u \, , \\
u &= f(h, \dot{h}, v) \, .
\end{aligned}
\tag{3.1}
$$

The first two equations model the longitudinal dynamics while the last equation specifies the control law used by the human driver to realize car following. Below we provide a more descriptive model for vehicle dynamics and we will define control laws that can be used to regulate the longitudinal motion of automated vehicles.

The second equation in (3.1) assigns the acceleration of the vehicle by the control input while neglecting the rolling resistance and the air drag. This model can be refined as follows. Consider the vehicle in Fig. 3.1a traveling with velocity $v \geq 0$ on a straight road with grade ϕ. Using the Newtonian approach [50] one may derive the equation of motion

$$
m_{\text{eff}}\dot{v} = -mg \sin\phi - mg\gamma \cos\phi - k(v + v_w)^2 + \frac{T_w}{R} \, . \tag{3.2}
$$

Here

$$
m_{\text{eff}} = m + \frac{J}{R^2} \tag{3.3}
$$

denotes the effective mass, which contains the mass m of the vehicle, the mass moment of inertia J of the rotating elements and the wheel radius R. Moreover, $g = 9.81\,[\text{m/s}^2]$ is the gravitational acceleration, γ is the rolling resistance coefficient, k is the air drag coefficient determined by the shape and frontal area of the vehicle, and v_w is the velocity of the wind acting on the front of the vehicle. Finally, T_w denotes the total torque exerted by the powertrain and the brakes on the wheels.

By dividing (3.2) with the effective mass m_{eff} we obtain

$$
\dot{v} = -p(v) + \text{sat}(u) \, , \tag{3.4}
$$

where

$$
\text{sat}(u) = \frac{T_w}{m_{\text{eff}} R} \, , \tag{3.5}
$$

indicates the effect of the torque T_w, while $p(v)$ describes the effect of resistance forces. The latter is a quadratic function of the velocity v, given by

$$p(v) = a + bv + cv^2 . \tag{3.6}$$

From now on, we neglect the effect of the wind for simplicity ($v_w = 0$), which yields the coefficients

$$a = \frac{m}{m_{\text{eff}}}(g \sin\phi + g\gamma \cos\phi) , \quad b = 0 , \quad c = \frac{k}{m_{\text{eff}}} , \tag{3.7}$$

and we will use the approximation $m/m_{\text{eff}} \approx 1$ in the numerical examples of this book. Note that $b > 0$ would correspond to having viscous damping in the system which is neglected here for simplicity.

In (3.4) u denotes the commanded acceleration of the vehicle, which will be regarded as a control input when designing longitudinal controllers for automated vehicles. In (3.5) we highlight that the engines and brakes of vehicles are not capable of generating arbitrarily large torque T_w on the wheels, hence vehicles are not able to realize arbitrarily large commanded acceleration u. We model these physical limits by the saturation function sat as

$$\text{sat}(u) = \begin{cases} -a_{\min} & \text{if } \; u \le -a_{\min} , \\ u & \text{if } \; -a_{\min} < u < \tilde{a}_{\max}(v) , \\ \tilde{a}_{\max}(v) & \text{if } \; u \ge \tilde{a}_{\max}(v) . \end{cases} \tag{3.8}$$

Here $-a_{\min}$ indicates the braking limit of the vehicle. Furthermore, $\tilde{a}_{\max}(v)$ denotes the acceleration limit, which is velocity-dependent and is determined by the maximum torque associated with $a_{\max}$ and the maximum power $P_{\max}$ of the powertrain:

$$\tilde{a}_{\max}(v) = \min\left\{a_{\max}, \frac{P_{\max}}{m_{\text{eff}}v}\right\} . \tag{3.9}$$

The saturation sat(u) is illustrated in Fig. 3.2a, while the acceleration capability $\tilde{a}_{\max}(v)$ is shown together with the resistance term $p(v)$ in Fig. 3.2b. Note that these functions make the longitudinal dynamics (3.4) nonlinear. This nonlinear model captures the physical limits of vehicles. On one hand, it provides the maximum achievable acceleration $a^*_{\max}$ under resistance forces (i.e., the maximum of $\dot{v}$) as $a^*_{\max} = -p(0) + a_{\max}$, that may be realized at $v = 0$ since p is a monotonically increasing function of v. On the other hand, the maximum achievable equilibrium velocity $v^*_{\max}$ (i.e., when $\dot{v} = 0$) is given by the solution of $p(v^*_{\max}) = \tilde{a}(v^*_{\max})$. This is often referred to as the top speed of the vehicle in laymen terms. These properties are illustrated schematically in Fig. 3.2b, while Fig. 3.3 depicts the numerical values of the resistance term and the acceleration limits for a passenger car in panel (a) and for a heavy-duty truck with a fully loaded trailer in panel (b); see Table 3.1

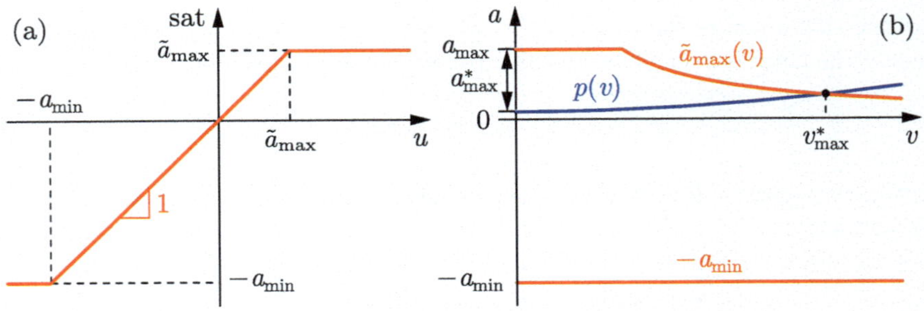

Fig. 3.2 Nonlinearities in the longitudinal dynamics and control of automated vehicles. Panel **a** shows the saturation (3.8) of the commanded acceleration u at the acceleration limits of the vehicle. Panel **b** indicates the acceleration limits (3.9) and the resistance term (3.6) as a function of the velocity

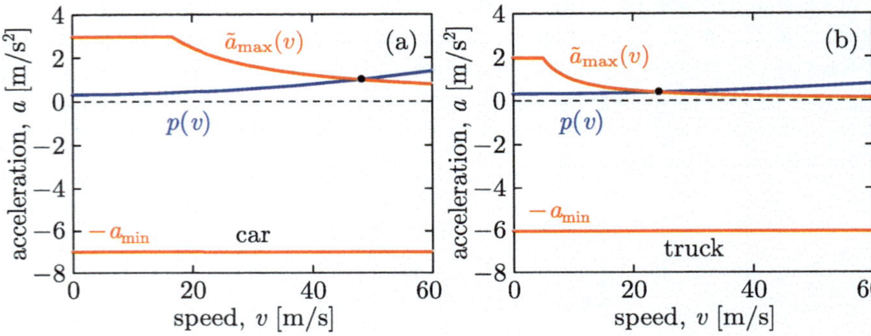

Fig. 3.3 Resistance term (3.6) and acceleration limits (3.9) in longitudinal vehicle dynamics. Panel **a** is associated with a passenger vehicle (car), and panel **b** corresponds to a heavy-duty vehicle with fully loaded trailer (truck). The underlying parameter values are listed in Table 3.1 and the road grade is assumed to be $\phi = 0.025$ [rad]

for the parameter values used. Notice that for the truck the nonlinearities are more significant, which makes longitudinal control more challenging.

Finally, we remark that (3.4) neglected the powertrain dynamics and brake dynamics, since it assumed that the torque demanded at the accelerator or brake pedals are realized at the wheels instantaneously. In reality, it may take significant time to transfer torque demand to the wheels. This can be captured by incorporating an additional time delay σ into the model:

$$\dot{v}(t) = -p(v(t)) + \mathrm{sat}(u(t-\sigma)) \,. \tag{3.10}$$

The value of σ depends on the type of the powertrain (internal combustion engines, electric motors, hybrid powertrains), on the type of the brake system (hydraulic brakes, air brakes) and also on the vehicle class; see Table 3.1 for some typical values we will use for passenger cars and heavy-duty trucks in the remainder of the book.

Table 3.1 Typical parameter values for longitudinal vehicle dynamics considering a passenger vehicle (car) and a heavy-duty vehicle with trailer (truck)

Description	Parameter	Value for car	Value for truck	Unit
Effective mass	m_{eff}	1500	29500	kg
Rolling resistance coefficient	γ	0.01	0.01	–
Air drag coefficient	k	0.45	3.85	kg/m
Braking limit	$a_{\min}$	7	6	m/s^2
Acceleration limit	$a_{\max}$	3	2	m/s^2
Power limit	$P_{\max}$	75	300	kW
Powertrain delay	σ	0.6	0.8	s

The parameter σ can also be used to incorporate the sensory and feedback delays for automated vehicles. We also remark that low-level controllers may be designed to compensate for the effect of resistance forces. Then, one may omit resistance terms when designing the high-level control algorithms and synthesize the input u by taking $p(v) = 0$.

In what follows, we design control laws for automated vehicles in order to execute desired longitudinal motions. We address how to command the acceleration u to achieve *cruise control (CC)* and *adaptive cruise control (ACC)*. First we neglect the time delay σ in the system (cf. (3.4)), and later we add back the delay for completeness (cf. (3.10)).

3.2 Cruise Control

Here we discuss cruise control (CC), in which the automated vehicle in Fig. 3.1a regulates its longitudinal motion described by (3.4) such that its velocity v approaches a—potentially time-varying—reference velocity v_{r}. This is accomplished via the control loop illustrated in Fig. 3.1b. The controller commands the acceleration u such that it drives the velocity error $v_{\mathrm{r}} - v$ to zero. In particular, we consider the proportional-integral (PI) controller given by

$$u = \alpha(v_{\mathrm{r}} - v) + \eta \int_0^t (v_{\mathrm{r}} - v)\, \mathrm{d}\tilde{t}\,, \tag{3.11}$$

with proportional gain $\alpha > 0$ and integral gain $\eta > 0$ to be designed.

As guidance to select the control gains α and η, let us analyze the closed-loop dynamics. For $\eta \neq 0$, the closed-loop system (3.4), (3.11) becomes

$$\begin{aligned} \dot{z} &= v_{\mathrm{r}} - v\ , \\ \dot{v} &= -p(v) + \operatorname{sat}\big(\alpha(v_{\mathrm{r}} - v) + \eta z\big)\ , \end{aligned} \tag{3.12}$$

where z indicates the integral term

$$z = \int_0^t (v_{\mathrm{r}} - v)\,\mathrm{d}\tilde{t}\ . \tag{3.13}$$

First, let us consider the scenario where the reference velocity v_{r} is constant in time: $v_{\mathrm{r}}(t) \equiv v_{\mathrm{r}}^* < v_{\mathrm{max}}^*$. Then, in steady state the cruise controller drives the automated vehicle at constant velocity v^*, given by the equilibrium

$$\begin{aligned} z(t) &\equiv z^*\ , \\ v(t) &\equiv v^*\ , \end{aligned} \tag{3.14}$$

that is associated with constant control input $u(t) \equiv u^*$. After substitution into (3.12), we obtain

$$\begin{aligned} 0 &= v_{\mathrm{r}}^* - v^*\ , \\ 0 &= -p(v^*) + \operatorname{sat}\big(\alpha(v_{\mathrm{r}}^* - v^*) + \eta z^*\big)\ . \end{aligned} \tag{3.15}$$

For $v_{\mathrm{r}}^* < v_{\mathrm{max}}^*$ the saturation can be dropped, and the equations can be rearranged to

$$\begin{aligned} v^* &= v_{\mathrm{r}}^*\ , \\ z^* &= \frac{p(v^*)}{\eta}\ , \end{aligned} \tag{3.16}$$

where the second equation corresponds to $u^* = p(v^*)$. This concludes that the controller is able to match the equilibrium velocity v^* to the reference v_{r}^* by using a control input u^* that overcomes the resistance term $p(v^*)$.

We remark that the desired equilibrium velocity ($v^* = v_{\mathrm{r}}^*$) is no longer achieved if one uses a proportional (P) controller and omits the integral term by taking $\eta = 0$. Then, the closed-loop dynamics become

$$\dot{v} = -p(v) + \operatorname{sat}\big(\alpha(v_{\mathrm{r}} - v)\big)\ , \tag{3.17}$$

with the equilibrium

$$v(t) \equiv v^*\ . \tag{3.18}$$

This equilibrium satisfies

$$v^* = v_{\mathrm{r}}^* - \frac{p(v^*)}{\alpha}\ , \tag{3.19}$$

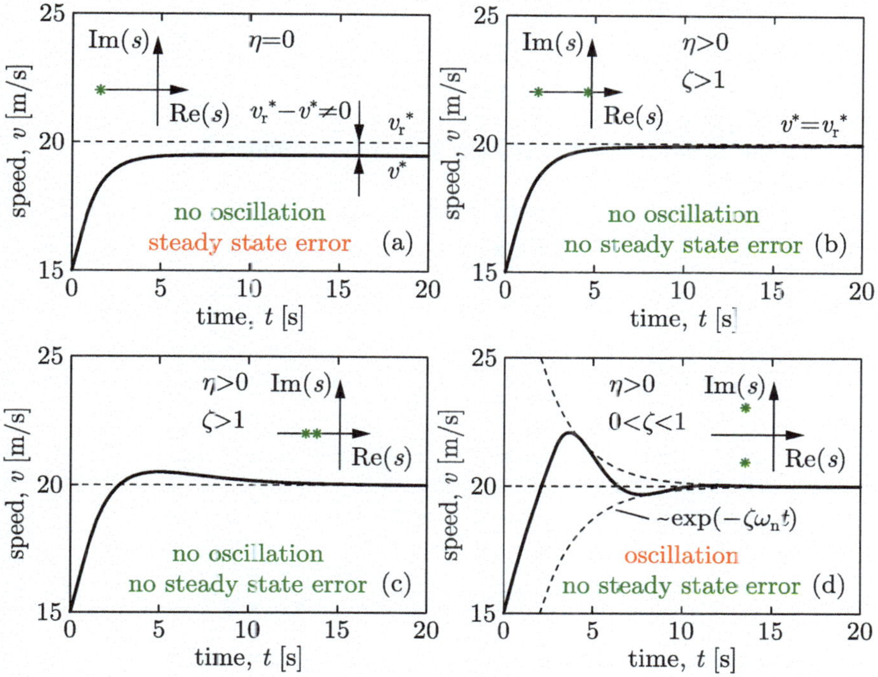

Fig. 3.4 Step response of an automated vehicle executing the cruise controller (3.11). The longitudinal motion is described by (3.4) using the passenger vehicle parameters in the first column of Table 3.1. The rest of the parameters are $v(0) = 15\,[\text{m/s}]$, $v_r(t) \equiv 20\,[\text{m/s}]$, $\phi = 0.025\,[\text{rad}]$, $\alpha = 0.9\,[1/\text{s}]$ and **a** $\eta = 0\,[1/\text{s}^2]$, **b** $\eta = 0.05\,[1/\text{s}^2]$, **c** $\eta = 0.2\,[1/\text{s}^2]$, **d** $\eta = 0.8\,[1/\text{s}^2]$. A steady-state velocity error exists for $\eta = 0$, and it is eliminated when $\eta > 0$. As η is increased, the response of the system gets faster, while overshoot and oscillations arise

that can be solved for v^*. For example, when $p(v) = a > 0$ (that is, $b = 0$ and $c = 0$ in (3.6)), we obtain

$$v^* = v_r^* - \frac{a}{\alpha} \neq v_r^* \,. \tag{3.20}$$

This indicates the existence of a steady-state velocity error a/α.

The P and PI controllers (with $\eta = 0$ and $\eta \neq 0$, respectively) are compared in Fig. 3.4, where the performance of the cruise controller (3.11) is illustrated when executed on a passenger vehicle with parameters given in the first column of Table 3.1. A steady-state error $v_r^* - v^* \approx 0.5\,[\text{m/s}]$ is clearly present when $\eta = 0$, while the steady-state error is eliminated when $\eta > 0$.

Now let us consider time-varying reference velocity $v_r(t)$ that fluctuates around a value v_r^*. We analyze the linearized dynamics around the equilibrium (3.14). By introducing the perturbations

$$\begin{aligned}\tilde{v}_{\rm r}(t) &= v_{\rm r}(t) - v_{\rm r}^* ,\\ \tilde{z}(t) &= z(t) - z^* ,\\ \tilde{v}(t) &= v(t) - v^* ,\end{aligned} \tag{3.21}$$

we can formulate the linearized dynamics within the saturation limits as

$$\begin{aligned}\dot{\mathbf{x}} &= \mathbf{a}\,\mathbf{x} + \mathbf{b}\,\tilde{v}_{\rm r} + \text{h.o.t.} ,\\ y &= \mathbf{c}\,\mathbf{x} ,\end{aligned} \tag{3.22}$$

with

$$\mathbf{x} = \begin{bmatrix}\tilde{z}\\ \tilde{v}\end{bmatrix}, \quad \mathbf{a} = \begin{bmatrix}0 & -1\\ \eta & -(\alpha+d)\end{bmatrix}, \quad \mathbf{b} = \begin{bmatrix}1\\ \alpha\end{bmatrix}, \quad \mathbf{c} = \begin{bmatrix}0 & 1\end{bmatrix}, \quad y = \tilde{v} , \tag{3.23}$$

where $d = p'(v^*)$ is the derivative of the resistance term $p(v)$ evaluated at v^* and this is typically very small ($d \approx 0.01\,[1/\text{s}]$).

The linearized dynamics are associated with the transfer function

$$T(s) = \mathbf{c}\,(s\mathbf{I} - \mathbf{a})^{-1}\,\mathbf{b} = \frac{\alpha s + \eta}{s^2 + (\alpha + d)s + \eta} , \tag{3.24}$$

cf. (2.78), and the characteristic equation

$$\det(s\mathbf{I} - \mathbf{a}) = s^2 + (\alpha + d)s + \eta = 0 , \tag{3.25}$$

cf. (2.79), with $\mathbf{I}$ denoting the two-dimensional identity matrix. This can be re-written to the well-known form of the characteristic equation of second-order systems:

$$s^2 + 2\zeta\omega_{\rm n}s + \omega_{\rm n}^2 = 0 , \tag{3.26}$$

cf. (2.36), where we introduced

$$\omega_{\rm n} = \sqrt{\eta} , \quad \zeta = \frac{\alpha + d}{2\sqrt{\eta}} \approx \frac{\alpha}{2\sqrt{\eta}} , \tag{3.27}$$

and the approximation holds since d is typically very small.

When designing the proportional gain α and the integral gain η in the PI controller (3.11), one may observe a trade-off. In particular, the constant α/η characterizes the time needed to eliminate the steady-state error. In order to make this time short, one may want to choose η to be large. This in fact places the zero $-\eta/\alpha$ of the transfer function (3.24) far to the left in the complex plane. On the other hand, the poles of (3.24), that is, the characteristic roots of (3.26) are given by

$$s_{1,2} = -\zeta\omega_{\rm n} \pm {\rm j}\omega_{\rm d} , \quad \omega_{\rm d} = \omega_{\rm n}\sqrt{1 - \zeta^2} , \tag{3.28}$$

cf. (2.37). These are complex conjugates for $\zeta < 1$ and real for $\zeta \geq 1$. The former case may lead to overshoots and oscillations, and thus, to achieve large values of $\zeta \approx \alpha/(2\sqrt{\eta})$, one may want to choose η to be small.

This trade-off is illustrated in Fig. 3.4 where the response of the cruise controller to a constant reference speed (i.e., a step response) is shown together with the eigenvalues for various values of η and fixed α. Observe that increasing η, increases the chance of overshoot in the system's response and leads to oscillations when $\zeta < 1$. At the same time, as η gets larger, the time constant of the integral controller decreases, making the time to reach the equilibrium shorter. Consequently, one needs to tune η and α carefully to achieve a fast response without overshoot and oscillations. Different car manufacturers use different gain setups giving different characteristics to their automobiles and trucks.

Finally, the response of a passenger vehicle (car) and a heavy-duty vehicle (truck) with trailer is compared in Fig. 3.5. The parameters of these vehicles are given in Table 3.1 while the powertrain delays are omitted ($\sigma = 0$). The cruise controller

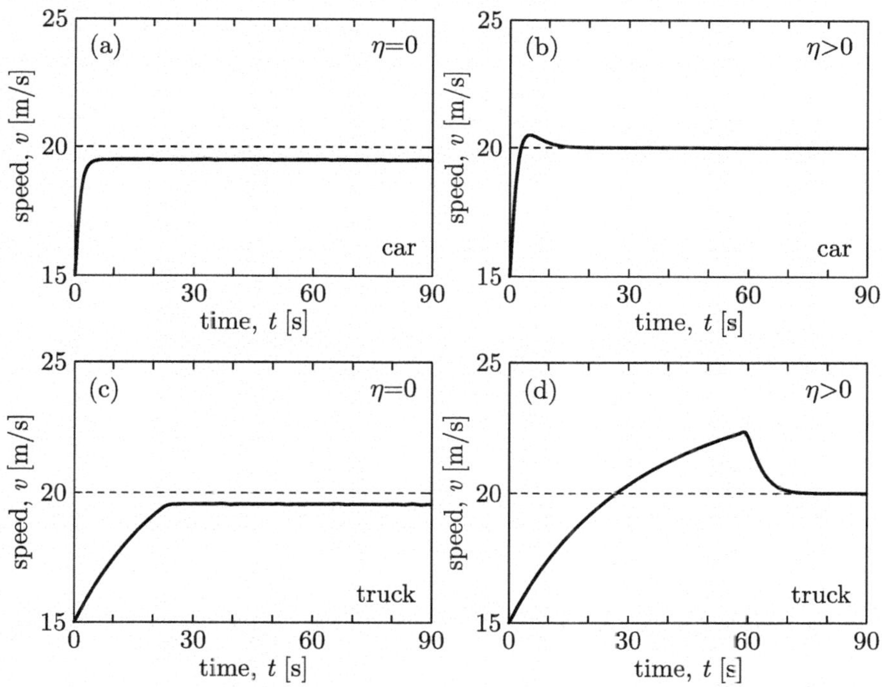

Fig. 3.5 Step response of the cruise controller (3.11). Panels **a**, **b** are associated with a passenger vehicle (car), while panels **c**, **d** correspond to a heavy-duty vehicle with fully loaded trailer (truck). The longitudinal motion is described by (3.4) using the parameters in Table 3.1. The rest of the parameters are $v(0) = 15\,[\text{m/s}]$, $v_r(t) \equiv 20\,[\text{m/s}]$, $\phi = 0.025\,[\text{rad}]$, $\alpha = 0.9\,[1/\text{s}]$ and **a**, **c** $\eta = 0\,[1/\text{s}^2]$, **b**, **d** $\eta = 0.2\,[1/\text{s}^2]$. The response of the truck is significantly slower and it is dominated by the nonlinearities presented in Fig. 3.2

parameters α and η are selected to be the same for the two vehicles. As expected, the response of the truck (bottom) is significantly slower than that of the car (top) due to its higher mass and limited acceleration capability, while it is also more susceptible to overshoot. Furthermore, the effect of the nonlinearities shown in Fig. 3.2 is more pronounced for the truck, which makes controller tuning more challenging.

3.3 Adaptive Cruise Control

In this section, we discuss adaptive cruise control (ACC) that enables automated vehicles to perform car following. Consider the scenario shown in Fig. 3.6 where an automated vehicle (AV) follows a preceding vehicle, which is, in this case, a human-driven vehicle (HV). The AV can measure its own velocity v and use on-board range sensors (such as radar, lidar, camera, or sonar) to detect the distance headway (range) h and the relative velocity (range rate) $\dot{h}$. In this book we omit detailed discussions on the advantages and disadvantages of different sensor kits, but simply mention that each of these comes with a range limit and operates at some finite sampling frequency. More discussion on the effects of sampling will be presented in Sect. 3.5.

The goals of ACC are to match the speed v of the vehicle to that v_1 of the preceding vehicle while keeping a safe and comfortable headway h. These are indeed the same as the aims of human drivers discussed in Chap. 2. ACC also relies on the same information: v, h and $\dot{h}$, or equivalently, v, h and v_1. Thus, we consider control laws in the form

$$u = F(h, v, v_1)\,, \tag{3.29}$$

where the nonlinear function F can be designed to achieve the above goals. The advantages of ACC over human car following are that AVs may have more consistent and faster response than humans, and that ACC can be tuned for optimal performance.

A particular control law motivated by the optimal velocity model (2.4) is given by

$$u = \alpha\big(V(h) - v\big) + \beta\big(W(v_1) - v\big)\,. \tag{3.30}$$

In order to compensate for resistance terms as in (3.4) and (3.10) one may add integral action similarly as in (3.11) and obtain

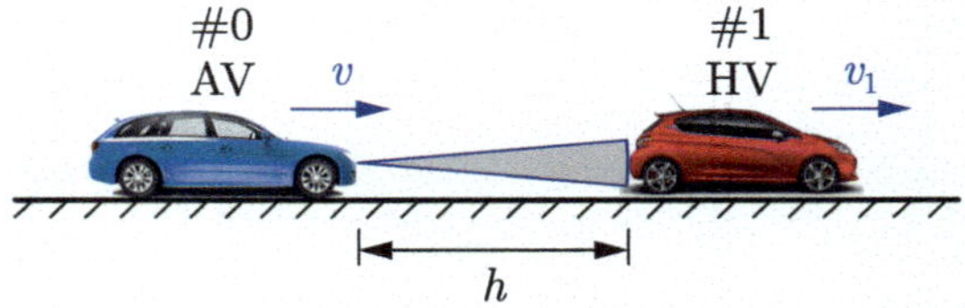

Fig. 3.6 Adaptive cruise control (ACC) where an automated vehicle (AV) follows a human-driven vehicle (HV) on a single lane. The AV responds to the HV by relying on its on-board range sensors

$$u = \alpha\big(V(h) - v\big) + \eta \int_0^t \big(V(h) - v\big)\,\mathrm{d}\tilde{t} + \beta\big(W(v_1) - v\big)\,, \qquad (3.31)$$

see [155]. This is, strictly speaking, not in the form of (3.29), but below we show that the same techniques can be used to analyze both (3.30) and (3.31).

The first term of the control law (3.30), and first two terms in the control law (3.31) are to ensure that the AV responds to the headway h while regulating its velocity v. To this end, the desired velocity V is prescribed as a function of the headway h through the range policy

$$V(h) = \begin{cases} 0 & \text{if} \quad h \le h_{\text{st}}\,, \\ \kappa(h - h_{\text{st}}) & \text{if} \quad h_{\text{st}} < h < h_{\text{go}}\,, \\ v_{\max} & \text{if} \quad h \ge h_{\text{go}}\,, \end{cases} \qquad (3.32)$$

where $\kappa = v_{\max}/(h_{\text{go}} - h_{\text{st}})$. The range policy is illustrated in Fig. 3.7a. This is inspired by the range policy (2.5), (2.6) of the optimal velocity model (2.4) which is often used to describe human drivers. When the headway drops below h_{st} the AV shall stop, and when the headway exceeds h_{go} the AV shall travel at the speed limit $v_{\max}$. Between h_{st} and h_{go}, the larger the headway is the faster the AV shall travel, according to an affine velocity-headway relationship associated with gradient κ, cf. (2.6), which gives the time headway $t_{\text{h}} = 1/\kappa$. While this parameter was already used in Chap. 2 to describe human drivers, now it is a tunable parameter that can be chosen when designing ACC. In commercial ACC systems, often the user can select how large the time headway shall be, which corresponds to selecting κ in (3.32). We remark that a nonlinear function could also be chosen for range policy, cf. (2.5).

The last term in control laws (3.30) and (3.31) is also motivated by the optimal velocity model (2.4), though it is formulated slightly differently. In particular, it contains the speed policy

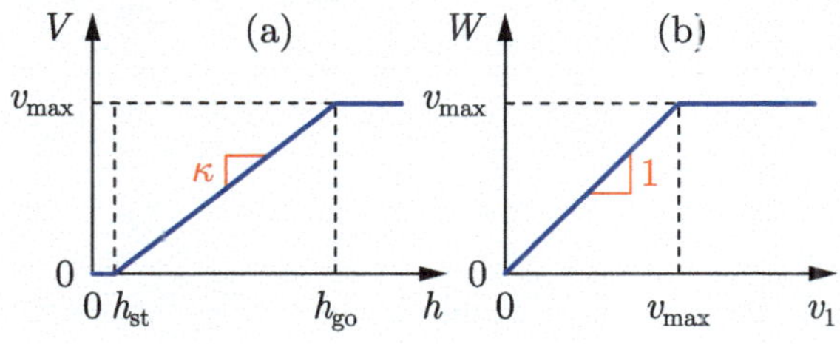

Fig. 3.7 Nonlinearities in the longitudinal control of automated vehicles. Panel **a** shows the range policy (3.32) describing the desired velocity versus the headway. Panel **b** shows the speed policy (3.33) addressing the speed limit imposed by traffic rules

$$W(v_1) = \begin{cases} v_1 & \text{if } v_1 \le v_{\max}\,, \\ v_{\max} & \text{if } v_1 > v_{\max}\,, \end{cases} \tag{3.33}$$

which is visualized in Fig. 3.7b. This is introduced to avoid following a speeding HV. In case of $v_1 \le v_{\max}$ the last term in the controllers becomes $\beta(v_1 - v) = \beta\dot{h}$ as in (2.4), while for $v_1 > v_{\max}$ we obtain $\beta(v_{\max} - v)$.

Note that when using radar, the range rate $\dot{h}$ is obtained from the Doppler effect, and then the velocity of the preceding vehicle is computed using $v_1 = v + \dot{h}$. We also remark that the control laws (3.30) and (3.31) are often made more complicated in practice in order to be able to handle scenarios when the AV is approaching a slow vehicle ahead or when another vehicle cuts in front of the AV. Some of these cases will be discussed in the context of experimental results in Chap. 5.

In summary, the controller (3.30) involves three tunable parameters α, β and κ, while the controller (3.31) involves four parameters α, η, β and κ, where η is added to handle the disturbances caused by the resistance term $p(v)$; see (3.4) and (3.10). Below we discuss how changing these parameters influences the control performance and highlight the differences between the two controllers. To do so, first we omit the powertrain delay σ and use the simplified model (3.4). Then we address the effects of the delay while using the model (3.10).

3.3.1 Adaptive Cruise Control Without Time Delay

Here, we analyze the closed-loop dynamics of ACC systems without response delay. Using the controller (3.29) in the model (3.4) yields

$$\begin{aligned} \dot{h} &= v_1 - v\,, \\ \dot{v} &= -p(v) + \mathrm{sat}\big(F(h, v, v_1)\big)\,. \end{aligned} \tag{3.34}$$

When the resistance term is omitted (or assumed to be compensated at the lower level) one may use the controller (3.30) and obtain

$$\begin{aligned} \dot{h} &= v_1 - v\,, \\ \dot{v} &= \mathrm{sat}\Big(\alpha\big(V(h) - v\big) + \beta\big(W(v_1) - v\big)\Big)\,, \end{aligned} \tag{3.35}$$

while once the resistance term is taken into account, using the controller (3.31) with integral action leads to

$$\begin{aligned} \dot{h} &= v_1 - v\,, \\ \dot{v} &= -p(v) + \mathrm{sat}\Big(\alpha\big(V(h) - v\big) + \eta z + \beta\big(W(v_1) - v\big)\Big)\,, \\ \dot{z} &= V(h) - v\,, \end{aligned} \tag{3.36}$$

where z denotes the integral term

$$z = \int_0^t \big(V(h) - v\big)\,\mathrm{d}\tilde{t}\ . \tag{3.37}$$

To simplify the analysis, we assume that the HV's velocity v_1 is below the speed limit $v_{\max}$ and that the commanded acceleration u is within the saturation limits $-a_{\min}$ and $\tilde{a}_{\max}(v)$. That is, we drop sat and W in (3.35) and (3.36). Then, (3.35) is identical to the dynamics of the optimal velocity model (2.21), (2.4). This implies that for special choices of ACC design AVs may drive such that they mimic human driver behavior. Below we summarize the stability results for the simpler case (3.35) before presenting the results for the more general case of (3.36).

First, consider the equilibrium of the closed-loop dynamics (3.35) when the HV is driving with constant velocity v_1^* and the AV follows it with constant velocity v^* and constant headway h^*, that is,

$$\begin{aligned} v_1(t) &\equiv v_1^*\ , \\ h(t) &\equiv h^*\ , \\ v(t) &\equiv v^*\ . \end{aligned} \tag{3.38}$$

Substituting this into (3.35) while dropping sat and W yields

$$\begin{aligned} v^* &= v_1^*\ , \\ v^* &= V(h^*)\ , \end{aligned} \tag{3.39}$$

cf. (2.15), (2.16), and the equilibrium headway is determined by the range policy $h^* = V^{-1}(v^*)$ when $0 < v^* < v_{\max}$.

The dynamics around the equilibrium can be described by the perturbations

$$\begin{aligned} \tilde{v}_1(t) &= v_1(t) - v_1^*\ , \\ \tilde{h}(t) &= h(t) - h^*\ , \\ \tilde{v}(t) &= v(t) - v^*\ , \end{aligned} \tag{3.40}$$

and linearizing (3.35) around (3.38) leads to

$$\begin{aligned} \dot{\mathbf{x}} &= \mathbf{a}\,\mathbf{x} + \mathbf{b}\,\tilde{v}_1 + \text{h.o.t.}\ , \\ y &= \mathbf{c}\,\mathbf{x}\ , \end{aligned} \tag{3.41}$$

with

$$\mathbf{x} = \begin{bmatrix} \tilde{h} \\ \tilde{v} \end{bmatrix}, \quad \mathbf{a} = \begin{bmatrix} 0 & -1 \\ \alpha\kappa & -(\alpha+\beta) \end{bmatrix}, \quad \mathbf{b} = \begin{bmatrix} 1 \\ \beta \end{bmatrix}, \quad \mathbf{c} = \begin{bmatrix} 0 & 1 \end{bmatrix}, \quad y = \tilde{v}\ , \tag{3.42}$$

and $\kappa = V'(h^*)$; cf. (2.24), (2.25), and (2.29). This leads to the transfer function

$$T(s) = \mathbf{c}(s\mathbf{I} - \mathbf{a})^{-1}\mathbf{b} = \frac{\beta s + \alpha\kappa}{s^2 + (\alpha + \beta)\, s + \alpha\kappa}\,, \tag{3.43}$$

cf. (2.78), and the characteristic equation

$$\det(s\mathbf{I} - \mathbf{a}) = s^2 + (\alpha + \beta)\, s + \alpha\kappa = 0\,, \tag{3.44}$$

cf. (2.79), which enables us to evaluate *plant stability* and *string stability* at the linear level as introduced in Chap. 2.

Plant stability is a fundamental requirement that every longitudinal control system must satisfy to be able to operate in practice. A plant stable ACC system enables the AV to approach the constant velocity determined by the preceding vehicle. Formally, this means that the linearized dynamics (3.41) are asymptotically stable, cf. Definition 2.1. Based on Theorem 2.1, the condition for plant stability is that the roots s_1 and s_2 of the characteristic equation (3.44) are located in the left-half plane:

$$\mathrm{Re}(s_1) < 0\,, \quad \mathrm{Re}(s_2) < 0\,. \tag{3.45}$$

This, according to the Routh–Hurwitz criteria, holds when

$$\begin{aligned} \alpha &> 0\,, \\ \alpha + \beta &> 0\,, \end{aligned} \tag{3.46}$$

cf. (2.80). If either of these conditions is not satisfied, the velocity of the AV may diverge from a desired equilibrium in a non-oscillatory or oscillatory manner.

String stability, on the other hand, is not essential for ACC systems to operate. In fact, some of the commercially available ACC systems are not string stable [34, 76]. Yet string stability can significantly enhance ACC performance in terms of achieving smooth driving behavior. For a string stable ACC system the velocity fluctuations of the AV are smaller than those of the HV with respect to a selected norm, cf. Definition 2.4. For example, string stability in the $\mathcal{L}_2$ norm is guaranteed (at the linear level) when the magnitude of the transfer function is less than one:

$$|T(\mathrm{j}\omega)| < 1\,, \quad \forall \omega > 0\,, \tag{3.47}$$

cf. Theorem 2.2. This condition can be directly verified for harmonic velocity perturbations

$$\tilde{v}_1(t) = v_{\mathrm{amp}} \sin(\omega t)\,, \tag{3.48}$$

to which the steady-state response of the AV is given by

$$\tilde{v}_{\mathrm{ss}}(t) = |T(\mathrm{j}\omega)|\, v_{\mathrm{amp}} \sin\big(\omega t + \angle T(\mathrm{j}\omega)\big)\,, \tag{3.49}$$

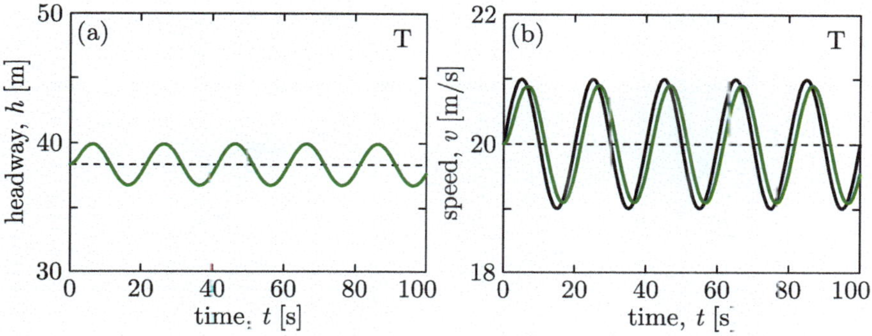

Fig. 3.8 Simulation results of system (3.35) for a passenger vehicle using the parameters in the first column of Table 3.1, the piecewise linear range policy (3.32) with $h_{st} = 5$ [m], $h_{go} = 55$ [m], $v_{max} = 30$ [m/s], and the control gains $\alpha = 0.4$ [1/s], $\beta = 0.5$ [1/s] corresponding to point T in Fig. 3.9b. The initial conditions are $v(0) = v^* = 20$ [m/s], $h(0) = V^{-1}(v^*) = 38.33$ [m] and the lead vehicle's velocity is $v_1(t) = v^* + v_{amp} \sin(\omega t)$, $v_{amp} = 1$ [m/s], $\omega = 0.1\pi$ [rad/s]

cf. (2.67), (2.69).

Substituting (3.43) into (3.47) and following the calculations (2.81)–(2.85) one can show that the string stable parameter domain is given by

$$\begin{aligned} &\alpha > 0\,, \\ &\alpha > 2(\kappa - \beta)\,, \end{aligned} \tag{3.50}$$

cf. (2.86).

Figure 3.8 demonstrates the performance of the ACC system (3.35) for a scenario where the HV's velocity fluctuates harmonically, as given by (3.48). The ACC design is both plant and string stable: the AV is able to follow the HV (keep a desired headway) in a stable manner while the velocity fluctuations of the AV are smaller than those of the HV. That is, the driving behavior of the AV is smoother than that of the HV thanks to string stability. This may be beneficial in terms of passenger comfort, energy efficiency, and keeping the flow of traffic smooth.

Figure 3.9 illustrates the stability charts of the ACC system (3.41), (3.42) in the (β, α) plane. The domains (3.46) of plant stable controller parameters are shaded as light gray and bounded by red curves, while the string stable domains (3.50) are shaded as dark gray and bounded by blue curves. We remark that Fig. 3.9b is identical to Fig. 2.7 describing the dynamics of human drivers. However, while in Chap. 2 parameters α, β and κ were characterizing human drivers, now they indicate the controller parameters of ACC, hence they can be tuned. Accordingly, Fig. 3.9 depicts stability charts for different values of κ. While the plant stable domain is independent of κ, the string stable domain shifts to the right as κ is increased; see point A located at $(\kappa, 0)$. That is, when one uses a range policy in ACC to keep shorter time headway $t_h = 1/\kappa$, larger control gains are required to maintain string stability.

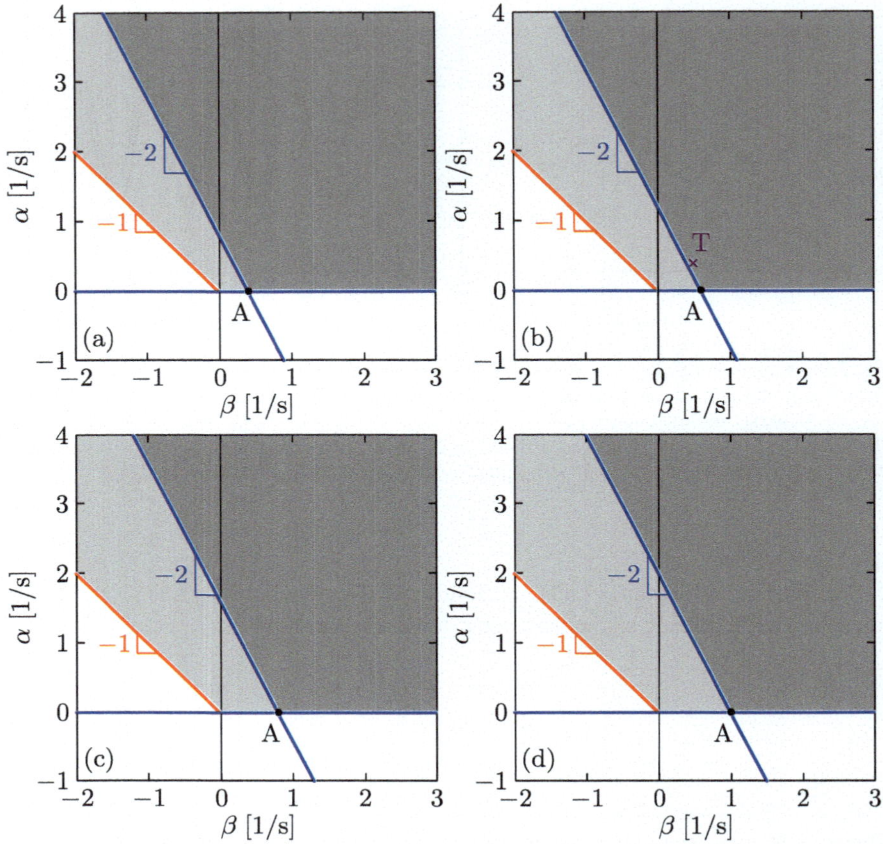

Fig. 3.9 Stability charts of system (3.41), (3.42). Panels **a–d** are associated with range policy gradient $\kappa = 0.4\,[1/\mathrm{s}]$, $\kappa = 0.6\,[1/\mathrm{s}]$, $\kappa = 0.8\,[1/\mathrm{s}]$ and $\kappa = 1\,[1/\mathrm{s}]$, respectively. The plant and string stable regions are shaded by light and dark gray, respectively. Point A is located at $(\kappa, 0)$, at the intersection of the string stability boundaries

Now, we analyze the closed-loop dynamics (3.36) where disturbances are compensated by integral action that was published in [155]. Assume again that the HV drives with constant velocity v_1^* and that the AV follows with constant velocity v^* and constant headway h^*. This yields the equilibrium

$$\begin{aligned} v_1(t) &\equiv v_1^*\,, \\ h(t) &\equiv h^*\,, \\ v(t) &\equiv v^*\,, \\ z(t) &\equiv z^*\,. \end{aligned} \tag{3.51}$$

Substituting this equilibrium solution into (3.36) while dropping sat and W leads to

$$\begin{aligned} 0 &= v_1^* - v^* \,, \\ 0 &= -p(v^*) + \alpha\left(V(h^*) - v^*\right) + \eta z^* + \beta\left(v_1^* - v^*\right). \\ 0 &= V(h^*) - v^* \,, \end{aligned} \tag{3.52}$$

which can be solved as

$$\begin{aligned} v^* &= v_1^* \,, \\ z^* &= \frac{p(v^*)}{\eta} \,, \\ v^* &= V(h^*) \,, \end{aligned} \tag{3.53}$$

Here the first and the last equation are equivalent to (3.39) while the second equation corresponds to the constant control input $u^* = p(v^*)$, that is, the controller compensates the resistance forces and ensures that the velocity v^* of the AV matches the velocity v_1^* of the HV. The equilibrium headway is given by $h^* = V^{-1}(v^*)$ when $0 < v^* < v_{\max}$.

Defining the perturbations

$$\begin{aligned} \tilde{v}_1(t) &= v_1(t) - v_1^* \,, \\ \tilde{h}(t) &= h(t) - h^* \,, \\ \tilde{v}(t) &= v(t) - v^* \,, \\ \tilde{z}(t) &= z(t) - z^* \,, \end{aligned} \tag{3.54}$$

and linearizing (3.36) around (3.51) leads to the dynamics

$$\begin{aligned} \dot{\mathbf{x}} &= \mathbf{a}\,\mathbf{x} + \mathbf{b}\,\tilde{v}_1 + \text{h.o.t.} \,, \\ y &= \mathbf{c}\,\mathbf{x} \,, \end{aligned} \tag{3.55}$$

with

$$\mathbf{x} = \begin{bmatrix} \tilde{h} \\ \tilde{v} \\ \tilde{z} \end{bmatrix}, \quad \mathbf{a} = \begin{bmatrix} 0 & -1 & 0 \\ \alpha\kappa & -(\alpha+\beta+d) & \eta \\ \kappa & -1 & 0 \end{bmatrix}, \quad \mathbf{b} = \begin{bmatrix} 1 \\ \beta \\ 0 \end{bmatrix}, \quad \mathbf{c} = \begin{bmatrix} 0 & 1 & 0 \end{bmatrix}, \quad y = \tilde{v} \,, \tag{3.56}$$

where $\kappa = V'(h^*)$ and $d = p'(v^*)$. These linearized dynamics are associated with transfer function

$$T(s) = \mathbf{c}\,(s\mathbf{I} - \mathbf{a})^{-1}\,\mathbf{b} = \frac{\beta s^2 + \alpha\kappa s + \eta\kappa}{s^3 + (\alpha+\beta+d)s^2 + (\alpha\kappa + \eta)\,s + \eta\kappa} \,, \tag{3.57}$$

and with characteristic equation

$$\det(s\mathbf{I} - \mathbf{a}) = s^3 + (\alpha + \beta + d)s^2 + (\alpha\kappa + \eta)\, s + \eta\kappa = 0\,, \tag{3.58}$$

which can be used to evaluate plant and string stability at the linear level.

Plant stability is achieved when all three roots s_1, s_2, and s_3 of the characteristic equation (3.58) are located in the left half of the complex plane, that is,

$$\mathrm{Re}(s_k) < 0\,, \quad k = 1, 2, 3\,. \tag{3.59}$$

Applying the Routh–Hurwitz criteria (see (A.10) in Appendix A we obtain

$$\begin{aligned}
&\alpha + \beta + d > 0\,,\\
&\alpha\kappa + \eta > 0\,,\\
&\eta\kappa > 0\,,\\
&(\alpha\kappa + \eta)\,(\alpha + \beta + d) - \eta\kappa > 0\,,
\end{aligned} \tag{3.60}$$

which, considering $\kappa > 0$, can be re-written as

$$\begin{aligned}
&\eta > 0\,,\\
&\alpha > \frac{1}{2}\left(-\beta - \frac{\eta}{\kappa} - d + \sqrt{\left(\beta - \frac{\eta}{\kappa} + d\right)^2 + 4\eta}\right).
\end{aligned} \tag{3.61}$$

The string stability condition (3.47) becomes

$$|T(\mathrm{j}\omega)|^2 = \frac{\left(\eta\kappa - \beta\omega^2\right)^2 + \left(\alpha\kappa\right)^2\omega^2}{\left(\eta\kappa - (\alpha + \beta + d)\omega^2\right)^2 + \left(\alpha\kappa + \eta - \omega^2\right)^2\omega^2} < 1, \quad \forall\omega > 0\,, \tag{3.62}$$

and this, according to (2.141), (2.142), can be re-written as

$$P(\omega) = \omega^4 + a\omega^2 + b > 0\,, \quad \forall\omega > 0\,, \tag{3.63}$$

where

$$\begin{aligned}
a &= (\alpha + \beta + d)^2 - \beta^2 - 2\,(\alpha\kappa + \eta)\,,\\
b &= \eta\,(\eta - 2\kappa d)\,.
\end{aligned} \tag{3.64}$$

Below we determine conditions for a and b to satisfy (3.63) and then turn them into conditions for α, β, κ and η.

We separate the cases $a \geq 0$ and $a < 0$. When $a \geq 0$ the polynomial $P(\omega)$ is strictly monotonously increasing in ω. That is, if $P(0) > 0$ holds then $P(\omega) > 0$, $\forall\omega > 0$ holds too. Thus,

$$b > 0\,, \tag{3.65}$$

ensures string stability. Using (3.64) and $\eta > 0$ (cf. (3.61)) we obtain

$$\eta > 2\kappa d \,. \tag{3.66}$$

When $a < 0$ then $P(\omega)$ has a minimum at $\hat{\omega} > 0$. That is, if $P(\hat{\omega}) > 0$ holds then $P(\omega) > 0, \forall \omega > 0$ holds too. To find the frequency $\hat{\omega}$ we calculate the derivative of $P(\omega)$:

$$\frac{\mathrm{d}}{\mathrm{d}\omega} P(\omega) = 4\omega^3 + 2a\omega = 0 \quad \Rightarrow \quad \hat{\omega} = \sqrt{-\frac{a}{2}}\,, \tag{3.67}$$

which yields the stability condition

$$P(\hat{\omega}) = b - \frac{a^2}{4} > 0 \,. \tag{3.68}$$

Note that this implies that $b > 0$ must hold, i.e., it is necessary for string stability regardless of the sign of a. By substituting (3.64) and considering $a < 0$ we obtain

$$\alpha > -(\beta - \kappa + d) + \sqrt{(\beta - \kappa)^2 - 2\kappa d + 2\eta - 2\sqrt{\eta\,(\eta - 2\kappa d)}}\,, \tag{3.69}$$

and

$$\alpha < -(\beta - \kappa + d) - \sqrt{(\beta - \kappa)^2 - 2\kappa d + 2\eta - 2\sqrt{\eta\,(\eta - 2\kappa d)}}\,, \tag{3.70}$$

where only condition (3.69) ensures string stability since for (3.70) the system is plant unstable; cf. (3.61).

We remark that in the special case of $d = 0$ the string stability conditions (3.69), (3.70) become independent of η and one obtains

$$\begin{aligned} \beta < \kappa \quad &\Rightarrow \quad \alpha > 2(\kappa - \beta)\,, \\ \beta > \kappa \quad &\Rightarrow \quad \alpha > 0\,, \end{aligned} \tag{3.71}$$

which is equivalent to (3.50) that was derived for the case without integral control. Since $d \approx 0$ one shall expect the string stability boundaries to be close to these boundaries.

Figure 3.10 shows the plant and string stability charts of the ACC system (3.55), (3.56) in the (β, α) plane for different values of η; cf. (3.61), (3.69) and (3.70). The plant/string stability boundaries are shown by red/blue curves, while the plant/string stable domains are shaded as light/dark gray. Figure 3.10a shows the $\eta = 0$ case which demonstrates the effect of the resistance term compared to Fig. 3.9b. As η is increased the plant stable domain changes significantly and the origin becomes plant unstable. In the meantime the string stability domains vary very little with η as suggested by the calculations above.

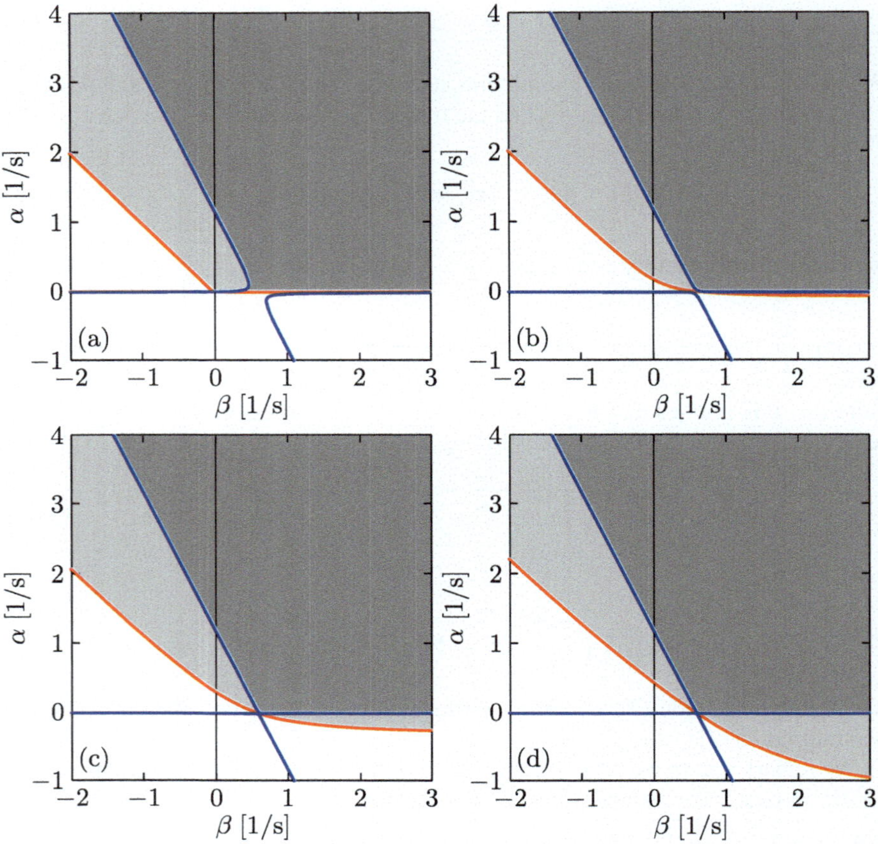

Fig. 3.10 Stability charts of system (3.55), (3.56) for $\kappa = 0.6\,[1/\mathrm{s}]$ and $d = 0.012\,[1/\mathrm{s}]$ corresponding to $v^* = 20\,[\mathrm{m/s}]$ and the parameters given in the first column of Table 3.1. Panels **a–d** correspond to the integral gains $\eta = 0$, $\eta = 0.05\,[1/\mathrm{s}^2]$, $\eta = 0.2\,[1/\mathrm{s}^2]$, $\eta = 0.8\,[1/\mathrm{s}^2]$, respectively. The plant and string stable regions are shaded by light and dark gray, respectively

3.3.2 Adaptive Cruise Control with Time Delay

Here, we discuss the plant and string stability of ACC when the response delay σ of the AV is taken into account. Utilizing the model (3.10) with the controller (3.29) results in

$$\begin{aligned}\dot{h}(t) &= v_1(t) - v(t)\,,\\ \dot{v}(t) &= -p\big(v(t)\big) + \mathrm{sat}\big(u(t-\sigma)\big)\,,\\ u &= F(h, v, v_1)\,,\end{aligned} \tag{3.72}$$

cf. (3.34). In particular, by dropping the resistance term and using the controller (3.30) we study the closed-loop dynamics

$$\begin{aligned}\dot{h}(t) &= v_1(t) - v(t)\,,\\ \dot{v}(t) &= \mathrm{sat}\big(u(t-\sigma)\big)\,,\\ u &= \alpha\big(V(h) - v\big) + \beta\big(W(v_1) - v\big)\,,\end{aligned} \tag{3.73}$$

cf. (3.35). We also study the case with disturbances and integral action while using the controller (3.31):

$$\begin{aligned}\dot{h}(t) &= v_1(t) - v(t)\,,\\ \dot{v}(t) &= -p\big(v(t)\big) + \mathrm{sat}\big(u(t-\sigma)\big)\,,\\ \dot{z}(t) &= V\big(h(t)\big) - v(t)\,,\\ u &= \alpha\big(V(h) - v\big) + \eta z + \beta\big(W(v_1) - v\big)\,,\end{aligned} \tag{3.74}$$

cf. (3.36).

Again, for simplicity, we drop sat and W in (3.73) and (3.74). In this case, (3.73) is identical to the optimal velocity model (2.129), (2.4), except that here we denote the delay with σ rather than τ. We first discuss the simpler system (3.73) before presenting the more general system (3.74).

It can be shown that the closed-loop dynamics (3.73) possesses the equilibrium (3.38), (3.39) where the HV and the AV drive with the same constant velocity and the AV keeps a constant headway. Defining the perturbations (3.40), the linearized dynamics around the equilibrium are given by

$$\begin{aligned}\dot{\mathbf{x}}(t) &= \mathbf{a}_0\,\mathbf{x}(t) + \mathbf{a}_\sigma\,\mathbf{x}(t-\sigma) + \mathbf{b}_0\,\tilde{v}_1(t) + \mathbf{b}_\sigma\,\tilde{v}_1(t-\sigma) + \text{h.o.t.}\,,\\ y(t) &= \mathbf{c}\,\mathbf{x}(t)\,.\end{aligned} \tag{3.75}$$

with

$$\mathbf{a}_0 = \begin{bmatrix} 0 & -1 \\ 0 & 0 \end{bmatrix}, \quad \mathbf{a}_\sigma = \begin{bmatrix} 0 & 0 \\ \alpha\kappa & -(\alpha+\beta) \end{bmatrix}, \quad \mathbf{b}_0 = \begin{bmatrix} 1 \\ 0 \end{bmatrix}, \quad \mathbf{b}_\sigma = \begin{bmatrix} 0 \\ \beta \end{bmatrix}, \quad \mathbf{c} = \begin{bmatrix} 0 & 1 \end{bmatrix}, \tag{3.76}$$

and $\kappa = V'(h^*)$; cf. (2.24), (2.131), (2.135) and (3.41), (3.42). This leads to the transfer function

$$T(s) = \mathbf{c}\big(s\mathbf{I} - \mathbf{a}_0 - \mathbf{a}_\sigma\, \mathrm{e}^{-s\sigma}\big)^{-1}\big(\mathbf{b}_0 + \mathbf{b}_\sigma\, \mathrm{e}^{-s\sigma}\big) = \frac{\beta s + \alpha\kappa}{s^2 \mathrm{e}^{s\sigma} + (\alpha+\beta)\, s + \alpha\kappa}, \tag{3.77}$$

cf. (2.133), (2.136) and (3.43), and to the characteristic equation

$$s^2 \mathrm{e}^{s\sigma} + (\alpha+\beta)\, s + \alpha\kappa = 0\,, \tag{3.78}$$

cf. (2.134), (2.137) and (3.44).

To ensure plant stability, all the infinitely many characteristic roots shall be located in the left-half complex plane according to Theorem 2.6, that is,

$$\mathrm{Re}(s_k) < 0, \quad k = 1, \ldots, \infty\,. \tag{3.79}$$

Equivalent to the calculations (2.138)–(2.139) in Chap. 2, the plant stability boundary corresponding to $s = 0$ is given by

$$\alpha = 0\,, \tag{3.80}$$

while the boundary corresponding to $s = \mathrm{j}\Omega$, $\Omega > 0$ is given by

$$\begin{aligned} \alpha &= \frac{\Omega^2 \cos(\Omega\sigma)}{\kappa}\,, \\ \beta &= \Omega \sin(\Omega\sigma) - \alpha\,. \end{aligned} \tag{3.81}$$

String stability is still given by the condition (3.47) and following the calculations (2.140)–(2.148) in Chap. 2, one can obtain the $\omega = 0$ string stability condition as

$$\begin{aligned} \alpha &> 0\,, \\ \alpha &> 2(\kappa - \beta)\,, \end{aligned} \tag{3.82}$$

and the $\omega > 0$ string stability boundary as

$$
\begin{aligned}
\alpha &= a \pm \sqrt{a^2+b}\,, & a &= \frac{\hat{\omega}(\kappa\sigma-1)+\kappa\sin(\hat{\omega}\sigma)\cos(\hat{\omega}\sigma)}{(2\kappa\sigma-1)\sin(\hat{\omega}\sigma)-\hat{\omega}\sigma\cos(\hat{\omega}\sigma)}\,,\\
& & b &= \frac{\hat{\omega}^2\big(\sin(\hat{\omega}\sigma)-\hat{\omega}\sigma\cos(\hat{\omega}\sigma)\big)}{(2\kappa\sigma-1)\sin(\hat{\omega}\sigma)-\hat{\omega}\sigma\cos(\hat{\omega}\sigma)}\,,\\
\beta &= \frac{\hat{\omega}+\alpha\kappa\sigma\sin(\hat{\omega}\sigma)}{\sin(\hat{\omega}\sigma)+\hat{\omega}\sigma\cos(\hat{\omega}\sigma)}-\alpha\,.
\end{aligned}
\tag{3.83}
$$

Moreover, following the calculations (2.149)–(2.151) one may show that for

$$
\sigma > \sigma_{\rm cr} = \frac{1}{2\kappa}\,, \tag{3.84}
$$

the string stable domain disappears.

Figure 3.11 depicts the stability charts of the ACC system (3.75), (3.76) in the (β,α) plane with delay in the AV's control loop. The boundaries (3.80), (3.81) of the plant stable domains are shown by red curves, while the boundaries (3.82), (3.83) of the string stable domains are shown by blue curves. The plant stable domain is shaded light gray, whereas the string stable domain is shaded dark gray. Observe that due to the delay, the stable domains are much smaller compared to the non-delayed system shown in Fig. 3.9. The stability chart in Fig. 3.11b is identical to Fig. 2.17 which describes the dynamics of human drivers. Panels (a)–(d) depict stability charts for different values of κ and demonstrate that as κ is increased the stable domains shrink. In particular, the points A, B, C move closer to each other and for $\kappa > 1/(2\sigma) \approx 0.83\,[1/{\rm s}]$ the string stable domain disappears; cf. (3.84).

One may conclude that choosing large gains no longer achieves string stability when the response delay σ of the AV is taken into account. This highlights a fundamental limitation of ACC design: setting the time headway $t_{\rm h} = 1/\kappa$ too small jeopardizes string stability. In Chap. 4 we will demonstrate that one may overcome this limitation by using connectivity that provides controllers with extra information in addition to instantaneous range and range rate data. This way string stability can be maintained even for smaller time headways than suggested by (3.84) and even the string unstable behavior of human-driven vehicles in the traffic flow can be compensated.

Now, we analyze the closed-loop dynamics (3.74) with delay in the control loop and with disturbances being compensated by integral action. It can be shown that the system possesses the equilibrium (3.51), (3.53), where the AV maintains the same constant velocity as the preceding vehicle and keeps a constant headway while in the meantime compensates for the disturbance. Defining the perturbations (3.54) one may linearize (3.74) around (3.51) and obtain

$$
\begin{aligned}
\dot{\mathbf{x}}(t) &= \mathbf{a}_0\,\mathbf{x}(t) + \mathbf{a}_\sigma\,\mathbf{x}(t-\sigma) + \mathbf{b}_0\,\tilde{v}_1(t) + \mathbf{b}_\sigma\,\tilde{v}_1(t-\sigma) + \text{h.o.t.}\,,\\
y(t) &= \mathbf{c}\,\mathbf{x}(t)\,,
\end{aligned}
\tag{3.85}
$$

with

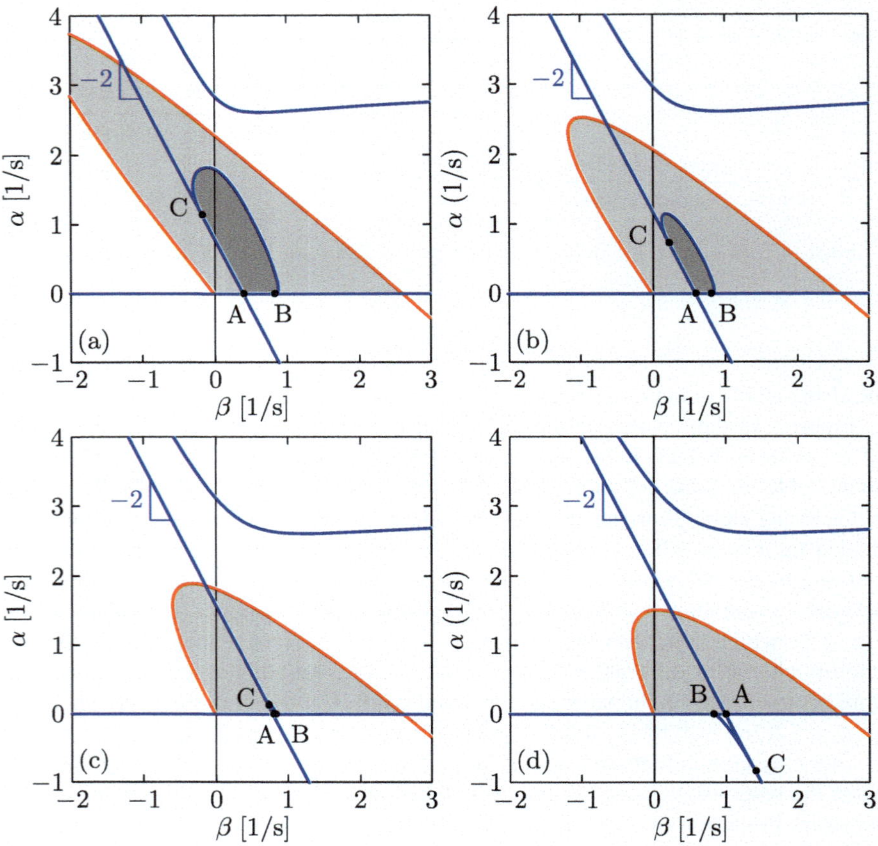

Fig. 3.11 Stability charts of system (3.75), (3.76) with delay $\sigma = 0.6$ [s]. Panels **a**–**d** are associated with range policy gradient $\kappa = 0.4$ [1/s], $\kappa = 0.6$ [1/s], $\kappa = 0.8$ [1/s] and $\kappa = 1$ [1/s], respectively. The plant and string stable regions are shaded by light and dark gray, respectively. Points A, B, and C are located at the intersections of the string stability boundaries

$$\mathbf{a}_0 = \begin{bmatrix} 0 & -1 & 0 \\ 0 & -d & 0 \\ \kappa & -1 & 0 \end{bmatrix}, \quad \mathbf{a}_\sigma = \begin{bmatrix} 0 & 0 & 0 \\ \alpha\kappa & -(\alpha+\beta) & \eta \\ 0 & 0 & 0 \end{bmatrix}, \quad \mathbf{b}_0 = \begin{bmatrix} 1 \\ 0 \\ 0 \end{bmatrix}, \quad \mathbf{b}_\sigma = \begin{bmatrix} 0 \\ \beta \\ 0 \end{bmatrix}, \quad \mathbf{c} = \begin{bmatrix} 0 & 1 & 0 \end{bmatrix}, \tag{3.86}$$

where $\kappa = V'(h^*)$ and $d = p'(v^*)$; cf. (3.55), (3.56). These linearized dynamics are associated with the transfer function

$$\begin{aligned} T(s) &= \mathbf{c}\big(s\mathbf{I} - \mathbf{a}_0 - \mathbf{a}_\sigma \mathrm{e}^{-s\sigma}\big)^{-1}\big(\mathbf{b}_0 + \mathbf{b}_\sigma \mathrm{e}^{-s\sigma}\big) \\ &= \frac{\beta s^2 + \alpha\kappa s + \eta\kappa}{\big(s^3 + ds^2\big)\mathrm{e}^{s\sigma} + (\alpha+\beta)\,s^2 + (\alpha\kappa + \eta)\,s + \eta\kappa}, \end{aligned} \tag{3.87}$$

cf. (3.57) and (3.77), and the characteristic equation

$$(s^3 + ds^2)e^{s\sigma} + (\alpha + \beta)\, s^2 + (\alpha\kappa + \eta)\, s + \eta\kappa = 0\,, \tag{3.88}$$

cf. (3.58) and (3.78), which are used to evaluate plant and string stability.

Due to the delay the characteristic equation (3.88) has infinitely many roots, and plant stability is ensured when these are all on the left-half complex plane; see (3.79). For $s = 0$, considering $\kappa > 0$, we obtain the plant stability boundary

$$\eta = 0\,, \tag{3.89}$$

which corresponds to the first condition in (3.61). For $s = \mathrm{j}\Omega$, $\Omega > 0$, by separating the real and imaginary parts of the characteristic equation, we obtain the plant stability boundary

$$\begin{aligned} \alpha &= \frac{1}{\kappa}\Big(\Omega^2\cos(\Omega\sigma) + d\Omega\sin(\Omega\sigma) - \eta\Big)\,, \\ \beta &= \Omega\sin(\Omega\sigma) - d\cos(\Omega\sigma) + \frac{\eta\kappa}{\Omega^2} - \alpha\,. \end{aligned} \tag{3.90}$$

Using the transformation (2.141), (2.142) the string stability condition (3.47) can be re-written as

$$\begin{aligned} P(\omega) &= \omega^4 + \big(\alpha^2 + 2\alpha\beta + d^2\big)\omega^2 + \eta^2 \\ &\quad - 2\Big(\big(\alpha\kappa + \eta - (\alpha + \beta)d\big)\omega^2 + \eta\kappa d\Big)\cos(\omega\sigma) \\ &\quad - 2\Big((\alpha + \beta)\omega^2 - \eta\kappa + (\alpha\kappa + \eta)d\Big)\omega\sin(\omega\sigma) > 0\,, \quad \forall\omega > 0\,. \end{aligned} \tag{3.91}$$

Then, the $\omega = 0$ string stability condition (2.144) yields

$$\eta > 2\kappa d\,, \tag{3.92}$$

which is the same as (3.66). Moreover, the $\omega > 0$ string stability boundary (2.147) can be obtained by differentiating (3.91) and solving for α and β. This results in

$$\begin{aligned} \alpha &= a \pm \sqrt{a^2 + b}\,, \\ \beta &= \frac{c_0 + c_1\alpha}{c_2} - \alpha\,, \\ a &= \frac{1}{2c_1 - c_2}\left(-c_0 + (c_2\kappa - c_1 d)\cos(\hat{\omega}\sigma) + \left(c_1 + \frac{c_2\kappa d}{\hat{\omega}^2}\right)\hat{\omega}\sin(\hat{\omega}\sigma)\right), \\ b &= \frac{1}{2c_1 - c_2}\left(-c_2\left(\hat{\omega}^2 + d^2 + \frac{\eta^2}{\hat{\omega}^2}\right) + 2\left(c_2\eta - c_0 d + \frac{c_2\eta\kappa d}{\hat{\omega}^2}\right)\cos(\hat{\omega}\sigma)\right. \\ &\quad \left. + 2\left(c_0 - \frac{c_2\eta(\kappa - d)}{\hat{\omega}^2}\right)\hat{\omega}\sin(\hat{\omega}\sigma)\right), \end{aligned}$$

$$
\begin{aligned}
c_0 &= \hat{\omega}^4 - \eta^2 + \eta\left((\kappa - d)\hat{\omega}^2\sigma + 2\kappa d\right)\cos(\hat{\omega}\sigma) + \eta\left(\hat{\omega}^2\sigma - (\kappa - d) + \kappa d\sigma\right)\hat{\omega}\sin(\hat{\omega}\sigma)\,,\\
c_1 &= \kappa\left(\hat{\omega}^2\sigma + d\right)\hat{\omega}\sin(\hat{\omega}\sigma) - \kappa d\sigma\hat{\omega}^2\cos(\hat{\omega}\sigma)\,,\\
c_2 &= (d\sigma + 1)\hat{\omega}^3\sin(\hat{\omega}\sigma) + \hat{\omega}^4\sigma\cos(\hat{\omega}\sigma)\,.
\end{aligned}
\tag{3.93}
$$

Finally, one may prove that considering $d \approx 0$ the critical delay—above which there exists no string stable domain in the (β, α, η) space—is given by

$$
\sigma > \sigma_{\rm cr} \approx \frac{1}{2\kappa}\,,
\tag{3.94}
$$

cf. (3.84). That is, adding integral action does not change the critical delay.

Figure 3.12 shows the stability charts of the ACC system (3.85), (3.86) in the (β, α) plane for different values of η. The plant stable domains are bounded by red curves and shaded light gray (cf. (3.90)), while the string stable domains are bounded by blue curves and shaded dark gray (cf. (3.93)). Again, the delay makes the stable domains much smaller compared to the delay-free case shown in Fig. 3.10. For Fig. 3.12a we have $\eta = 0$ and this shows the effect of the resistance term compared to Fig. 3.11b. For the other panels in Fig. 3.12 we have $\eta > 0$ and they show that the stable domains shrink as η is increased. Also, notice that when $\eta > 0$, there are no $s = 0$ plant stability curves and there are no $\omega = 0$ string stability curves. The reminiscence of these is now embedded in the low frequency sections of the $s = \pm \mathrm{j}\Omega$ plant stability curves and in the low frequency sections of the $\omega > 0$ string stability curves. This highlights that such zero frequency stability losses are in fact structurally unstable, and are due to the simplified vehicle models and controllers used. Yet, these still provide us with a good guidance for the stability charts of the real physical systems as will also be demonstrated in the experimental Chap. 5.

In order to illustrate the nonlinear behavior of the closed-loop system (3.74) we select the control gains corresponding to point T in Fig. 3.12b and simulate the response of the system to a harmonic input. Figure 3.13a, b show the headway and the velocity of a car using the parameters in the first column of Table 3.1. Notice that, corresponding to the linear design, both plant an string stability are satisfied: the velocity amplitude of the AV is smaller than that of the HV. Also notice that the response of the AV is close to harmonic. Figure 3.13c, d depict the response of a heavy-duty truck (with a fully loaded trailer) using the parameters in the second column of Table 3.1. Observe that the power and torque limitations play a significant role in shaping the response of the truck, which is far from being harmonic.

Finally, we remark that the first-order lag approximation presented in Sect. 2.3.4 can also be applied to the control system (3.72) yielding

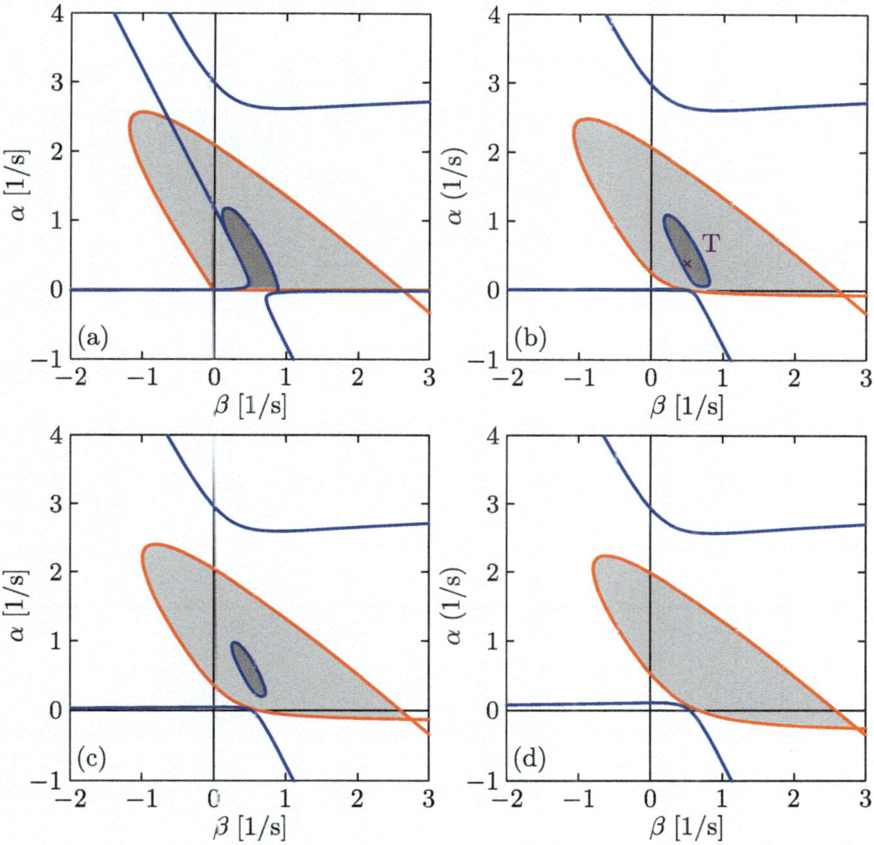

Fig. 3.12 Stability charts of system (3.85), (3.86) with $\sigma = 0.6$ [s], $\kappa = 0.6$ [1/s] and $d = 0.012$ [1/s] corresponding to $v^* = 20$ [m/s] and to the parameters given in the first column of Table 3.1. Panels **a–d** correspond to the integral gains $\eta = 0$, $\eta = 0.05$ [1/s^2], $\eta = 0.1$ [1/s^2], $\eta = 0.2$ [1/s^2], respectively. The plant and string stable regions are shaded by light and dark gray, respectively

$$
\begin{aligned}
\dot{h} &= v_1 - v\,, \\
\dot{v} &= -p(v) + \mathrm{sat}(u)\,, \\
\dot{u} &= \frac{1}{\sigma}\Big(F(h, v, v_1) - u\Big)\,,
\end{aligned}
\tag{3.95}
$$

cf. (2.176), which is often used in the literature [260, 262].

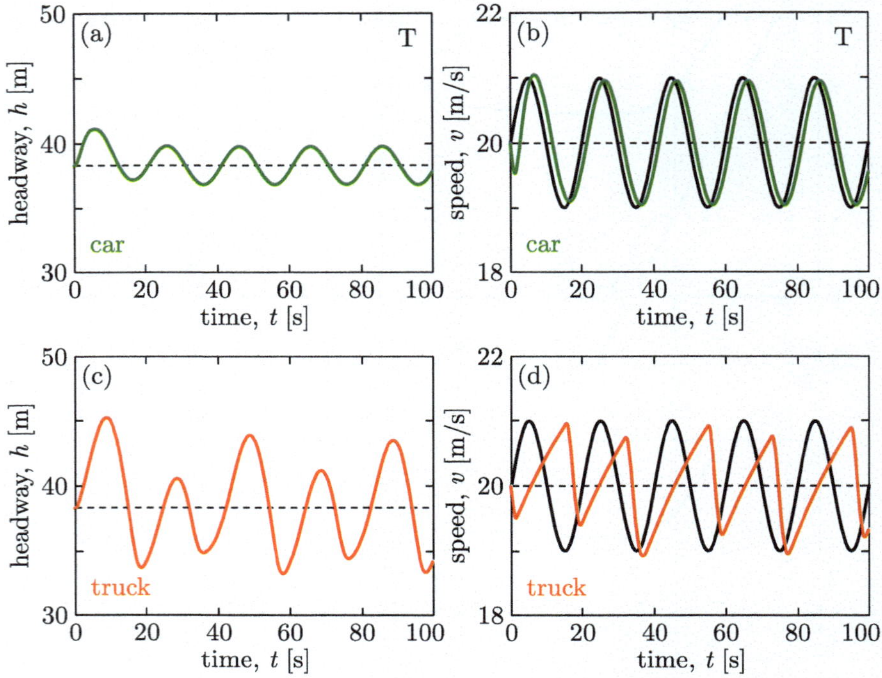

Fig. 3.13 Simulation results of system (3.74) for a passenger vehicle (car) and a heavy-duty vehicle (truck) using the parameters in Table 3.1, the road grade $\phi = 0.025$ [rad], the piecewise linear range policy (3.32) with $h_{\mathrm{st}} = 5$ [m], $h_{\mathrm{go}} = 55$ [m], $v_{\max} = 30$ [m/s], and the control gains $\alpha = 0.4$ [1/s], $\eta = 0.05$ [1/s^2], $\beta = 0.5$ [1/s] corresponding to point T in Fig. 3.12b. The initial conditions are $v(t) = v^* = 20$ [m/s], $h(t) = V^{-1}(v^*) = 38.33$ [m], $z(t) = 0$ [m] for $t \in [-\sigma, 0]$ and the lead vehicle's velocity is $v_1(t) = v^* + v_{\mathrm{amp}} \sin(\omega t)$, $v_{\mathrm{amp}} = 1$ [m/s], $\omega = 0.1\pi$ [rad/s]

Specifically, the approximation of (3.73) becomes

$$\begin{aligned} \dot{h} &= v_1 - v\,, \\ \dot{v} &= \mathrm{sat}(u)\,, \\ \dot{u} &= \frac{1}{\sigma}\Big(\alpha\big(V(h) - v\big) + \beta\big(W(v_1) - v\big) - u\Big)\,. \end{aligned} \tag{3.96}$$

When dropping sat and W, the equilibrium of this system is given by (3.38), (3.39) and $u(t) \equiv 0$. Linearizing about this equilibrium leads to

$$\begin{bmatrix} \dot{\tilde{h}} \\ \dot{\tilde{v}} \\ \dot{\tilde{u}} \end{bmatrix} = \begin{bmatrix} 0 & -1 & 0 \\ 0 & 0 & 1 \\ \dfrac{\alpha\kappa}{\sigma} & -\dfrac{\alpha+\beta}{\sigma} & -\dfrac{1}{\sigma} \end{bmatrix} \begin{bmatrix} \tilde{h} \\ \tilde{v} \\ \tilde{u} \end{bmatrix} + \begin{bmatrix} 1 \\ 0 \\ \dfrac{\beta}{\sigma} \end{bmatrix} \tilde{v}_1 + \text{h.o.t.}\,, \tag{3.97}$$

cf. (2.178), (2.179) and (2.181). Using $\tilde{v}$ as output, we obtain the transfer function

$$T(s) = \frac{\beta s + \alpha\kappa}{\sigma s^3 + s^2 + (\alpha+\beta)s + \alpha\kappa}\,, \tag{3.98}$$

cf. (2.182), and the stability charts are similar to the one in Fig. 2.26. Note that one may obtain this transfer function by substituting $e^{s\sigma} \approx 1 + s\sigma$ into (3.77).

Similarly, applying the first-order lag approximation to (3.74) yields

$$\begin{aligned}
\dot{h} &= v_1 - v\,,\\
\dot{v} &= -p(v) + \mathrm{sat}(u)\,,\\
\dot{z} &= V(h) - v\,,\\
\dot{u} &= \frac{1}{\sigma}\Big(\alpha\big(V(h) - v\big) + \eta z + \beta\big(W(v_1) - v\big) - u\Big)\,.
\end{aligned} \tag{3.99}$$

When dropping sat and W, the equilibrium of this system is given by (3.51), (3.53) and $u(t) \equiv u^* = p(v^*)$. Linearization about the equilibrium yields

$$\begin{bmatrix} \dot{\tilde{h}} \\ \dot{\tilde{v}} \\ \dot{\tilde{z}} \\ \dot{\tilde{u}} \end{bmatrix} = \begin{bmatrix} 0 & -1 & 0 & 0 \\ 0 & -d & 0 & 1 \\ \kappa & -1 & 0 & 0 \\ \frac{\alpha\kappa}{\sigma} & -\frac{\alpha+\beta}{\sigma} & \frac{\eta}{\sigma} & -\frac{1}{\sigma} \end{bmatrix} \begin{bmatrix} \tilde{h} \\ \tilde{v} \\ \tilde{z} \\ \tilde{u} \end{bmatrix} + \begin{bmatrix} 1 \\ 0 \\ 0 \\ \frac{\beta}{\sigma} \end{bmatrix} \tilde{v}_1 + \text{h.o.t.}\,, \tag{3.100}$$

and the transfer function

$$T(s) = \frac{\beta s^2 + \alpha\kappa s + \eta\kappa}{\sigma s^4 + (\sigma d + 1)s^3 + (\alpha + \beta + d)s^2 + (\alpha\kappa + \eta)s + \eta\kappa}\,, \tag{3.101}$$

which can also be obtained by substituting $e^{s\sigma} \approx 1 + s\sigma$ into (3.87). Here we do not analyze this system further but the analysis shall follow the same steps as laid out above.

To sum up, one may state that adaptive cruise control design shows similarities to the analysis of human car following. In particular, the time delay in the control loop of the automated vehicle plays a significant role in ACC design, and leads to a fundamental limitation about how close an AV can follow the preceding vehicle (in terms of time headway) without compromising string stability. On the other hand, while integral action can be added to eliminate disturbances, one may still use a vehicle model without disturbances (but with delay) when designing ACC. This simplified model will also be used in the Sect. 3.5 when analyzing the effect of digital sampling and used as the base for the connected cruise control design in the next chapter. In the next section, however, we use the model with disturbances and the controller with integral action in order to represent automated vehicles when those are mixed into the flow of human-driven vehicles.

3.4 Mitigating Congestion with Adaptive Cruise Control in Mixed Traffic Systems

Above we have laid out the principles of adaptive cruise control (ACC) design, and we discussed different performance metrics (e.g., plant stability and string stability) at the level of a single leader-follower pair. In this section, we insert automated vehicles (AVs) driven by ACC into the flow of human-driven vehicles (HVs) as illustrated in Fig. 3.14 and evaluate the performance of the whole vehicle chain using the analytical tools developed in Sect. 2.2.1. We also illustrate the results using numerical simulations.

We investigate open chains of mixed traffic flow for different AV penetration rates defined as

$$p = \frac{N_{\rm AV}}{N} = \frac{N_{\rm AV}}{N_{\rm HV} + N_{\rm AV}} \,. \tag{3.102}$$

Here $N_{\rm AV}$ and $N_{\rm HV}$ denotes the numbers of automated vehicles and human-driven vehicles, respectively, and $N = N_{\rm HV} + N_{\rm AV}$ gives the total number of follower vehicles. The key hypothesis is that the string unstable motion of HVs can be compensated by the string stable behavior of AVs, and N-to-0 string stability can be achieved once the penetration of AVs exceeds a critical value. For further discussions on stabilizing traffic flows with AVs, see [133].

In order to motivate the following analysis, we start with numerical simulations. To represent the human-driven vehicles we utilize the nonlinear car-following model (2.207), which combines (2.129) with the simple saturation function (2.208) that uses constant acceleration limits. In particular, we utilize the OVM (2.207), (2.4) with piecewise linear range policy (2.5), (2.6). The human parameters are chosen such that the reaction time delay is larger that the critical delay (2.151), that is, these vehicles are string unstable for any combination of the gain parameters. To model the automated vehicles, we utilize the physics-based model (3.10) with the higher-fidelity saturation function (3.8), (3.9) that incorporates the power limit, while we omit the grade ($\phi = 0$) and the headwind ($v_{\rm w} = 0$). We assume that the AVs utilize the range policy (3.32) and the speed policy (3.33). We apply the controller (3.31) with integral action yielding the closed-loop dynamics (3.74). That is, we take into account the resistance terms and use integral action in the controller in order to compensate those. We use the parameters from the first column of Table 3.1 and the gain parameters corresponding to point T in Fig. 3.12b.

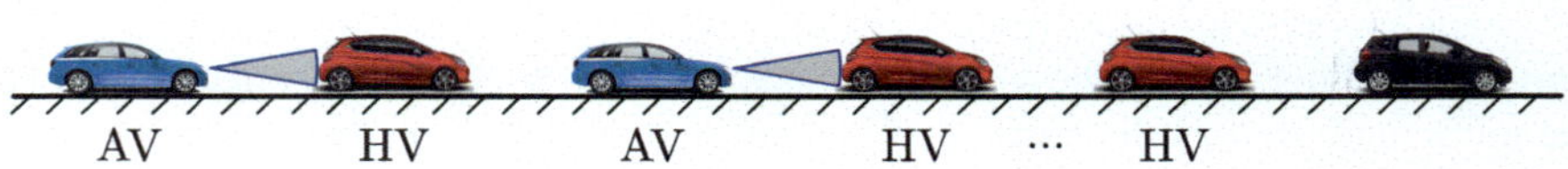

Fig. 3.14 Illustration of mixed traffic flow consisting of both automated vehicles (AVs) that use adaptive cruise control (ACC) and human-driven vehicles (HVs)

The simulation results are shown for 24 follower vehicles in Fig. 3.15, where the first row represents the case of human-driven vehicles, while the other rows correspond to $p = 16.67\%$, $p = 33.33\%$ and $p = 50\%$ AV penetration levels, cf. (3.102). These levels are achieved by making every 6th, 3rd, and 2nd vehicles to be AVs, respectively. Due to the string instability of human drivers the perturbation imposed by the lead vehicle increases as it propagates backward along the chain of human-driven vehicles, cf. Fig. 2.12a. For $p = 16.67\%$ the perturbation level is maintained at (approximately) the same level. Finally, for $p = 33.33\%$ and for $p = 50\%$ the perturbations decay along the vehicle chain indicating string stable behavior. These results demonstrate that having large enough penetration of AVs (with well-designed ACC algorithms) can be used to mitigate traffic congestion in mixed traffic.

Now we present the formal string stability analysis of vehicle chains consisting of N_{HV} human-driven vehicles and N_{AV} automated vehicles. While the analysis holds at the linear level only, it can be applied to evaluate the string stability of vehicle chains independent of the order of HVs and AVs in the chain. In particular, we utilize Theorem 2.3 which allows one to evaluate the $(i+1)$-to-$(i-k)$ string stability by using the product of the transfer functions. Having $N = N_{\mathrm{HV}} + N_{\mathrm{AV}}$ following vehicles we analyze the N-to-0 string stability which we call *head-to-tail string stability*. According to (2.72), (2.74), the vehicle chain is head-to-tail string stable is and only if

$$|G_{0N}(\mathrm{j}\omega)| < 1 \,, \quad \forall \omega > 0 \,, \tag{3.103}$$

where

$$G_{0N}(s) = T_{01}(s)\, T_{12}(s) \cdots T_{N-1,N}(s) \,, \tag{3.104}$$

is called the *head-to-tail transfer function*.

The transfer functions $T_{i,i+1}(s)$, $i = 0, \ldots, N-1$ describe how individual vehicles respond to the motion of their immediate predecessors, and these are referred to as *link transfer functions*. For example, using the OVM parametrization, human drivers are described by the link transfer function

$$T_{\mathrm{h}}(s) = \frac{\beta_{\mathrm{h}} s + \alpha_{\mathrm{h}} \kappa_{\mathrm{h}}}{s^2 \mathrm{e}^{s\tau} + (\alpha_{\mathrm{h}} + \beta_{\mathrm{h}}) s + \alpha_{\mathrm{h}} \kappa_{\mathrm{h}}} \,, \tag{3.105}$$

cf. (2.136), where we added the subscript h to refer to humans. Analogously, for automated vehicles one may use the link transfer functions

$$T(s) = \frac{\beta s + \alpha \kappa}{s^2 \mathrm{e}^{s\sigma} + (\alpha + \beta)\, s + \alpha \kappa} \,, \tag{3.106}$$

cf. (3.77), or

$$T(s) = \frac{\beta s^2 + \alpha \kappa s + \eta \kappa}{\left(s^3 + d s^2\right) \mathrm{e}^{s\sigma} + (\alpha + \beta)\, s^2 + (\alpha \kappa + \eta)\, s + \eta \kappa} \,, \tag{3.107}$$

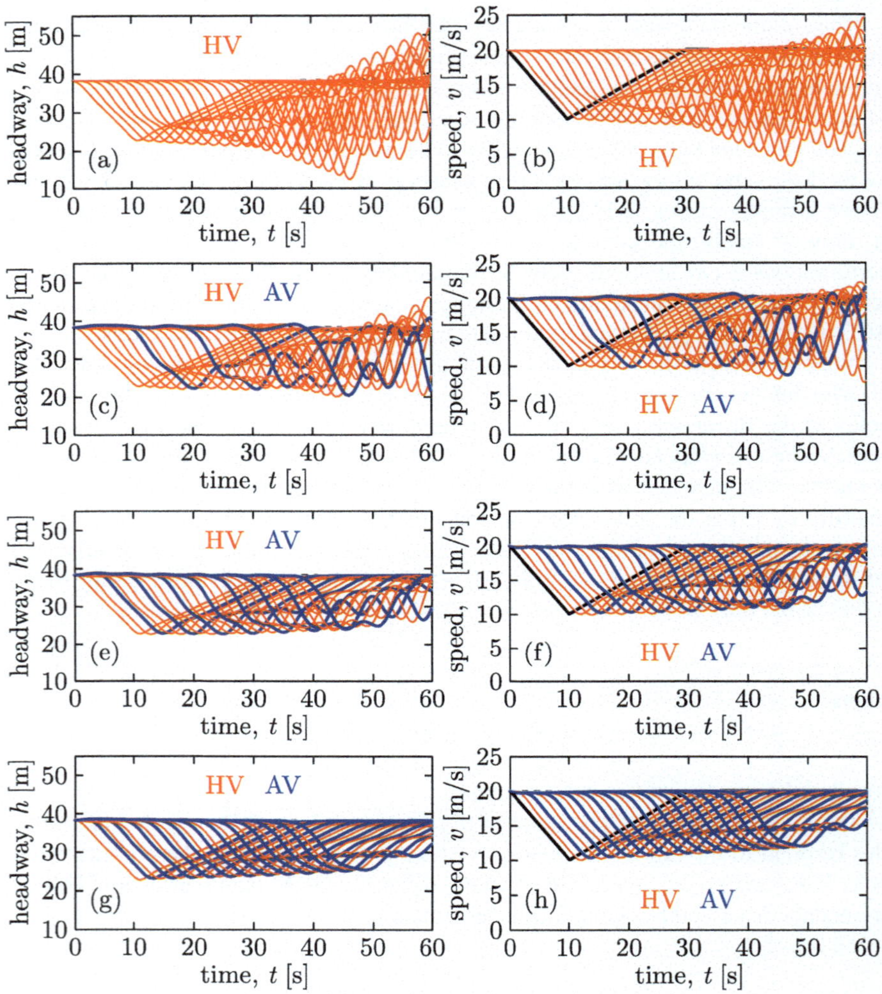

Fig. 3.15 Numerical simulations of mixed traffic for the case of 24 follower vehicles with AV penetration rates **a**, **b** $p = 0\%$, **c**, **d** $p = 16.67\%$, **e**, **f** $p = 33.33\%$, **g**, **h** $p = 50\%$. HVs are modeled by the OVM (2.207), (2.4) with piecewise linear range policy (2.5), (2.6), gains $\alpha_{\rm h} = 0.1\,[1/{\rm s}]$, $\beta_{\rm h} = 0.6\,[1/{\rm s}]$, and delay $\tau = 1\,[{\rm s}]$. AVs are modeled by (3.74) with range policy (3.32), speed policy (3.33), gains $\alpha = 0.4\,[1/{\rm s}]$, $\beta = 0.5\,[1/{\rm s}]$, $\eta = 0.05\,[1/{\rm s}^2]$, and delay $\sigma = 0.6\,[{\rm s}]$. Other AV parameters are drawn from the first column of Table 3.1, while $h_{\rm st} = 5\,[{\rm m}]$, $h_{\rm go} = 55\,[{\rm m}]$, $v_{\max} = 30\,[{\rm m/s}]$, $a_{\min} = 7\,[{\rm m/s}^2]$, $a_{\max} = 3\,[{\rm m/s}^2]$ are used for all vehicles. The initial conditions are $v_i(t) = v^* = 20\,[{\rm m/s}]$, $h_i(t) = V_{\rm h}^{-1}(v^*) = 38.33\,[{\rm m}]$ for $t \in [-\tau, 0]$ for HVs, and $v_i(t) = v^* = 20\,[{\rm m/s}]$, $h_i(t) = V^{-1}(v^*) = 38.33\,[{\rm m}]$, $z_i(t) = 0\,[{\rm m}]$ for $t \in [-\sigma, 0]$ for AVs. At the equilibrium speed $v^* = 20\,[{\rm m/s}]$, the range policy gradients are $\kappa_{\rm h} = 0.6\,[1/{\rm s}]$ and $\kappa = 0.6\,[1/{\rm s}]$. The lead vehicle uses constant $-1\,[{\rm m/s}^2]$ and $0.5\,[{\rm m/s}^2]$ acceleration to slow down from $20\,[{\rm m/s}]$ to $10\,[{\rm m/s}]$ and to speed back up

cf. (3.87). Here we use the latter one corresponding to the above-described scenario where the AVs compensate the disturbances by using integral actions.

Assuming that the mixed traffic chain consists of $N_{\rm HV}$ identical human-driven vehicles (with link transfer function $T_{\rm h}(s)$) and $N_{\rm AV}$ identical automated vehicles (with link transfer function $T(s)$), the head-to-tail transfer function (3.104) becomes

$$G_{0N}(s) = T_{\rm h}^{N_{\rm HV}}(s)\, T^{N_{\rm AV}}(s)\,. \tag{3.108}$$

Taking the N-th root and using the definition (3.102) of the AV penetration rate, we obtain

$$G(s) = \sqrt[N]{G_{0N}(s)} = T_{\rm h}^{1-p}(s)\, T^{p}(s)\,. \tag{3.109}$$

Correspondingly, the head-to-tail string stability condition (3.103) can be re-written as

$$|G(\mathrm{j}\omega)| < 1\,, \quad \forall \omega > 0\,. \tag{3.110}$$

For a chain of human-driven vehicles we have $p = 0$ yielding $G(s) = T_{\rm h}(s)$, while for a chain of automated vehicles we have $p = 1$ yielding $G(s) = T(s)$. Note that (3.109) allows one to compare vehicle chains of different lengths as it only depends on the link transfer functions and the penetration rate p.

These formulae emphasize that even if the HVs are string unstable (i.e., $|T_{\rm h}(\mathrm{j}\omega)| > 1$ for some angular frequency $\omega > 0$), head-to-tail string stability can be achieved with the help of string stable AVs (i.e., $|T(\mathrm{j}\omega)| < 1$ for all $\omega > 0$). Moreover, the higher the AV penetration is, the easier it is to achieve string stability. Note, however, that perturbations may still grow on certain sections of the chain; head-to-tail string stability only guarantees that the velocity fluctuations of the tail vehicle are smaller than those of the head vehicle. On the other hand, according to Theorem 2.4, mixed vehicle chains are plant stable if and only if all HVs and AVs are plant stable, because the characteristic equation includes the product of the denominators of the link transfer functions.

The frequency responses are shown in Fig. 3.16 for different AV penetration levels. The top panels show that when all vehicles in the chain are human driven (red curve) the system is head-to-tail string unstable, while when all vehicles are automated (blue curve) the system is head-to-tail string stable. As the AV penetration rate is increased, the maximum value of the frequency response curve decreases, and above a critical penetration rate head-to-tail string stability is achieved. The bottom panels illustrate this process by showing the value of the frequency response curve as a function of the AV penetration rate for a particular angular frequency ω^* (which is chosen where the human frequency response curve takes its maximum). It can be observed that the head-to-tail amplification is (approximately) a linear function of the penetration rate. The critical value, where the function goes below one, depends on the parameters, in particular on the human reaction time delay τ. This is highlighted by the difference between the left and right panels which show the frequency response for $\tau = 1$ [s] and $\tau = 0.9$ [s], respectively. More details about robustness against uncertainties in human parameters will be given in Sect. 4.7.

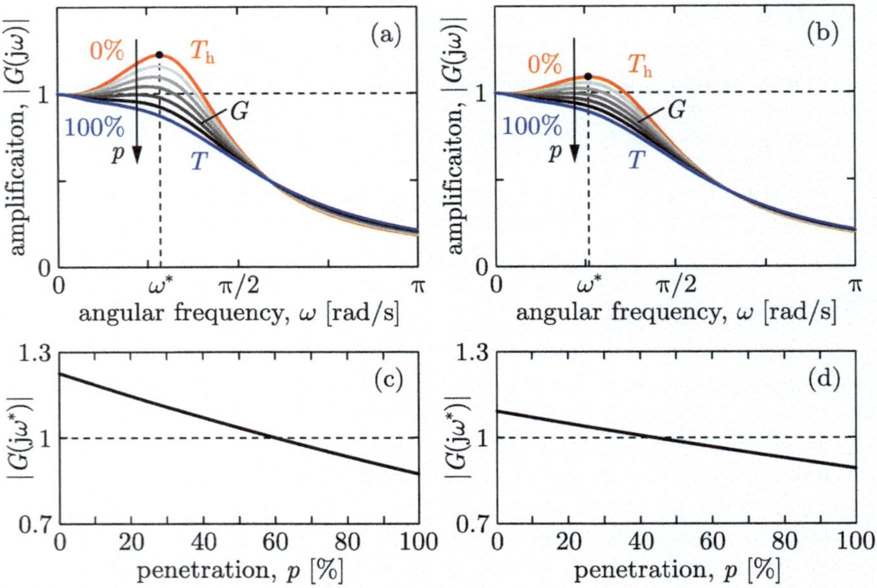

Fig. 3.16 Frequency responses of vehicle chains for mixed traffic scenarios based on the scaled head-to-tail transfer function (3.109) while using the link transfer functions (3.105) for HVs and (3.107) for AVs. Panels **a**, **b** depict the frequency response curves for penetration rates $p = 0\%$ (red), $p = 16.67, 33.33, 50, 66.67, 83.33\%$ (gray) and $p = 100\%$ (blue), while panels **c**, **d** show the head-to-tail amplification at angular frequency ω^* where the human frequency response is maximal. The HV parameters are $\alpha_h = 0.1$ [1/s], $\beta_h = 0.6$ [1/s], $\kappa_h = 0.6$ [1/s], and **a**, **c** $\tau = 1$ [s] and (b,d) $\tau = 0.9$ [s]. The AV parameters are $\alpha = 0.4$ [1/s], $\beta = 0.5$ [1/s], $\eta = 0.05$ [1/s^2], $\kappa = 0.6$ [1/s], $\sigma = 0.6$ [s], and $d = 0.012$ [1/s] corresponding to $v^* = 20$ [m/s] and to the parameters given in the first column of Table 3.1

The analysis above only holds at the linear level. As we illustrated in Sect. 2.4, nonlinearities in human behavior may lead to bistability phenomena. Here we omit the nonlinear analysis of mixed traffic but we will return to this problem in Sect. 4.5 after introducing connectivity in Chap. 4.

3.5 Effect of Sampling in Adaptive Cruise Control

In this section we discuss the application of digital control for adaptive cruise control and relate the effects of sampling to the effects of time delays in the control loop. In real hardware the ACC algorithms (3.30) and (3.31) are realized using digital control since range sensors provide information about the motion of vehicles with a given frequency. In other words they sample the motion signals with a sampling time Δt. This will remain the case once the information received via vehicle-to-everything (V2X)

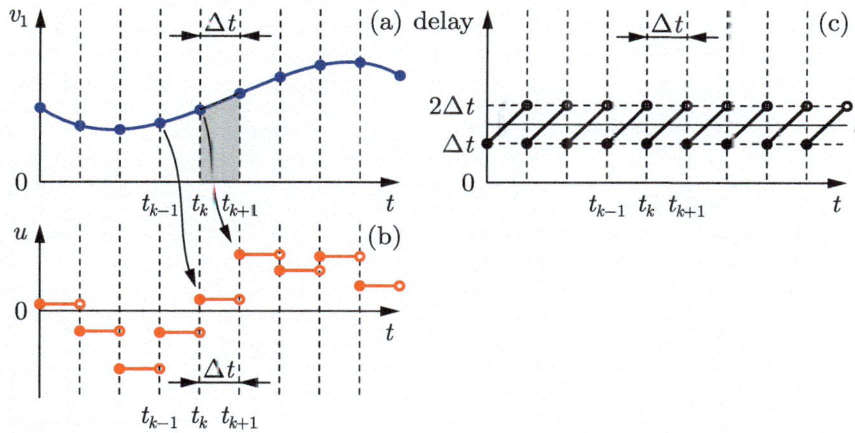

Fig. 3.17 Sample-and-hold strategy used for digital control. Panel **a** illustrates the sampling of the velocity of the preceding vehicle, panel **b** shows how the collected information is utilized by the controller, while panel **c** plots the time-varying time delay in the control loop

communication is utilized in the controllers, which will be discussed in Chap. 4. Thus, understanding the impact of digital effects is of high importance for automated vehicles (AVs) and for connected automated vehicles (CAVs).

The key features of digital control are highlighted in Fig. 3.17. Panel (a) indicates that the velocity of the preceding vehicle is sampled in Δt time intervals, that is, information is available at $t_k = k\Delta t$, $k \in \mathbb{Z}$. Panel (b) shows that the data collected at t_{k-1} is used to determine the control input u applied at t_k. The input is kept constant along the time interval $t \in [t_k, t_{k+1})$ using a zero-order hold (ZOH), and new input is applied at t_{k+1} (based on data collected at t_k). This is often referred to as sample-and-hold strategy. Panel (c) depicts the effective time delay arising as the result of this digital control implementation. At t_k the control input relies on data from t_{k-1}, that is, the information is Δt old. As the same data is being used, the age of information grows to $2\Delta t$ by t_{k+1}, when it drops back to Δt as new control input is applied based on new data. That is, digital control leads to time-varying time delays in the control loop with average delay value

$$\hat{\sigma} = \frac{3}{2}\Delta t \,. \tag{3.111}$$

Below we develop the necessary mathematical tools to analyze the dynamics of automated vehicles implementing ACC (3.73) with sample-and-hold strategy based on [138, 172, 173]. First, we focus on the scenario without powertrain delay ($\sigma = 0$); see also (3.35). Later we extend the analysis to the case of nonzero powertrain delay. Using $\sigma = 0$ in (3.73) the sample-and-hold strategy can be formulated as

$$\begin{aligned}\dot{h}(t) &= v_1(t) - v(t)\,,\\ \dot{v}(t) &= a(t_k)\,,\\ a(t_k) &= \mathrm{sat}\big(u(t_{k-1})\big)\,,\\ u &= \alpha\big(V(h) - v\big) + \beta\big(W(v_1) - v\big)\,,\end{aligned} \tag{3.112}$$

in the time domain $t \in [t_k, t_{k+1})$. Observe that the acceleration a on the right-hand side of the second equation is constant in this time domain. That is, integrating the second equation with respect to time yields

$$v(t) = v(t_k) + a(t_k)(t - t_k)\,. \tag{3.113}$$

At $t = t_{k+1}$ we obtain

$$v(t_{k+1}) = v(t_k) + a(t_k)\Delta t\,. \tag{3.114}$$

Substituting (3.113) into the first equation in (3.112) and integrating with respect to time yields

$$h(t) = h(t_k) - v(t_k)(t - t_k) - \frac{1}{2}a(t_k)(t - t_k)^2 + \int_{t_k}^{t} v_1(\tilde{t})\,\mathrm{d}\tilde{t}\,. \tag{3.115}$$

At $t = t_{k+1}$ this results in

$$h(t_{k+1}) = h(t_k) - v(t_k)\Delta t - \frac{1}{2}a(t_k)\Delta t^2 + \int_{t_k}^{t_{k+1}} v_1(\tilde{t})\,\mathrm{d}\tilde{t}\,. \tag{3.116}$$

Furthermore, the corresponding acceleration $a(t_{k+1})$ can be written as

$$a(t_{k+1}) = \mathrm{sat}\big(u(t_k)\big)\,, \tag{3.117}$$

where the control input $u(t_k)$ is given by

$$u(t_k) = \alpha\big(V(h(t_k)) - v(t_k)\big) + \beta\big(W(v_1(t_k)) - v(t_k)\big)\,. \tag{3.118}$$

The integral in (3.116) can be approximated by the trapezoidal rule:

$$\int_{t_k}^{t_{k+1}} v_1(\tilde{t})\,\mathrm{d}\tilde{t} \approx \frac{v_1(t_k) + v_1(t_{k+1})}{2}\Delta t\,, \tag{3.119}$$

and using the simplified notation

$$h(t_k) = h(k)\,, \quad v(t_k) = v(k)\,, \quad a(t_k) = a(k)\,, \tag{3.120}$$

the closed-loop system (3.114), (3.116)–(3.118) can be formulated as

$$\begin{bmatrix} h(k+1) \\ v(k+1) \\ a(k+1) \end{bmatrix} = \begin{bmatrix} 1 & -\Delta t & -\frac{\Delta t^2}{2} \\ 0 & 1 & \Delta t \\ 0 & 0 & 0 \end{bmatrix} \begin{bmatrix} h(k) \\ v(k) \\ a(k) \end{bmatrix} + \begin{bmatrix} \frac{\Delta t}{2} \\ 0 \\ 0 \end{bmatrix} v_1(k) + \begin{bmatrix} \frac{\Delta t}{2} \\ 0 \\ 0 \end{bmatrix} v_1(k+1)$$
$$+ \begin{bmatrix} 0 \\ 0 \\ \mathrm{sat}\Big(\alpha\big(V(h(k)) - v(k)\big) + \beta\big(W(v_1(k)) - v(k)\big)\Big) \end{bmatrix}. \tag{3.121}$$

Next, we analyze the equilibrium of the discrete-time dynamics (3.121) by dropping sat and W, and we show how its stability is affected by the sampling time Δt. To calculate the equilibrium, we assume that the HV drives with constant velocity v_1^*, while the AV follows with constant velocity v^* and constant headway h^*, that is,

$$\begin{aligned} v_1(k) &\equiv v_1^* , \\ h(k) &\equiv h^* , \\ v(k) &\equiv v^* , \\ a(k) &\equiv 0 . \end{aligned} \tag{3.122}$$

This leads to the equilibrium (3.39). That is, the sample-and-hold unit does not affect the equilibrium state, just like the powertrain delay had no effect on it.

To analyze plant and string stability, we consider perturbations around the equilibrium, in the form

$$\begin{aligned} \tilde{v}_1(k) &= v_1(k) - v_1^* , \\ \tilde{h}(k) &= h(k) - h^* , \\ \tilde{v}(k) &= v(k) - v^* , \\ \tilde{a}(k) &= a(k) , \end{aligned} \tag{3.123}$$

cf. (3.40). Linearizing the discrete-time dynamics (3.121) around the equilibrium (3.39), (3.122) leads to

$$\begin{bmatrix} \tilde{h}(k+1) \\ \tilde{v}(k+1) \\ \tilde{a}(k+1) \end{bmatrix} = \begin{bmatrix} 1 & -\Delta t & -\frac{\Delta t^2}{2} \\ 0 & 1 & \Delta t \\ \alpha\kappa & -(\alpha+\beta) & 0 \end{bmatrix} \begin{bmatrix} \tilde{h}(k) \\ \tilde{v}(k) \\ \tilde{a}(k) \end{bmatrix} + \begin{bmatrix} \frac{\Delta t}{2} \\ 0 \\ \beta \end{bmatrix} \tilde{v}_1(k) + \begin{bmatrix} \frac{\Delta t}{2} \\ 0 \\ 0 \end{bmatrix} \tilde{v}_1(k+1) + \text{h.o.t.} , \tag{3.124}$$

with $\kappa = V'(h^*)$.

By introducing the state, input and output as

$$\mathbf{x}(k) = \begin{bmatrix} \tilde{h}(k) \\ \tilde{v}(k) \\ \tilde{a}(k) \end{bmatrix} , \quad u(k) = \tilde{v}_1(k) , \quad y(k) = \tilde{v}(k) , \tag{3.125}$$

respectively, the linearized discrete-time dynamics of ACC is compactly written as

$$\begin{aligned}\mathbf{x}(k+1) &= \mathbf{A}\mathbf{x}(k) + \mathbf{B}u(k) + \mathbf{B}_1 u(k+1) + \text{h.o.t.}\,,\\ y(k) &= \mathbf{C}\mathbf{x}(k)\,,\end{aligned} \tag{3.126}$$

cf. (3.75). Here the coefficient matrices are

$$\mathbf{A} = \begin{bmatrix} 1 & -\Delta t & -\frac{\Delta t^2}{2} \\ 0 & 1 & \Delta t \\ \alpha\kappa & -(\alpha+\beta) & 0 \end{bmatrix}, \quad \mathbf{B} = \begin{bmatrix} \frac{\Delta t}{2} \\ 0 \\ \beta \end{bmatrix}, \quad \mathbf{B}_1 = \begin{bmatrix} \frac{\Delta t}{2} \\ 0 \\ 0 \end{bmatrix}, \quad \mathbf{C} = \begin{bmatrix} 0 & 1 & 0 \end{bmatrix}. \tag{3.127}$$

The linearized dynamics can be analyzed through Z-transform of (3.126). This leads to the transfer function

$$T(z) = \mathbf{C}\,(z\mathbf{I} - \mathbf{A})^{-1}\,(\mathbf{B} + z\mathbf{B}_1)\,, \tag{3.128}$$

which relates the Z-transforms $U(z)$ and $Y(z)$ of the input and output, respectively, with $z \in \mathbb{C}$; cf. (3.77). The transfer function can be expressed as follows by substituting the coefficient matrices in (3.127):

$$T(z) = \frac{\left(\alpha\kappa\frac{\Delta t^2}{2} + \beta\Delta t\right)z + \alpha\kappa\frac{\Delta t^2}{2} - \beta\Delta t}{z^3 - 2z^2 + \left(\alpha\kappa\frac{\Delta t^2}{2} + (\alpha+\beta)\Delta t + 1\right)z + \alpha\kappa\frac{\Delta t^2}{2} - (\alpha+\beta)\Delta t}\,. \tag{3.129}$$

We first discuss plant stability by considering the linearized dynamics with zero input as

$$\mathbf{x}(k+1) = \mathbf{A}\mathbf{x}(k)\,. \tag{3.130}$$

We adapt Definition 2.1 of plant stability to discrete-time systems as follows.

Definition 3.1 The automated vehicle is called plant stable if the trivial solution $\mathbf{x}(k) \equiv \mathbf{0}$ is asymptotically stable (i.e., it is Lyapunov stable and $\lim_{k\to\infty}\mathbf{x}(k) = \mathbf{0}$) for all initial conditions $\mathbf{x}(0) \in \mathbb{R}^3$.

The solution of system (3.130) can be sought for in the form $\mathbf{x}(k) = z^k\mathbf{x}(0)$ where z is an eigenvalue of matrix $\mathbf{A}$, satisfying

$$\det(z\mathbf{I} - \mathbf{A}) = 0\,. \tag{3.131}$$

Therefore, the condition of plant stability is given as follows.

Theorem 3.1 *The automated vehicle is plant stable at the linear level if and only if the eigenvalues of matrix* $\mathbf{A}$ *are located within the unit circle, that is,* $|z_i| < 1$, $i = 1, 2, 3$ *holds for the solutions of (3.131).*

Notice that the characteristic equation (3.131) gives the denominator of the transfer function (3.128). The eigenvalues z_i are also called as characteristic multipliers.

These must have modulus less than one, as opposed to the eigenvalues in Theorem 2.1 that must have negative real parts. Based on Theorem 3.1, plant stability may be lost through three different mechanisms: a characteristic multiplier crosses the unit circle at $z = 1$, $z = -1$ or $z = \mathrm{e}^{\pm \mathrm{j}\theta}$ with some $\theta \in (0, \pi)$. The latter can be re-written as $z = \mathrm{e}^{\pm \mathrm{j}\Omega \Delta t}$ where $\Omega \in (0, \pi/\Delta t)$ is a bounded angular frequency.

The plant stability boundaries associated with these three cases can be calculated as follows. Substituting (3.127) into (3.131) yields the characteristic equation

$$z^3 - 2z^2 + \left(\alpha\kappa\frac{\Delta t^2}{2} + (\alpha+\beta)\Delta t + 1\right)z + \alpha\kappa\frac{\Delta t^2}{2} - (\alpha+\beta)\Delta t = 0\,, \tag{3.132}$$

cf. the denominator of (3.129). By substituting $z = 1$ into (3.132), the first plant stability boundary becomes

$$\alpha = 0\,, \tag{3.133}$$

cf. (3.80). The second boundary is obtained from (3.132) with $z = -1$ in the form

$$\alpha = -\frac{2}{\Delta t} - \beta\,. \tag{3.134}$$

The third boundary is provided by substituting $z = \mathrm{e}^{\mathrm{j}\Omega \Delta t}$ into (3.132) and separating it into real and imaginary parts. This leads to the two equations

$$\begin{aligned}
\cos(3\Omega\Delta t) - 2\cos(2\Omega\Delta t) + \left(\alpha\kappa\frac{\Delta t^2}{2} + (\alpha+\beta)\Delta t + 1\right)\cos(\Omega\Delta t)& \\
+\alpha\kappa\frac{\Delta t^2}{2} - (\alpha+\beta)\Delta t = 0\,,& \\
\sin(3\Omega\Delta t) - 2\sin(2\Omega\Delta t) + \left(\alpha\kappa\frac{\Delta t^2}{2} + (\alpha+\beta)\Delta t + 1\right)\sin(\Omega\Delta t) = 0\,,&
\end{aligned} \tag{3.135}$$

which yield a curve parameterized by $\Omega \in (0, \pi/\Delta t)$:

$$\begin{aligned}
\alpha &= \frac{2}{\kappa\Delta t^2}\left(2\sin^2(\Omega\Delta t) + 3\cos(\Omega\Delta t) - 3\right)\,, \\
\beta &= \frac{1}{\Delta t}\left(2\sin^2(\Omega\Delta t) + \cos(\Omega\Delta t) - 1\right) - \alpha\,.
\end{aligned} \tag{3.136}$$

Figure 3.18 depicts the plant stability charts of the ACC system (3.126), (3.127). The plant stability boundaries given by (3.133), (3.134) and (3.136) are plotted as red curves. We remark that the $z = -1$ boundary is outside the frame of Fig. 3.18, thus it is less relevant in practice. The plant stable domain is shaded as light gray. In this domain, the condition in Theorem 3.1 holds. This can be verified through Fig. 3.19, which depicts the eigenvalues of matrix $\mathbf{A}$ for points P, T, and V in Fig. 3.18b. Point

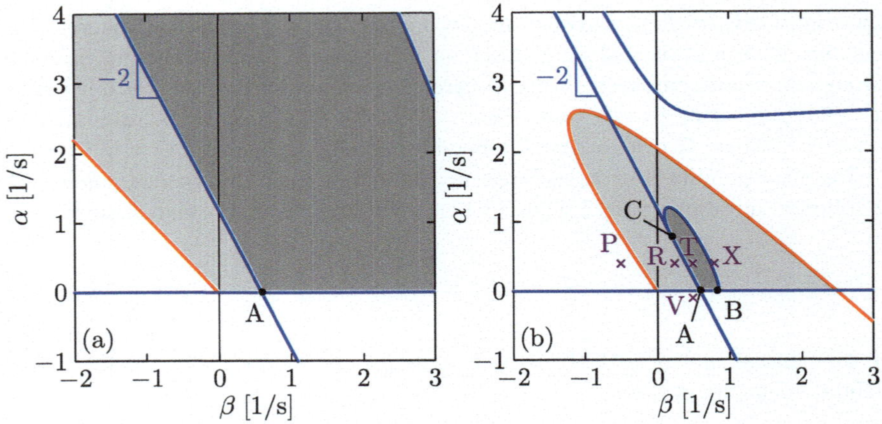

Fig. 3.18 Stability charts of system (3.126), (3.127) for $\kappa = 0.6$ [1/s]. Panels **a** and **b** are associated with sampling time $\Delta t = 0.1$ [s] and $\Delta t = 0.4$ [s], respectively. The plant and string stable regions are shaded by light and dark gray, respectively. Points A, B and C are located at the intersections of the string stability boundaries, while points P(−0.5,0.4), R(0.25,0.4), T(0.5,0.4), V(0.5,−0.1), and X(0.8,0.4) correspond to the spectra and frequency response plots in Figs. 3.19 and 3.20

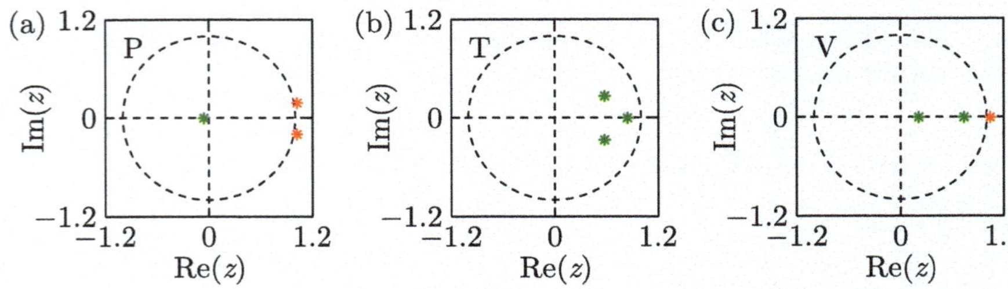

Fig. 3.19 Spectrum plots obtained from the characteristic equation (3.132) with $\kappa = 0.6$ [1/s] and $\Delta t = 0.4$ [s]. Panels **a**–**c** correspond to points P(−0.5,0.4), T(0.5,0.4) and V(0.5,−0.1) in Fig. 3.18, respectively

P lies beyond the $z = e^{\pm j\Omega\Delta t}$ plant stability boundary, hence a complex conjugate pair of characteristic multipliers is located outside the unit circle in Fig. 3.19a. Point T is plant stable, hence all eigenvalues are within the unit circle in Fig. 3.19b. Point V is located beyond the $z = 1$ plant stability boundary, which leads to the existence of a positive real eigenvalue outside the unit circle in Fig. 3.19c.

Next, we continue with string stability calculations. Again, we adapt the corresponding tools of continuous-time analysis to discrete-time systems. We consider a harmonic motion for the HV as

$$\tilde{v}_1(k) = v_{\text{amp}} \sin(\omega k \Delta t) , \tag{3.137}$$

cf. (3.48). This induces a harmonic response from the AV in steady state in case of linearized dynamics:

$$\tilde{v}_{ss}(k) = \left|T\left(e^{j\omega\Delta t}\right)\right| v_{amp} \sin\left(\omega k \Delta t + \angle T\left(e^{j\omega\Delta t}\right)\right), \tag{3.138}$$

cf. (3.49). Note that $T\left(e^{j\omega\Delta t}\right)$ is periodic in ω, hence we restrict ourselves to studying the response for the period $\omega \in [0, 2\pi/\Delta t)$ only.

We state the discrete-time counterpart of Theorem 2.2 to establish string stability.

Theorem 3.2 *If and only if*

$$\left|T\left(e^{j\omega\Delta t}\right)\right| < 1, \quad \forall \omega \in (0, 2\pi/\Delta t), \tag{3.139}$$

then the HV-AV pair are

- *string stable in the $\mathcal{L}_2$ norm for any $\tilde{v}_1 \in \mathcal{L}_2$;*
- *string stable in the $\mathcal{L}_\infty$ norm for the harmonic input (3.137) when $k \to \infty$.*

Based on the analogy with Theorem 2.2, the rest of the string stability calculations follow those in (2.141)–(2.151). First, according to (2.142), we formulate

$$P(\omega) = \frac{1}{\omega^2}\left(\mathrm{Den}\left(\left|T\left(e^{j\omega\Delta t}\right)\right|^2\right) - \mathrm{Num}\left(\left|T\left(e^{j\omega\Delta t}\right)\right|^2\right)\right), \tag{3.140}$$

that makes the string stability condition (3.139) equivalent to $P(\omega) > 0$, $\forall \omega \in (0, 2\pi/\Delta t)$. After substituting (3.129) into (3.140), the expression of $P(\omega)$ becomes

$$\begin{aligned} P(\omega) = \frac{2}{\omega^2}\Big((b-c)\cos(3\omega\Delta t) + (-b+3c+1)\cos(2\omega\Delta t) \\ - (a+b+3c+4)\cos(\omega\Delta t) + a+b+c+3\Big), \end{aligned} \tag{3.141}$$

where

$$a = \alpha(\alpha+2\beta)\Delta t^2, \quad b = \alpha\kappa\frac{\Delta t^2}{2}, \quad c = (\alpha+\beta)\Delta t. \tag{3.142}$$

Similar to plant stability, we analyze string stability by considering three cases: $\omega = 0$, $\omega = \pi/\Delta t$ and $\omega = \hat{\omega}$ with $0 < \hat{\omega} \neq \pi/\Delta t$ (that is, $e^{j\omega\Delta t} = 1$, $e^{j\omega\Delta t} = -1$ and $e^{j\omega\Delta t} = e^{j\hat{\omega}\Delta t} \in \mathbb{C}$, along the unit circle). By taking the limit $\omega \to 0$, we obtain

$$P(0) = (a-4b)\Delta t^2 = \alpha(\alpha+2\beta-2\kappa)\Delta t^4, \tag{3.143}$$

that leads to the $\omega = 0$ string stability condition

$$\alpha > 0, \quad \alpha > 2(\kappa - \beta), \tag{3.144}$$

cf. (3.82). For $\omega = \pi/\Delta t$, we get

$$P\left(\frac{\pi}{\Delta t}\right) = \frac{4\Delta t^2}{\pi^2}(a + 4c + 4) = \frac{4\Delta t^4}{\pi^2}\left(\alpha + \frac{2}{\Delta t}\right)\left(\alpha + 2\beta + \frac{2}{\Delta t}\right), \tag{3.145}$$

that yields the string stability condition

$$\alpha > -\frac{2}{\Delta t}, \quad \alpha > -2\left(\beta + \frac{1}{\Delta t}\right). \tag{3.146}$$

Finally, we can derive the boundaries of string stability for $\omega = \hat{\omega}$ from

$$\begin{aligned} P(\hat{\omega}) &= 0\,, \\ Q(\hat{\omega}) &= 0\,, \end{aligned} \tag{3.147}$$

where $Q(\omega) = \frac{1}{2\omega}\frac{\mathrm{d}P}{\mathrm{d}\omega}$ is given by

$$\begin{aligned} Q(\omega) = -\frac{1}{\omega^2}P(\omega) - \frac{\Delta t}{\omega^3}\Big(&3(b-c)\sin(3\omega\Delta t) + 2(-b+3c+1)\sin(2\omega\Delta t) \\ &- (a+b+3c+4)\sin(\omega\Delta t)\Big)\,. \end{aligned} \tag{3.148}$$

We omit the expressions of these boundaries due to their increased complexity.

The result of the string stability analysis is summarized in Fig. 3.18 as the stability charts of the ACC system (3.126), (3.127). The string stability boundaries given by (3.144), (3.146) and (3.147) are plotted as blue curves (and the $\omega = \pi/\Delta t$ boundary lies outside the frame). In the dark gray shaded string stable domain, the condition in Theorem 3.2 holds. This is demonstrated by Fig. 3.19, that shows the frequency response for points R, T, and X marked in Fig. 3.18b. The frequency responses reflect the conclusions from the stability chart: point R is $\omega = 0$ string unstable, point T is string stable, whereas point X is $\omega = \hat{\omega}$ string unstable (with $0 < \hat{\omega} \neq \pi/\Delta t$).

Notably, the stable domains of the sample-and-hold ACC (3.126), (3.127) in Fig. 3.18b are almost indistinguishable from those of ACC with delay (3.75), (3.76) in Fig. 3.11b. Likewise, the frequency response of sample-and-hold ACC in Fig. 3.20, that is shown by solid lines, closely matches the response of the corresponding continuous-time system with delay at low frequencies, as shown by the dashed lines repeated from Fig. 2.20. In fact, the sampling time Δt of the sample-and-hold example corresponds to the same average delay $\hat{\sigma}$ in (3.111) as the delay σ of the continuous-time case. Based on this observation, the stability properties of sample-and-hold ACC systems can be approximated by those of continuous-time ACC with the same average delay.

Moreover, notice that the stable regions in Fig. 3.18 shrink when the sampling time Δt is increased, similar to how the delay affects the corresponding continuous-time dynamics. If the sampling time Δt exceeds a critical value Δt_{cr}, the string stable domain ceases to exist. The critical sampling time can be found as follows.

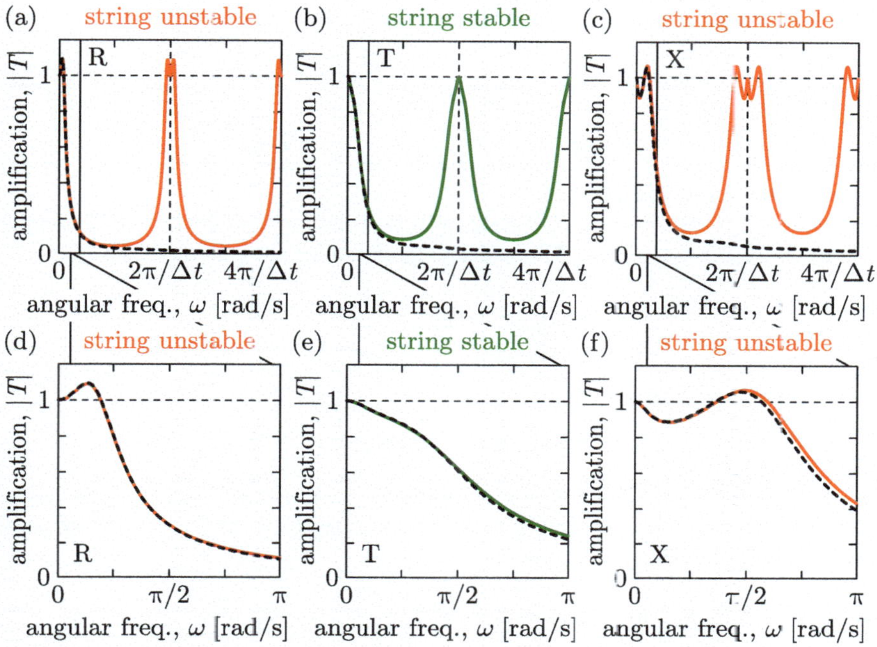

Fig. 3.20 Frequency response plots corresponding to the transfer function (3.129) with $\kappa = 0.6\,[1/\mathrm{s}]$ and $\Delta t = 0.4\,[\mathrm{s}]$. The bottom panels show a zoom of the top panels. The left, middle, and right panels are associated with points R(0.25,0.4), T(0.5,0.4), and X(0.8,0.4) in Fig. 3.18, respectively. For reference, the response of the corresponding continuous-time system with the same average delay is repeated from panels **a**, **c**, and **e** of Fig. 2.20 as dashed lines

If $\Delta t = \Delta t_{\mathrm{cr}}$, the string stable domain shrinks to a point, and the corner points A, B, and C coincide. Point A is found at the intersection of the boundaries in (3.144). Points B and C are the $\omega \to 0$ starting points of the boundary (3.147), hence they are given by $P(0) = 0$ and $Q(0) = 0$, where $P(0)$ is given by (3.143) and

$$Q(0) = \frac{\Delta t^4}{12}(-a + 64b - 36c + 12) = \frac{\Delta t^4}{12}\Big(-\alpha(\alpha+2\beta)\Delta t^2 + 32\alpha\kappa\Delta t^2 - 36(\alpha+\beta)\Delta t + 12\Big)\,. \tag{3.149}$$

These equations lead to the following coordinates of points A, B, and C:

$$A:(\kappa,0)\,,\quad B:\left(\frac{1}{3\Delta t},0\right),\quad C:\left(\frac{5\kappa^2\Delta t^2 - 6\kappa\Delta t + 1}{\Delta t(5\kappa\Delta t - 3)},\,\frac{2(3\kappa\Delta t - 1)}{\Delta t(5\kappa\Delta t - 3)}\right), \tag{3.150}$$

which coincide for $\Delta t = \Delta t_{\mathrm{cr}}$ with

$$\Delta t_{\mathrm{cr}} = \frac{1}{3\kappa}\,. \tag{3.151}$$

Remarkably, the average delay (3.111)

$$\hat{\sigma}_{\mathrm{cr}} = \frac{1}{2\kappa} , \tag{3.152}$$

corresponding to the critical sampling time is exactly the same as the critical delay of continuous-time ACC in (3.84).

Finally, we briefly discuss ACC systems by considering both the powertrain delay and the sample-and-hold unit. The underlying dynamics are governed by

$$\begin{aligned} \dot{h}(t) &= v_1(t) - v(t) , \\ \dot{v}(t) &= a(t_k) , \\ a(t_k) &= \mathrm{sat}\big(u(t_{k-1} - \sigma)\big) , \\ u &= \alpha\big(V(h) - v\big) + \beta\big(W(v_1) - v\big) , \end{aligned} \tag{3.153}$$

for $t \in [t_k, t_{k+1})$; cf. (3.73), (3.112). For simplicity, we illustrate results only for cases where the delay σ is integer multiples of the sampling time Δt, i.e., $\sigma = r\Delta t$ with some $r \in \mathbb{N}$. This yields the dynamics

$$\begin{aligned} \dot{h}(t) &= v_1(t) - v(t) , \\ \dot{v}(t) &= a(t_k) , \\ a(t_k) &= \mathrm{sat}\big(u(t_{k-1-r})\big) , \\ u &= \alpha\big(V(h) - v\big) + \beta\big(W(v_1) - v\big) . \end{aligned} \tag{3.154}$$

In discrete time, the expressions (3.114), (3.116) of the velocity and headway still hold, while the acceleration expression (3.117) changes to

$$a(t_{k+1}) = \mathrm{sat}\big(u(t_{k-r})\big) . \tag{3.155}$$

By using the trapezoidal rule (3.119) for integrating the preceding vehicle's velocity, the discrete-time dynamics become

$$\begin{aligned} \begin{bmatrix} h(k+1) \\ v(k+1) \\ a(k+1) \end{bmatrix} &= \begin{bmatrix} 1 & -\Delta t & -\frac{\Delta t^2}{2} \\ 0 & 1 & \Delta t \\ 0 & 0 & 0 \end{bmatrix} \begin{bmatrix} h(k) \\ v(k) \\ a(k) \end{bmatrix} + \begin{bmatrix} \frac{\Delta t}{2} \\ 0 \\ 0 \end{bmatrix} v_1(k) + \begin{bmatrix} \frac{\Delta t}{2} \\ 0 \\ 0 \end{bmatrix} v_1(k+1) \\ &+ \begin{bmatrix} 0 \\ 0 \\ \mathrm{sat}\Big(\alpha\big(V(h(k-r)) - v(k-r)\big) + \beta\big(W(v_1(k-r)) - v(k-r)\big)\Big) \end{bmatrix} , \end{aligned} \tag{3.156}$$

cf. (3.121).

Analogously to (3.124), the corresponding linearized system becomes

$$\begin{bmatrix} \tilde{h}(k+1) \\ \tilde{v}(k+1) \\ \tilde{a}(k+1) \end{bmatrix} = \begin{bmatrix} 1 & -\Delta t & -\frac{\Delta t^2}{2} \\ 0 & 1 & \Delta t \\ 0 & 0 & 0 \end{bmatrix} \begin{bmatrix} \tilde{h}(k) \\ \tilde{v}(k) \\ \tilde{a}(k) \end{bmatrix} + \begin{bmatrix} 0 & 0 & 0 \\ 0 & 0 & 0 \\ \alpha\kappa & -(\alpha+\beta) & 0 \end{bmatrix} \begin{bmatrix} \tilde{h}(k-r) \\ \tilde{v}(k-r) \\ \tilde{a}(k-r) \end{bmatrix}$$
$$+ \begin{bmatrix} 0 \\ 0 \\ \beta \end{bmatrix} \tilde{v}_1(k-r) + \begin{bmatrix} \frac{\Delta t}{2} \\ 0 \\ 0 \end{bmatrix} \tilde{v}_1(k) + \begin{bmatrix} \frac{\Delta t}{2} \\ 0 \\ 0 \end{bmatrix} \tilde{v}_1(k+1) + \text{h.o.t.}\,, \tag{3.157}$$

with $\kappa = V'(h^*)$. This can be written compactly as

$$\begin{aligned} \mathbf{x}(k+1) &= \mathbf{A}_0\mathbf{x}(k) + \mathbf{A}_{-r}\mathbf{x}(k-r) \\ &\quad + \mathbf{B}_{-r}u(k-r) + \mathbf{B}_0 u(k) + \mathbf{B}_1 u(k+1) + \text{h.o.t.}\,, \\ y(k) &= \mathbf{C}\mathbf{x}(k)\,, \end{aligned} \tag{3.158}$$

where the state $\mathbf{x}(k)$, input $u(k)$ and output $y(k)$ are defined in (3.125), and

$$\begin{aligned} \mathbf{A}_0 &= \begin{bmatrix} 1 & -\Delta t & -\frac{\Delta t^2}{2} \\ 0 & 1 & \Delta t \\ 0 & 0 & 0 \end{bmatrix}, \quad \mathbf{A}_{-r} = \begin{bmatrix} 0 & 0 & 0 \\ 0 & 0 & 0 \\ \alpha\kappa & -(\alpha+\beta) & 0 \end{bmatrix}, \\ \mathbf{B}_{-r} &= \begin{bmatrix} 0 \\ 0 \\ \beta \end{bmatrix}, \quad \mathbf{B}_0 = \mathbf{B}_1 = \begin{bmatrix} \frac{\Delta t}{2} \\ 0 \\ 0 \end{bmatrix}, \quad \mathbf{C} = \begin{bmatrix} 0 & 1 & 0 \end{bmatrix}, \end{aligned} \tag{3.159}$$

cf. (3.127). Here $\mathbf{A}_0$ represents double integrator dynamics with piecewise constant acceleration (zero-order hold), $\mathbf{A}_{-r}$ and $\mathbf{B}_{-r}$ come from the adaptive cruise control law, while $\mathbf{B}_0$ and $\mathbf{B}_1$ are related to the integral of the preceding vehicle's velocity that was approximated by the trapezoidal rule. Notice that $\mathbf{A} = \mathbf{A}_0 + \mathbf{A}_{-r}$ and $\mathbf{B} = \mathbf{B}_0 + \mathbf{B}_{-r}$ in (3.127).

System (3.158) is associated with the transfer function

$$T(z) = \mathbf{C}\left(z\mathbf{I} - \mathbf{A}_0 - \mathbf{A}_{-r}z^{-r}\right)^{-1}\left(\mathbf{B}_{-r}z^{-r} + \mathbf{B}_0 + z\mathbf{B}_1\right), \tag{3.160}$$

cf. (3.128). By substituting the matrices in (3.159), the transfer function can be expressed as

$$T(z) = \frac{\left(\alpha\kappa\frac{\Delta t^2}{2} + \beta\Delta t\right)z + \alpha\kappa\frac{\Delta t^2}{2} - \beta\Delta t}{(z^3 - 2z^2 + z)z^r + \left(\alpha\kappa\frac{\Delta t^2}{2} + (\alpha+\beta)\Delta t\right)z + \alpha\kappa\frac{\Delta t^2}{2} - (\alpha+\beta)\Delta t}\,, \tag{3.161}$$

that simplifies to (3.129) for $r = 0$. The stability of this system can be analyzed through the same procedure as discussed for $r = 0$.

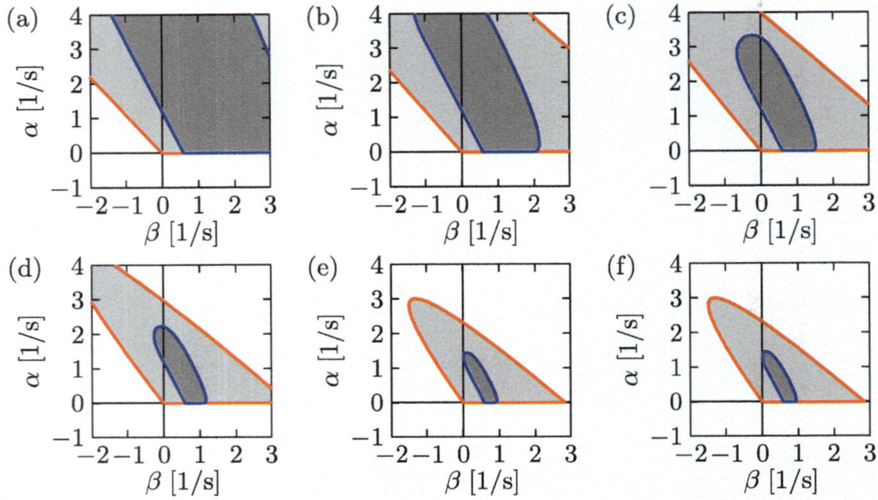

Fig. 3.21 Stability charts of system (3.158), (3.159) for $\kappa = 0.6\,[1/\mathrm{s}]$. Panels **a**–**e** correspond to sampling time $\Delta t = 0.1\,[\mathrm{s}]$ and time delays $\sigma = 0\,[\mathrm{s}]$, $\sigma = 0.1\,[\mathrm{s}]$, $\sigma = 0.2\,[\mathrm{s}]$, $\sigma = 0.3\,[\mathrm{s}]$ and $\sigma = 0.4\,[\mathrm{s}]$, respectively. The dark gray shaded string stable region shrinks significantly as the delay is increased. Panel **f** is associated with sampling time $\Delta t = 0\,[\mathrm{s}]$ and time delay $\sigma = 0.55\,[\mathrm{s}]$. Since the average delay $\sigma + \frac{3}{2}\Delta t$ is the same in panels **e** and **f**, the stable regions closely match

The resulting stability charts are depicted in Fig. 3.21 for various combinations of the delay σ and the sampling time Δt. Based on panels (a)–(e), we can again conclude that increasing the delay negatively impacts stability. Furthermore, panels (e) and (f) highlight that the stable regions are approximately the same for different delay and sampling time combinations if the corresponding average delay $\sigma + \frac{3}{2}\Delta t$ is the same. As such, the effect of sampling can be approximated well by considering continuous-time control systems with the same average delay. Therefore, in the rest of this book we restrict ourselves to continuous-time systems with delays, and we omit further discussions on the effects of sampling.

Chapter 4
Connected Cruise Control

In this chapter we discuss how wireless connectivity can be utilized in order to improve the longitudinal control of automated vehicles. We use the models of human driving introduced in Chap. 2 and extend the tools developed for the control design in Chap. 3 by incorporation of the motion information from multiple vehicles ahead. We evaluate the impacts of connectivity on string stability of mixed vehicle chains that contain human-driven and automated vehicles. We also study the effects of nonlinearities on connected vehicle systems. Finally, we highlight the benefits of communication for energy consumption and investigate the effects of uncertainties in human parameters.

4.1 Controlling Vehicles Using Vehicle-to-Everything Connectivity

In the last few decades there has been significant efforts to make vehicles connected to each other and to the infrastructure. Multiple different communication technologies have been developed that are based on "broadcast and catch" protocols. These can facilitate low-latency peer-to-peer communication between road participants and can lead to flexible connectivity topologies. For example, a connected vehicle crossing an intersection can communicate with nearby connected vehicles and the infrastructure, while being within the communication range, and improve its situation awareness through the received information.

Depending on who communicates with whom, the communication may be referred to as vehicle-to-vehicle (V2V), vehicles-to-infrastructure (V2I), vehicle-to-pedestrian (V2P) communication, etc. These can be merged under the name vehicle-to-everything (V2X) communication. Currently there are two different technologies supporting V2X. One is called dedicated short range communication (DSRC). It is based on WiFi and has been developed since the early 2000s. The other is referred to

G. Orosz and T. G. Molnár, *Dynamics and Control of Connected Vehicles*,
Surveys and Tutorials in the Applied Mathematical Sciences 15,
https://doi.org/10.1007/978-3-031-94555-7_4

as cellular vehicle-to-everything (CV2X) communication. It is based on cellular LTE and has been developed since the late 2010s. Both technologies operate at 5.9 GHz with the update rate 10 Hz and the range being a few hundred meters. That is, a vehicle may send and receive V2X messages to and from other road participants within the communication range in about every 100 ms. We remark that the results presented in this book are agnostic to which V2X communication technology is utilized and they shall be generalizable for new, currently non-existing wireless communication technologies that satisfy some minimal latency requirements.

Initially, V2X has been established to support human drivers by providing them warnings based on the received V2X messages. In the meantime, it became evident that automated vehicles may also benefit from V2X information. In particular, V2X may allow them to receive information that is difficult to extract from sensory information, e.g., acceleration and yaw rate. Moreover, V2X may also reach beyond the line of sight of sensors and receive information from vehicles beyond the corner or multiple vehicles ahead in the traffic flow.

In order to distinguish vehicles with and without connectivity and with and without automation, we will use the notation which was introduced in Chap. 1 and was utilized in Sect. 3.4 of Chap. 3. Namely, those vehicles that are neither automated nor connected are called human-driven vehicles (HVs); those vehicles that are automated but not connected are called automated vehicles (AVs); those vehicles that are not automated but connected are called connected human-driven vehicles (CHVs); and those vehicles that are connected and automated are called connected automated vehicles (CAVs). Figures 4.1, 4.7 and 4.11 illustrate mixed traffic scenarios where a CAV responds to the motion of HVs and CHVs. We refer to the corresponding algorithms as connected cruise control (CCC) to differentiate it from cooperative adaptive cruise control (CACC) that requires all vehicles in the chain to be CAVs; cf. the literature review in Chap. 1. Finally, we remark that when there are no CHVs in the communication range of the CAV, CCC degrades to ACC which was discussed in Chap. 3. The control algorithms described below are designed to allow graceful degradation.

Consider the scenario shown in Fig. 4.1 where a connected automated vehicle (CAV) responds to the motion of a single preceding vehicle, a connected human-

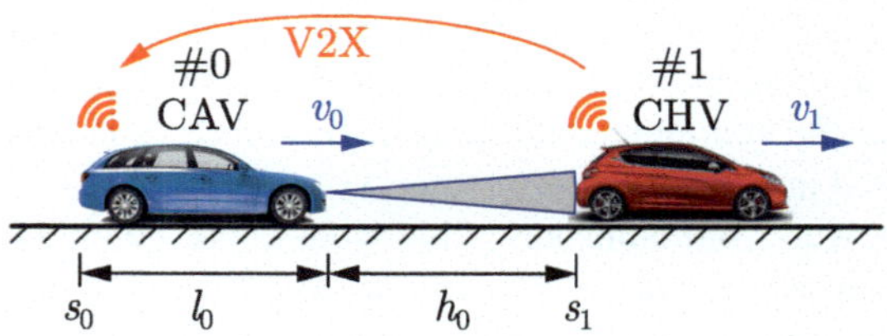

Fig. 4.1 Connected cruise control (CCC) where a connected automated vehicle (CAV) follows a connected human-driven vehicle (CHV) on a single lane. The CAV is able to respond to the CHV by relying both on its on-board range sensors and the information shared by the CHV via vehicle-to-everything (V2X) connectivity

driven vehicle (CHV). The CAV may detect the distance and the speed difference from the CHV via on-board range sensors such as radar or lidar as discussed for adaptive cruise control (ACC) in Chap. 3. Alternatively, both vehicles may measure their own GPS position and speed, and the CHV may transmit this information to the CAV via vehicle-to-everything (V2X) communication. Based on the received information, the CAV can compute the distance and the speed difference and use these to control its longitudinal motion. This leads to the simplest realization of connected cruise control (CCC).

Hereinafter we use index 0 to refer to the CAV and index 1 for the CHV ahead. That is, the headway ahead of the CAV is denoted by h_0, the CAV's position is s_0 and its speed is v_0, while the CHV's position is s_1 and its speed is v_1. The headway can be measured by radar or calculated from GSP coordinates according to

$$h_0 = s_1 - s_0 - l_0 \,, \tag{4.1}$$

where l_0 denotes the length of the CAV; cf. (2.1).

We remark that in practice s_0 and s_1 are not determined individually but rather the distance $s_1 - s_0$ is calculated using the Haversine formula [92]:

$$s_1 - s_0 = 2R \arcsin \sqrt{\sin^2\left(\frac{\phi_0 - \phi_1}{2}\right) + \cos\phi_0 \cos\phi_1 \sin^2\left(\frac{\lambda_0 - \lambda_1}{2}\right)} \,. \tag{4.2}$$

This provides the great circle distance between two points of GPS coordinates (ϕ_0, λ_0) and (ϕ_1, λ_1). Here ϕ and λ denote the latitude and longitude, respectively, while R denotes the radius of Earth (estimated at the location of the experiment).

Similarly, the range rate $\dot{h}_0$ can be obtained by radar (using the Doppler effect) or calculated as

$$\dot{h}_0 = v_1 - v_0 \,, \tag{4.3}$$

where the speed v_0 is measured on board by the CAV while v_1 is received via V2X communication.

For simplicity, let us approximate the dynamics of the CAV's motion by the delayed double integrator model:

$$\begin{aligned} \dot{h}_0(t) &= v_1(t) - v_0(t) \,, \\ \dot{v}_0(t) &= \text{sat}(u(t-\sigma)) \,, \end{aligned} \tag{4.4}$$

cf. (3.72) when the resistance term is neglected. Accordingly, the CAV's acceleration is directly prescribed by the control input u that is subject to a time delay σ. This delay incorporates the effects of communication latency arising from scheduling and sending V2X messages, the computation time of sample-and-hold units, and the actuation (powertrain) delays. The commanded control input saturates outside $[-a_{\min}, \tilde{a}_{\max}(v)]$. This is given by (3.8), where we often neglect the effect of the power limit $P_{\max}$ for simplicity, i.e., we use $\tilde{a}_{\max}(v) = a_{\max}$.

4.2 Connected Cruise Control with Acceleration Feedback

Connectivity may also allow the CHV to share additional pieces of information apart from position and speed. In what follows, we analyze how acceleration data transmitted through V2X connectivity can be utilized by the CAV to improve its longitudinal control performance [62]. We analyze the stability and frequency response of CCC algorithms with acceleration feedback and demonstrate control performance via simulations, first for an open chain of vehicles, then for the ring configuration.

4.2.1 Open Chain of Vehicles

Consider that apart from the headway h_0 and speeds v_0, v_1, the CAV also has access to the acceleration $\dot{v}_1$ of the CHV and it utilizes this information in its controller:

$$u = F(h_0, v_0, v_1, \dot{v}_1) , \tag{4.5}$$

where the expression of F depends on the choice of control law. In particular, we will focus on a specific control law proposed in [62] that is similar to the optimal velocity model (2.4) in Chap. 2 and the ACC law (3.30) in Chap. 3, supplemented by an additional acceleration feedback term with gain $\gamma \geq 0$:

$$u = \alpha\big(V(h_0) - v_0\big) + \beta\big(W(v_1) - v_0\big) + \gamma\, \dot{v}_1 . \tag{4.6}$$

Here V indicates the range policy (3.32) and W is the speed policy (3.33) that enforces the speed limit v_{max} even when the CHV's speed v_1 exceeds it; see Fig. 3.7.

The potential benefits of acceleration feedback are demonstrated in Fig. 4.2, where an emergency braking scenario is simulated. The CHV makes an excessive braking and comes to a full stop to which the CAV must respond to. If the CHV transmits its acceleration to the CAV via V2X connectivity, it enables the CAV to recognize the upcoming danger of collision and respond to it faster. In this example, the CAV is unable to avoid collision (indicated by $h < 0$) when acceleration feedback is not used (top panels), while the extra acceleration feedback term allows it to stop behind the CHV safely (bottom panels).

Now let us analyze the dynamics (4.4) subject to the general controller with acceleration feedback (4.5). The equilibrium of the car-following scenario is given by (2.15), (2.16), i.e., the CAV maintains the same constants speed v^* as the one given by the CHV's motion while keeping constant headway h^* according to the range policy V. Stability and string stability can be analyzed by considering perturbations $\tilde{h}_0$, $\tilde{v}_0$ and $\tilde{v}_1$ around the equilibrium in the form (2.22).

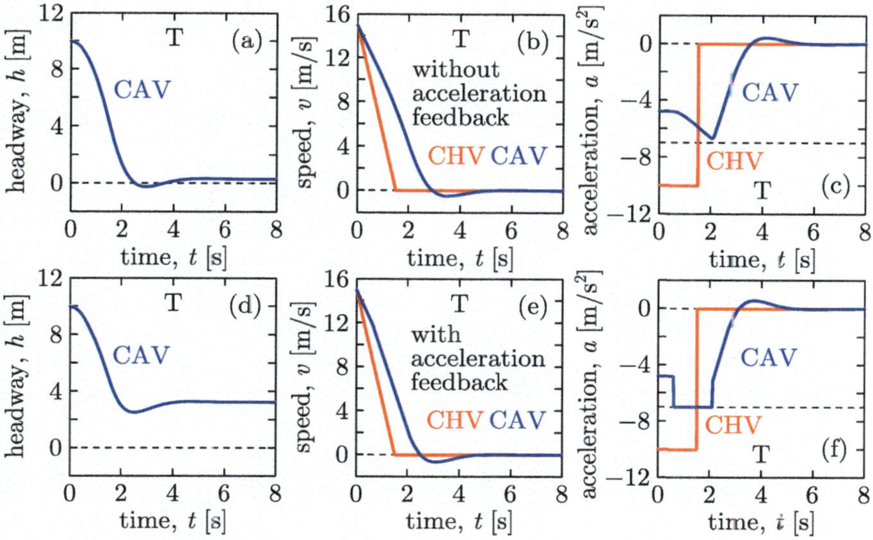

Fig. 4.2 Numerical simulation of CCC (4.4) with acceleration feedback (4.6) during an emergency braking scenario. The lead vehicle reduces its speed from 15 to 0 [m/s] with constant $-10\,[\mathrm{m/s^2}]$ deceleration. The initial conditions are given by $v_0(t) = 15\,[\mathrm{m/s}]$ and $h_0(t) = 10\,[\mathrm{m}]$ over $t \in [-\sigma, 0]$. The range policy is piecewise linear as in (3.32) with $h_{\mathrm{st}} = 5\,[\mathrm{m}]$, $h_{\mathrm{go}} = 55\,[\mathrm{m}]$ and $v_{\max} = 30\,[\mathrm{m/s}]$, while the rest of the parameters are $a_{\min} = 7\,[\mathrm{m/s^2}]$, $a_{\max} = 3\,[\mathrm{m/s^2}]$ and $\sigma = 0.6\,[\mathrm{s}]$. Panels **a–c** correspond to point T(0.5,0.4) in Fig. 4.4a where acceleration feedback is not used, i.e., $\gamma = 0$. Panels **d–f** are associated with point T(0.5,0.4) in Fig. 4.4b with $\gamma = 0.25$

This leads to the linearized system

$$
\begin{aligned}
\dot{\tilde{h}}_0(t) &= \tilde{v}_1(t) - \tilde{v}_0(t)\,, \\
\dot{\tilde{v}}_0(t) &= F^*_{h_0}\tilde{h}_0(t-\sigma) + F^*_{v_0}\tilde{v}_0(t-\sigma) + F^*_{v_1}\tilde{v}_1(t-\sigma) + F^*_{\dot{v}_1}\dot{\tilde{v}}_1(t-\sigma) + \text{h.o.t.}\,,
\end{aligned}
\tag{4.7}
$$

where we utilized the notation $F^*_{h_0} = \partial_{h_0} F(h^*, v^*, v^*, 0)$, $F^*_{v_0} = \partial_{v_0} F(h^*, v^*, v^*, 0)$, $F^*_{v_1} = \partial_{v_1} F(h^*, v^*, v^*, 0)$, $F^*_{\dot{v}_1} = \partial_{\dot{v}_1} F(h^*, v^*, v^*, 0)$, cf. (2.130). With the state, input and output defined by (2.24), we obtain the form

$$
\begin{aligned}
\dot{\mathbf{x}}_0(t) &= \mathbf{a}_0\,\mathbf{x}_0(t) + \mathbf{a}_\sigma\,\mathbf{x}_0(t-\sigma) + \mathbf{b}_0\,u_0(t) + \mathbf{b}_\sigma\,u_0(t-\sigma) + \hat{\mathbf{b}}_\sigma\,\dot{u}_0(t-\sigma) + \text{h.o.t.}\,, \\
y_0(t) &= \mathbf{c}\,\mathbf{x}_0(t)\,,
\end{aligned}
\tag{4.8}
$$

cf. (2.131), where the coefficient matrices become

$$
\mathbf{a}_0 = \begin{bmatrix} 0 & -1 \\ 0 & 0 \end{bmatrix}, \quad \mathbf{a}_\sigma = \begin{bmatrix} 0 & 0 \\ F^*_{h_0} & F^*_{v_0} \end{bmatrix}, \quad \mathbf{b}_0 = \begin{bmatrix} 1 \\ 0 \end{bmatrix}, \quad \mathbf{b}_\sigma = \begin{bmatrix} 0 \\ F^*_{v_1} \end{bmatrix}, \quad \hat{\mathbf{b}}_\sigma = \begin{bmatrix} 0 \\ F^*_{\dot{v}_1} \end{bmatrix}, \quad \mathbf{c} = \begin{bmatrix} 0 & 1 \end{bmatrix}, \tag{4.9}
$$

cf. (2.132). Similarly to (2.133), the transfer function between the CHV's and the CAV's speed fluctuations can be given as

$$T(s) = \mathbf{c}\Big(s\mathbf{I} - \mathbf{a}_0 - \mathbf{a}_\sigma\, \mathrm{e}^{-s\sigma}\Big)^{-1}\Big(\mathbf{b}_0 + \mathbf{b}_\sigma\, \mathrm{e}^{-s\sigma} + \hat{\mathbf{b}}_\sigma\, s\mathrm{e}^{-s\sigma}\Big)\,. \tag{4.10}$$

Specifically, for control law (4.6), the matrices (4.9) are

$$\mathbf{a}_0 = \begin{bmatrix} 0 & -1 \\ 0 & 0 \end{bmatrix},\quad \mathbf{a}_\sigma = \begin{bmatrix} 0 & 0 \\ \alpha\kappa & -(\alpha+\beta) \end{bmatrix},\quad \mathbf{b}_0 = \begin{bmatrix} 1 \\ 0 \end{bmatrix},\quad \mathbf{b}_\sigma = \begin{bmatrix} 0 \\ \beta \end{bmatrix},\quad \hat{\mathbf{b}}_\sigma = \begin{bmatrix} 0 \\ \gamma \end{bmatrix},\quad \mathbf{c} = \begin{bmatrix} 0 & 1 \end{bmatrix}, \tag{4.11}$$

cf. (2.135). Here we still have $\kappa = V'(h^*)$, while we dropped the saturation functions sat and W by assuming that they are not effective around the equilibrium. In this case the transfer function (4.10) becomes

$$T(s) = \frac{\gamma s^2 + \beta s + \alpha\kappa}{s^2 \mathrm{e}^{s\sigma} + (\alpha+\beta)s + \alpha\kappa}\,. \tag{4.12}$$

Notice that this transfer function is similar to that of the optimal velocity model (OVM) given in (2.136) that was used in Chap. 2 to characterize the response of human drivers, and to that of the adaptive cruise controller given in (3.77) that was designed in Chap. 3 to regulate the longitudinal motion of automated vehicles. However, apart from the position and velocity gains α and β, we also have the acceleration gain γ that can be tuned to achieve the desired control performance.

Let us determine the control gains for which plant and string stability are achieved. Since the denominator of the transfer function (4.12) is the same as that of (2.136) that characterizes human drivers (apart from the delay being σ instead of τ), the plant stability boundaries are the same as the ones in (2.138), (2.139). That is, for $s = 0$ we have

$$\alpha = 0\,, \tag{4.13}$$

while for $s = \mathrm{j}\Omega$, $\Omega > 0$ we obtain

$$\begin{aligned} \alpha &= \frac{\Omega^2 \cos(\Omega\sigma)}{\kappa}\,, \\ \beta &= \Omega \sin(\Omega\sigma) - \alpha\,. \end{aligned} \tag{4.14}$$

The plant stability chart is therefore the same as the one presented in Fig. 2.17 for the OVM parametrization of human car-following, irrespective of the acceleration gain γ.

String stability, on the other hand, is affected by acceleration feedback. As established in Chap. 2, string stability requires the magnitude $|T(\mathrm{j}\omega)|$ of the frequency response to be smaller than 1 according to (2.81), which may be violated at $\omega = 0$ or at $\omega > 0$ in case of string instability. The frequency response corresponding to (4.12) is plotted in Fig. 4.3 for various values of the acceleration gain γ. In this example, the

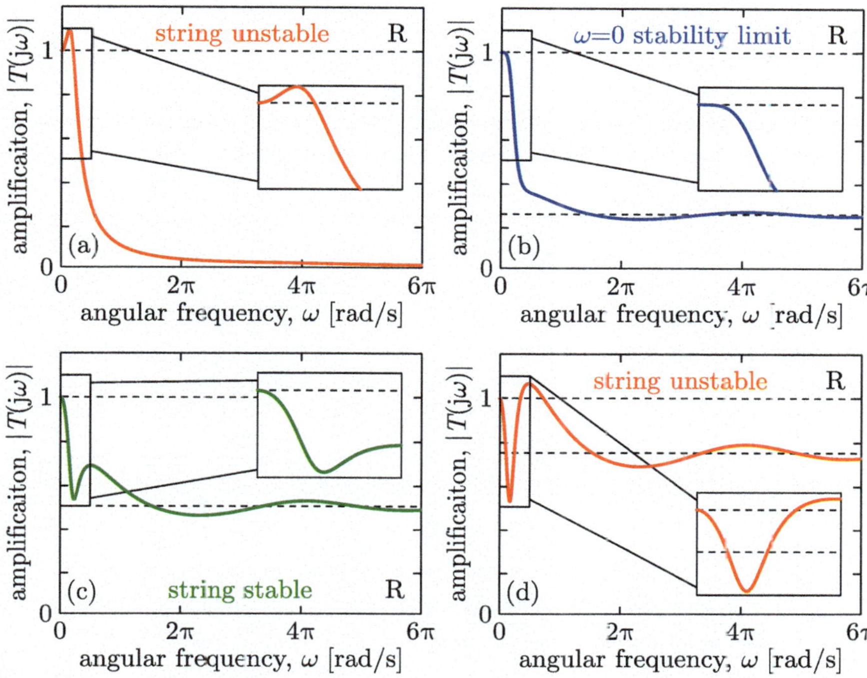

Fig. 4.3 Frequency response plots corresponding to the transfer function (4.12) with $\kappa = 0.6\,[1/\mathrm{s}]$ and $\sigma = 0.6\,[\mathrm{s}]$. Panels **a–d** are associated with point R(0.25,0.4) in Fig. 4.4 and $\gamma = 0$, $\gamma = 0.25$, $\gamma = 0.5$, and $\gamma = 0.75$, respectively

$\omega = 0$ string unstable case in panel (a) is stabilized by increasing γ: panel (b) corresponds to the string stability boundary and string stability is achieved in panel (c). However, selecting even larger γ values leads to $\omega > 0$ string instability as depicted in panel (d). In fact, the magnitude of the frequency response approaches γ at high frequencies (see dashed lines):

$$\lim_{\omega\to\infty} |T(\mathrm{j}\omega)| = \gamma\,, \tag{4.15}$$

which restricts the acceleration gain to $\gamma < 1$.

Now let us derive the $\omega = 0$ and $\omega > 0$ string stability boundaries. For algebraic convenience, we convert the condition $|T(\mathrm{j}\omega)| < 1$ into the form $P(\omega) > 0$ through (2.141) and (2.142), where $P(\omega)$ now becomes

$$P(\omega) = (1-\gamma^2)\omega^2 + \alpha^2 + 2\alpha\beta - 2(\alpha+\beta)\omega\sin(\omega\sigma) - 2\alpha\kappa\cos(\omega\sigma) + 2\gamma\alpha\kappa\,, \tag{4.16}$$

cf. (2.143) which is recovered for $\gamma = 0$ and $\sigma = \tau$. The $\omega = 0$ string stability condition in (2.144) gives

$$
\begin{aligned}
&\alpha > 0\,,\\
&\alpha > 2\big(\kappa(1-\gamma)-\beta\big),
\end{aligned}
\tag{4.17}
$$

considering that $\alpha > 0$ is required for plant stability. The $\omega > 0$ string stability boundaries can be derived from (2.147), where again we can notice that $\frac{\mathrm{d}P}{\mathrm{d}\omega}(\omega) = 2\omega Q(\omega)$ with

$$
Q(\omega) = 1-\gamma^2-(\alpha+\beta)\sigma\frac{\sin(\omega\sigma)}{\omega\sigma}-(\alpha+\beta)\sigma\cos(\omega\sigma)+\alpha\kappa\sigma^2\frac{\sin(\omega\sigma)}{\omega\sigma}\,,
\tag{4.18}
$$

cf. (2.149).

Solving (2.147) leads to the form

$$
\begin{aligned}
\alpha &= a \pm \sqrt{a^2+b}\,,\\
\beta &= \frac{(1-\gamma^2)\hat{\omega}+\alpha\kappa\sigma\sin(\hat{\omega}\sigma)}{\sin(\hat{\omega}\sigma)+\hat{\omega}\sigma\cos(\hat{\omega}\sigma)}-\alpha\,,\\
a &= \frac{\hat{\omega}(\kappa\sigma-1+\gamma^2)+\kappa\sin(\hat{\omega}\sigma)\big(\cos(\hat{\omega}\sigma)-\gamma\big)-\gamma\kappa\hat{\omega}\sigma\cos(\hat{\omega}\sigma)}{(2\kappa\sigma-1)\sin(\hat{\omega}\sigma)-\hat{\omega}\sigma\cos(\hat{\omega}\sigma)}\,,\\
b &= \frac{(1-\gamma^2)\hat{\omega}^2\big(\sin(\hat{\omega}\sigma)-\hat{\omega}\sigma\cos(\hat{\omega}\sigma)\big)}{(2\kappa\sigma-1)\sin(\hat{\omega}\sigma)-\hat{\omega}\sigma\cos(\hat{\omega}\sigma)}\,,
\end{aligned}
\tag{4.19}
$$

which reduces to expression (2.148) for $\gamma = 0$ and $\sigma = \tau$.

The stability charts of CCC with acceleration feedback are shown in Fig. 4.4 for various acceleration gains. Red curves indicate the plant stability boundaries that enclose the light gray shaded plant stable region, whereas blue curves mark the string stability boundaries surrounding the dark gray shaded string stable domain. According to the figure, the acceleration gain γ can help enlarging the string stable domain as long as it is not selected to be too large. The delay σ negatively impacts stability, as it makes the stable domains bounded and they shrink with increasing delay.

The string stable domain vanishes if the delay exceeds a critical value σ_{cr} that depends on γ. The critical delay can be found by locating points A, B and C in Fig. 4.4, since these points coincide when the delay reaches the critical value. Point A is given by the intersection of the two lines in (4.17), while points B and C are obtained by taking the limit $\hat{\omega} \to 0$ along the string stability boundaries (4.19).

Again, this limit can be calculated by solving $P(0) = 0$ and $Q(0) = 0$, resulting in

$$
\begin{aligned}
&\mathrm{A}: \big(\kappa(1-\gamma), 0\big)\,, \quad \mathrm{B}: \left(\frac{1-\gamma^2}{2\sigma}, 0\right),\\
&\mathrm{C}: \left((1-\gamma)\frac{2\kappa^2\sigma^2-4\kappa\sigma+1+\gamma}{2\sigma(\kappa\sigma-1)}, (1-\gamma)\frac{2\kappa\sigma-1-\gamma}{\sigma(\kappa\sigma-1)}\right),
\end{aligned}
\tag{4.20}
$$

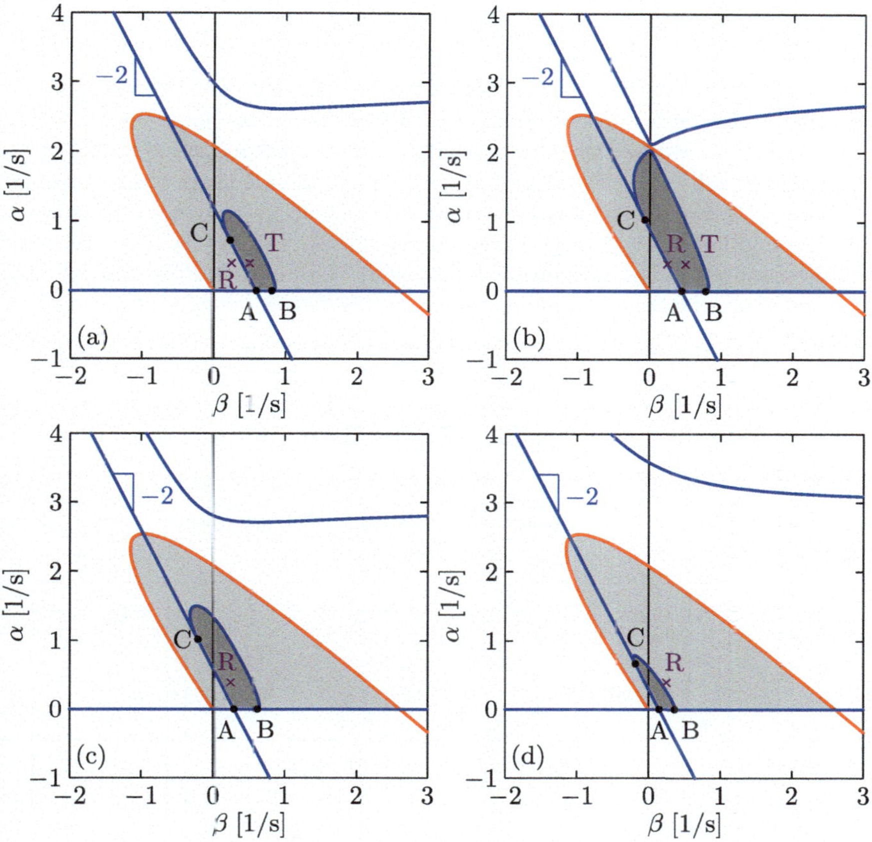

Fig. 4.4 Stability charts of system (4.8), (4.11) for $\kappa = 0.6\,[1/\mathrm{s}]$ and $\sigma = 0.6\,[\mathrm{s}]$. Panels **a–d** are associated with acceleration gains $\gamma = 0$, $\gamma = 0.25$, $\gamma = 0.5$ and $\gamma = 0.75$, respectively. The plant and string stable regions are shaded by light and dark gray, respectively. Point A is located at the intersection $(\kappa, 0)$, while points B and C lie at the starting points of the string stability boundaries. Point R(0.25,04) corresponds to the frequency response plots in Fig. 4.3, while point T(0.5,0.4) is associated with the numerical simulations in Fig. 4.2

which coincide for the critical delay

$$\sigma_{\mathrm{cr}} = \frac{1+\gamma}{2\kappa} \,. \tag{4.21}$$

The critical delay is $(1+\gamma)$ times larger than the one without acceleration feedback, cf. (2.151) and (3.84), i.e., it can be increased at most two-fold if one takes $\gamma \to 1$. Acceleration feedback is thus capable of achieving string stability even for larger delays, although the string stable domains may be small for high acceleration gains; see the illustration in Fig. 4.4d.

4.2.2 Vehicles on a Ring

Here we address the stability analysis for the case of CCC with acceleration feedback in the ring configuration illustrated by Fig. 2.13. Consider a ring of net length L where N identical CAVs follow each other using the acceleration feedback controller (4.5), or more specifically (4.6). Regardless of the CAVs' behavior, the periodic boundary condition (2.91) and the headway constraint (2.92) hold for the ring configuration, while the equilibrium is given by the uniform flow (2.15), (2.93). The linearized dynamics around this equilibrium are described by (4.8) for each CAV. Utilizing that the response of vehicle $i+1$ is input to vehicle i, that is, $u_i(t) = y_{i+1}(t) = \mathbf{c}\,\mathbf{x}_{i+1}(t)$, we obtain

$$
\begin{aligned}
\dot{\mathbf{x}}_i(t) &= \mathbf{a}_0\,\mathbf{x}_i(t) + \mathbf{a}_\sigma\,\mathbf{x}_i(t-\sigma) \\
&\quad + \mathbf{b}_0\mathbf{c}\,\mathbf{x}_{i+1}(t) + \mathbf{b}_\sigma\mathbf{c}\,\mathbf{x}_{i+1}(t-\sigma) + \hat{\mathbf{b}}_\sigma\mathbf{c}\,\dot{\mathbf{x}}_{i+1}(t-\sigma)\,, \quad i = 1, \ldots, N\,,
\end{aligned}
\tag{4.22}
$$

cf. (2.152).

Combined with the periodic boundary condition (2.97), this implies the dynamical system

$$
\begin{aligned}
\begin{bmatrix} \dot{\mathbf{x}}_1(t) \\ \dot{\mathbf{x}}_2(t) \\ \vdots \\ \dot{\mathbf{x}}_N(t) \end{bmatrix}
&= \begin{bmatrix} \mathbf{a}_0 & \mathbf{b}_0\mathbf{c} & & \\ & \ddots & \ddots & \\ & & \mathbf{a}_0 & \mathbf{b}_0\mathbf{c} \\ \mathbf{b}_0\mathbf{c} & & & \mathbf{a}_0 \end{bmatrix}
\begin{bmatrix} \mathbf{x}_1(t) \\ \mathbf{x}_2(t) \\ \vdots \\ \mathbf{x}_N(t) \end{bmatrix}
+ \begin{bmatrix} \mathbf{a}_\sigma & \mathbf{b}_\sigma\mathbf{c} & & \\ & \ddots & \ddots & \\ & & \mathbf{a}_\sigma & \mathbf{b}_\sigma\mathbf{c} \\ \mathbf{b}_\sigma\mathbf{c} & & & \mathbf{a}_\sigma \end{bmatrix}
\begin{bmatrix} \mathbf{x}_1(t-\sigma) \\ \mathbf{x}_2(t-\sigma) \\ \vdots \\ \mathbf{x}_N(t-\sigma) \end{bmatrix} \\
&\quad + \begin{bmatrix} \mathbf{0} & \hat{\mathbf{b}}_\sigma\mathbf{c} & & \\ & \ddots & \ddots & \\ & & \mathbf{0} & \hat{\mathbf{b}}_\sigma\mathbf{c} \\ \hat{\mathbf{b}}_\sigma\mathbf{c} & & & \mathbf{0} \end{bmatrix}
\begin{bmatrix} \dot{\mathbf{x}}_1(t-\sigma) \\ \dot{\mathbf{x}}_2(t-\sigma) \\ \vdots \\ \dot{\mathbf{x}}_N(t-\sigma) \end{bmatrix}.
\end{aligned}
\tag{4.23}
$$

In short form, one can write

$$
\dot{\mathbf{X}}(t) = \mathbf{A}_0\mathbf{X}(t) + \mathbf{A}_\sigma\mathbf{X}(t-\sigma) + \hat{\mathbf{A}}_\sigma\dot{\mathbf{X}}(t-\sigma)\,, \tag{4.24}
$$

with the coefficient matrices

$$
\begin{aligned}
\mathbf{A}_0 &= \mathbf{I}_N \otimes \mathbf{a}_0 + \mathbf{R}_N \otimes \mathbf{b}_0\mathbf{c}\,, \\
\mathbf{A}_\sigma &= \mathbf{I}_N \otimes \mathbf{a}_\sigma + \mathbf{R}_N \otimes \mathbf{b}_\sigma\mathbf{c}\,, \\
\hat{\mathbf{A}}_\sigma &= \mathbf{R}_N \otimes \hat{\mathbf{b}}_\sigma\mathbf{c}\,,
\end{aligned}
\tag{4.25}
$$

similarly to the case of human car-following with delay in (2.153)–(2.155), where $\mathbf{I}_N$ and $\mathbf{R}_N$ are the N-dimensional identity and circulant matrices.

Notice that the linearized system (4.24) involves the state derivative with delay. Equations of this kind are called neutral delay differential equations [94, 130, 254]. As shown in Appendix B, for neutral systems, a necessary condition for stability is the stability of the so-called difference part constructed from the derivative terms:

$$\mathbf{X}(t) = \hat{\mathbf{A}}_\sigma \mathbf{X}(t-\sigma)\,. \tag{4.26}$$

The difference part is associated with the characteristic equation

$$\det\left(\mu\mathbf{I} - \hat{\mathbf{A}}_\sigma\right) = 0\,, \tag{4.27}$$

where $\mu = \mathrm{e}^{\lambda\sigma}$. The solutions μ_ℓ, $\ell = 1, \ldots, 2N$ of (4.27) are the eigenvalues of matrix $\hat{\mathbf{A}}_\sigma$, also called as characteristic multipliers. To ensure stability, μ_ℓ must be located within the unit circle of the complex plane, since $\mathrm{Re}(\lambda_\ell) < 0$ is equivalent to $|\mu_\ell| < 1$.

By exploiting the circulant structure of the coefficient matrix $\hat{\mathbf{A}}_\sigma$ via the transformation (2.103), (2.104), the characteristic equation (4.27) can be decoupled into

$$\det\left(\mu\mathbf{I} - \hat{\mathbf{b}}_\sigma \mathbf{c}\,\mathrm{e}^{\mathrm{j}\frac{2k\pi}{N}}\right) = 0\,, \tag{4.28}$$

for $k = 0, 1, \ldots, N-1$. Substituting the matrices from (4.11) simplifies (4.28) to

$$\mu\left(\mu - \gamma\mathrm{e}^{\mathrm{j}\frac{2k\pi}{N}}\right) = 0\,. \tag{4.29}$$

The two characteristic multipliers associated with mode k are therefore $\mu_{2k+1} = 0$ and $\mu_{2k+2} = \gamma\,\mathrm{e}^{\mathrm{j}\frac{2k\pi}{N}}$, implying $|\mu_{2k+1}| = 0$ and $|\mu_{2k+2}| = \gamma$, $k = 0, 1, \ldots, N-1$. That is, considering all N modes, there are N multipliers lying in the origin of the complex plane and N multipliers are located along a circle of radius γ. To ensure the stability of the neutral system (4.24), the roots of (4.29) must be within the unit circle of the complex plane. This results in the condition $\gamma < 1$. Recall that the same condition was required as a necessity for the string stability of the open chain of vehicles when analyzing the high-frequency response in (4.15).

The analysis above only provides a necessary condition for stability. Necessary and sufficient conditions are achieved according to Theorem 2.7 by analyzing the roots of the characteristic equation

$$\det\left(\lambda\mathbf{I} - \mathbf{A}_0 - \mathbf{A}_\sigma\mathrm{e}^{-\lambda\sigma} - \lambda\hat{\mathbf{A}}_\sigma\mathrm{e}^{-\lambda\sigma}\right) = 0\,, \tag{4.30}$$

corresponding to (4.24), which is equivalent to

$$\det\left(\lambda\mathbf{I} - \left(\mathbf{a}_0 + \mathbf{b}_0\mathbf{c}\,\mathrm{e}^{\mathrm{j}\frac{2k\pi}{N}}\right) - \left(\mathbf{a}_\sigma + \mathbf{b}_\sigma\mathbf{c}\,\mathrm{e}^{\mathrm{j}\frac{2k\pi}{N}} + \lambda\hat{\mathbf{b}}_\sigma\mathbf{c}\,\mathrm{e}^{\mathrm{j}\frac{2k\pi}{N}}\right)\mathrm{e}^{-\lambda\sigma}\right) = 0\,, \tag{4.31}$$

for $k = 0, 1, \ldots, N-1$; cf. (2.156), (2.159). Substituting the coefficient matrices from (4.11) gives

$$\lambda^2 e^{\lambda\sigma} + (\alpha+\beta)\lambda + \alpha\kappa - (\gamma\lambda^2 + \beta\lambda + \alpha\kappa)e^{j\frac{2k\pi}{N}} = 0 \,. \tag{4.32}$$

Then, the stability boundaries can be obtained by considering the cases $\lambda = 0$ and $\lambda = j\omega$, $\omega > 0$. Note that similar to (2.160) for $k = 0$ a root $\lambda = 0$ exists regardless the parameters of the CAV's controller. However, this root is an artifact of the translational symmetry along the ring, and therefore it can be disregarded. The rest of the characteristic equation $\lambda e^{\lambda\sigma} - \gamma\lambda + \alpha = 0$ leads to the stability condition

$$0 < \alpha < \frac{\arccos\gamma}{\sigma}\sqrt{1-\gamma^2} \,, \tag{4.33}$$

for $\gamma < 1$, cf. (2.161) for $\gamma = 0$.

For $k > 0$, substituting $\lambda = 0$ into (4.32), one gets the stability boundary

$$\alpha = 0 \,, \tag{4.34}$$

cf. (2.162). For $\lambda = j\omega$, $\omega > 0$, decomposing (4.32) into real and imaginary parts and solving for α and β lead to

$$\begin{aligned}
\alpha &= \frac{\omega^2\left(\cos(\omega\sigma) - \cos\left(\omega\sigma - \frac{2k\pi}{N}\right) + \gamma\left(1 - \cos\frac{2k\pi}{N}\right)\right)}{-\omega\sin\frac{2k\pi}{N} + 2\kappa\left(1 - \cos\frac{2k\pi}{N}\right)} \,, \\
\beta &= \frac{-\omega^2\left(\cos(\omega\sigma) - \gamma\cos\frac{2k\pi}{N}\right) + \kappa\omega\left(\sin(\omega\sigma) - \sin\left(\omega\sigma - \frac{2k\pi}{N}\right) - \gamma\sin\frac{2k\pi}{N}\right)}{-\omega\sin\frac{2k\pi}{N} + 2\kappa\left(1 - \cos\frac{2k\pi}{N}\right)} \,,
\end{aligned} \tag{4.35}$$

$k = 1, \ldots, N-1$. These stability boundaries simplify to the ones given by (2.163) for the OVM parametrization of human car-following models, if one takes $\gamma = 0$ and $\sigma = \tau$.

Again, one may use the trigonometric identities $\left(1 - \cos\frac{2k\pi}{N}\right) = 2\sin^2\frac{k\pi}{N}$ and $\sin\frac{2k\pi}{N} = 2\sin\frac{k\pi}{N}\cos\frac{k\pi}{N}$ to obtain

$$\begin{aligned}
\alpha &= \frac{\omega^2\left(\left(\cos(\omega\sigma) + \gamma\right)\tan\frac{k\pi}{N} - \sin(\omega\sigma)\right)}{2\kappa\tan\frac{k\pi}{N} - \omega} \,, \\
\beta &= \frac{\omega\left(\cos(\omega\sigma) - \gamma + \sin(\omega\sigma)\tan\frac{k\pi}{N}\right)}{2\tan\frac{k\pi}{N}} - \frac{\alpha}{2} \,,
\end{aligned} \tag{4.36}$$

cf. (2.164). Indeed, (4.35) and (4.36) are equivalent.

The stability chart of CCC with acceleration feedback in the ring configuration is depicted in Fig. 4.5. The stability boundaries (4.34) and (4.35) are indicated by different colors for different modes k. Similarly to Chap. 2, the stable region of the

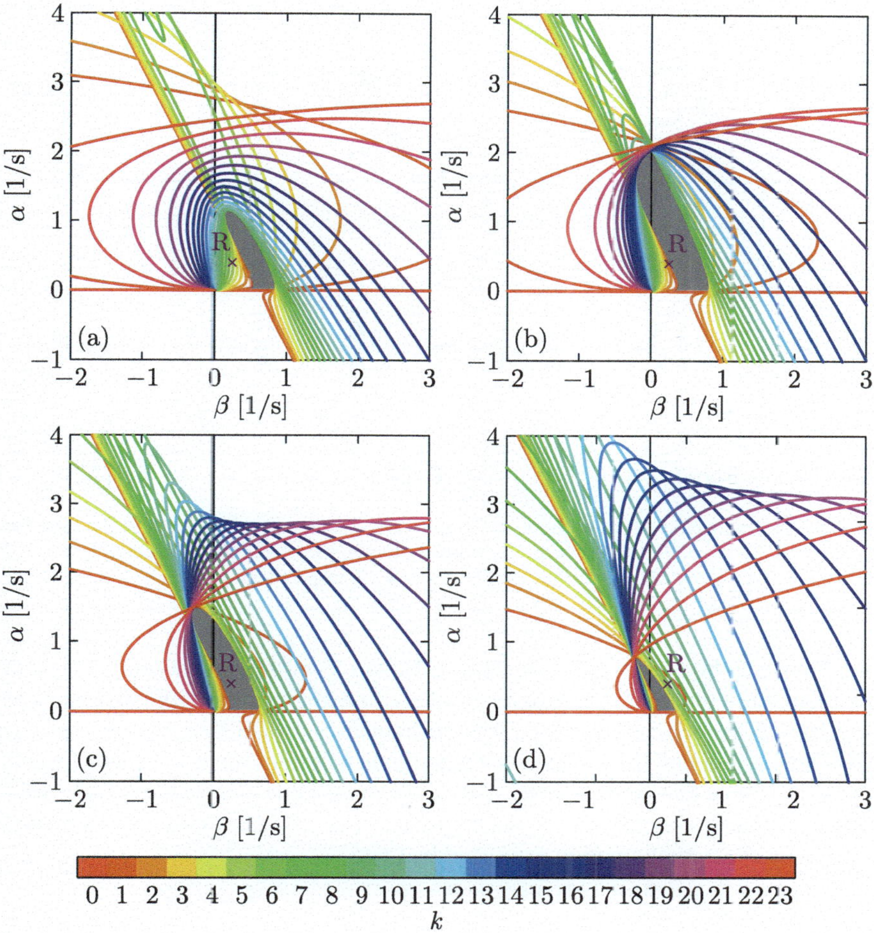

Fig. 4.5 Stability chart of system (4.23), (4.11) for $N = 24$ vehicles, $\kappa = 0.6$ [1/s] and $\sigma = 0.6$ [s]. Panels **a–d** are associated with acceleration gains $\gamma = 0$, $\gamma = 0.25$, $\gamma = 0.5$, and $\gamma = 0.75$, respectively. The wave numbers k are indicated using color bar

ring configuration (indicated by dark gray shading) closely approximates the string stable domain of the open chain scenario if the number N of vehicles is high enough; cf. Fig. 4.4.

For the ring configuration, stability is determined by the roots of the characteristic equation (4.30), which are plotted in Fig. 4.6 for point R marked in Fig. 4.5. Panel (a) corresponds to $\gamma = 0$; this plot identical to Fig. 2.25a. For $\gamma = 0$ the system is unstable: there are eigenvalues with small imaginary parts located in the right-half plane (corresponding to low-frequency instability). By adding the acceleration feedback term ($\gamma > 0$) and increasing γ, the system gets close to the stability boundary

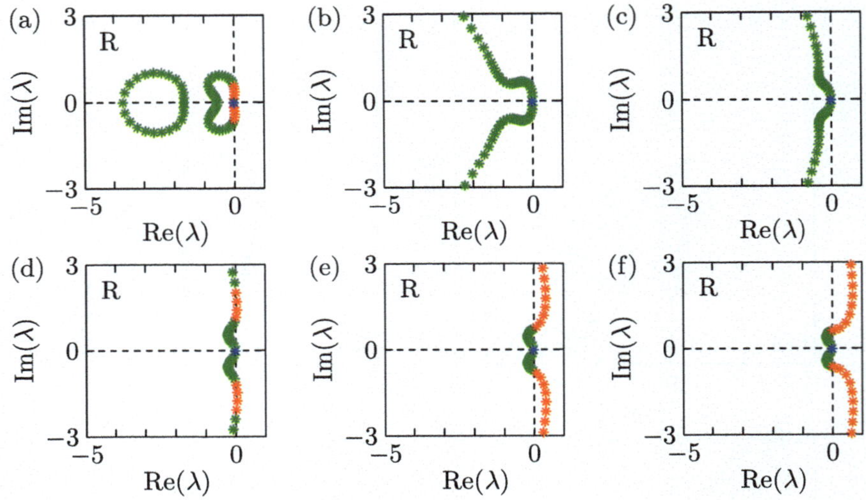

Fig. 4.6 Spectrum plots obtained from the characteristic equation (4.30) for $N = 24$ vehicles, $\kappa = 0.6\,[1/\mathrm{s}]$ and $\sigma = 0.6\,[\mathrm{s}]$. Panels **a–f** correspond to point R(0.25, 0.4) marked in Fig. 4.5 and $\gamma = 0$, $\gamma = 0.25$, $\gamma = 0.5$, $\gamma = 0.75$, $\gamma = 1$ and $\gamma = 1.25$, respectively

(panel (b)) and becomes stable (panel (c)). However, further increase in γ leads to loss of stability (panel (d)), where eigenvalues with larger imaginary parts move to the right-half plane (corresponding to high-frequency instability). As γ exceeds 1, infinitely many eigenvalues are located in the right half of the complex plane (panels (e) and (f)). Note that the imaginary parts of these unstable eigenvalues can be arbitrarily large, which corresponds to the high-frequency string instability phenomenon of the open chain configuration expressed by (4.15).

4.3 Connected Vehicle Networks

The acceleration feedback controller in Sect. 4.2 vividly illustrates an example where V2X connectivity improves control performance by providing additional information to the connected automated vehicle about the motion of the preceding vehicle, which would be hard to obtain by on-board sensors. Applications of connectivity, however, are not limited to obtaining information from the immediate preceding vehicle. Connectivity reaches beyond the range of on-board sensors, and thus, it can extend perception to beyond the line of sight. In particular, V2X connectivity allows vehicles to receive information from distant vehicles in their lane even when their optical sensors are blocked by nearby vehicles; see Fig. 4.7.

This enables a connected automated vehicle to respond to multiple vehicles ahead of it as long as these vehicles are equipped with communication devices [251].

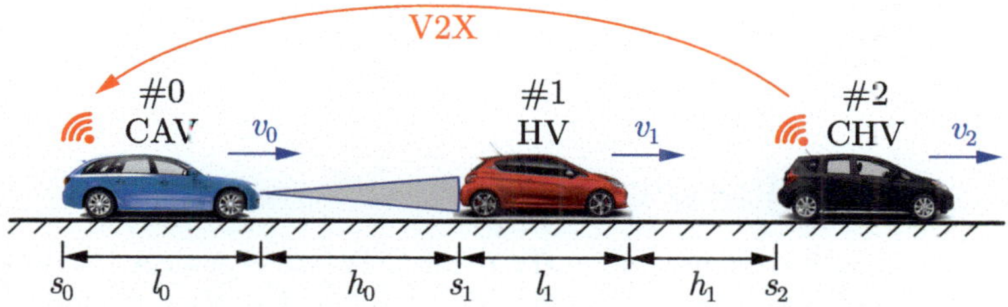

Fig. 4.7 A simple connected vehicle network where a connected automated vehicle (CAV) responds to a human-driven vehicle (HV) and a connected human-driven vehicle (CHV) via connected cruise control (CCC) on a single lane. The CAV is able to detect the preceding HV by on-board sensors and it can also obtain information from the CHV via vehicle-to-everything (V2X) connectivity

Note that the vehicles ahead are only required to be connected, but not necessarily automated: a connected automated vehicle may respond both to connected human-driven and connected automated vehicles with different levels automation. In this section we consider connected vehicle networks where multiple (but not necessarily all) vehicles are equipped with connectivity, and we design longitudinal controllers for connected automated vehicles that allow them to respond to multiple vehicles ahead. These controllers do not assume full penetration of connectivity, they simply respond to the vehicle immediately ahead and to other vehicles farther ahead that happen to be connected and traveling within communication range.

4.3.1 A Simple Three-Vehicle Network

We begin our discussion on connected vehicle networks with a small-scale simple example [251]. Consider the network of three vehicles shown in Fig. 4.7, where a connected automated vehicle (CAV) responds to two vehicles: a human-driven vehicle (HV) immediately ahead and a connected human-driven vehicle (CHV) two vehicles ahead. Since the CAV can potentially detect the immediate preceding HV by range sensors, connectivity is not necessarily required for this vehicle, although connectivity can support the sensory data with additional information as explained above. However, obtaining information from the CHV two vehicles ahead would be challenging without connectivity.

Let us investigate the car-following dynamics by considering the motion of the lead CHV as given and modeling the response of the middle HV and the CAV. We number the vehicles by using index 0 for the CAV, 1 for the HV, and 2 for the CHV (i.e., with indices increasing in the direction of motion). We denote the lengths, positions and speeds of the vehicles by l_i, s_i, and v_i, $i = 0, 1, 2$, respectively. The longitudinal dynamics can be modeled by delayed double integrators according to

$$\begin{aligned}
\dot{s}_1(t) &= v_1(t) , \\
\dot{v}_1(t) &= \text{sat}\big(u_{12}(t-\sigma_{12})\big) , \\
\dot{s}_0(t) &= v_0(t) , \\
\dot{v}_0(t) &= \text{sat}\big(u_{01}(t-\sigma_{01}) + u_{02}(t-\sigma_{02})\big) .
\end{aligned} \tag{4.37}$$

Here u_{12} indicates the response of the HV to the CHV that is subject to a delay σ_{12} caused by human reaction time and powertrain dynamics. The control inputs u_{01} and u_{02} stand for the response of the CAV to the HV and the CHV, respectively, with delays σ_{01} and σ_{02} originating from sensing, communication, feedback and actuation. Note that these two delays can potentially be different and may even be tuned (above some minimum feasible values) as design parameters [191]. The desired accelerations u_{12} and $u_{01} + u_{02}$ saturate at the acceleration capabilities of the vehicles. Although saturation limits can be different for each vehicle, we consider the same saturation function (sat) for simplicity.

The desired acceleration of the HV can be captured by the human driver models discussed in Chap. 2, which take the form

$$u_{12} = F_{12}(s_1, s_2, v_1, v_2) . \tag{4.38}$$

For example, the OVM (2.4) now becomes

$$u_{12} = \alpha_{12}\big(V_1(s_2 - s_1 - l_1) - v_1\big) + \beta_{12}(v_2 - v_1) , \tag{4.39}$$

where $h_1 = s_2 - s_1 - l_1$ is the headway ahead of the HV, α_{12} and β_{12} are driver parameters, and V_1 is the range policy of the HV.

The desired acceleration of the CAV, on the other hand, can be designed by prescribing control laws of the form

$$\begin{aligned}
u_{01} &= F_{01}(s_0, s_1, v_0, v_1) , \\
u_{02} &= F_{02}(s_0, s_2, v_0, v_2) .
\end{aligned} \tag{4.40}$$

For example, one can utilize a controller that resembles the OVM:

$$\begin{aligned}
u_{01} &= \alpha_{01}\big(V_0(s_1 - s_0 - l_0) - v_0\big) + \beta_{01}\big(W(v_1) - v_0\big) , \\
u_{02} &= \alpha_{02}\big(V_0\left((s_2 - s_0 - l_0 - l_1)/2\right) - v_0\big) + \beta_{02}\big(W(v_2) - v_0\big) ,
\end{aligned} \tag{4.41}$$

where α_{01}, α_{02}, β_{01} and β_{02} are control gains to be designed, V_0 is the range policy of the CAV evaluated at the headway

$$h_{01} = h_0 = s_1 - s_0 - l_0 , \tag{4.42}$$

cf. (4.1) and at the average headway

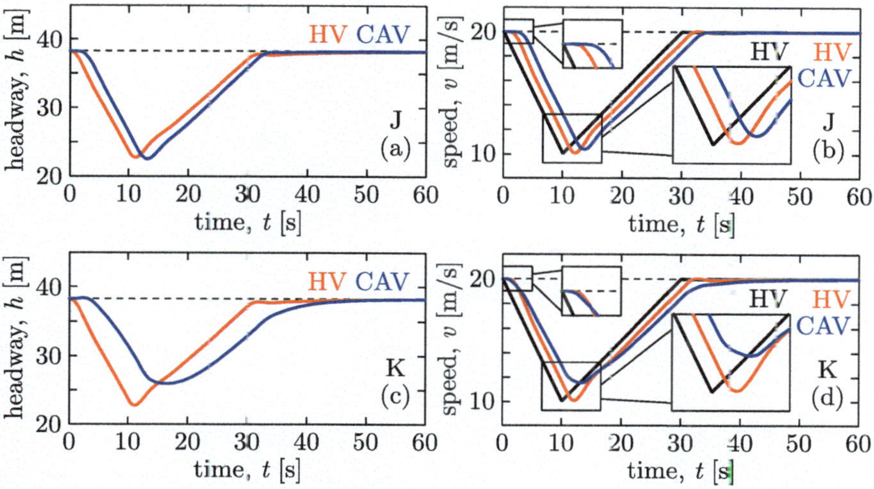

Fig. 4.8 Numerical simulation of the connected vehicle network in Fig. 4.7 where the HV and CAV are governed by (4.37), (4.39) and (4.41). The lead vehicle uses constant -1 and $0.5\,[\mathrm{m/s^2}]$ acceleration to slow down from 20 to $10\,[\mathrm{m/s}]$ and to speed back up. The range policy is piecewise linear (2.5), (2.6) and (3.32) for both the HV and the CAV with $h_{\mathrm{st}} = 5\,[\mathrm{m}]$, $h_{\mathrm{go}} = 55\,[\mathrm{m}]$ and $v_{\max} = 30\,[\mathrm{m/s}]$. The initial conditions over $t \le 0$ are $v_i(t) = v^* = 20\,[\mathrm{m/s}]$, $h_i(t) = V_i^{-1}(v^*) = 38.33\,[\mathrm{m}]$, $i = 0, 1$. The human driver parameters are $\alpha_{12} = 0.1\,[1/\mathrm{s}]$, $\beta_{12} = 0.6\,[1/\mathrm{s}]$ and $\tau_1 = 1\,[\mathrm{s}]$. The CAV's control parameters are $\alpha_{01} = 0.4\,[1/\mathrm{s}]$, $\alpha_{02} = 0$ and $\sigma_{01} = \sigma_{02} = 0.6\,[\mathrm{s}]$. The rest of the parameters are $l = 5\,[\mathrm{m}]$, $a_{\min} = 7\,[\mathrm{m/s^2}]$ and $a_{\max} = 3\,[\mathrm{m/s^2}]$. The two rows of the figure correspond to points J(0.5,0) and K(0.5,0.5) in the (β_{01}, β_{02}) plane in Fig. 4.9b, that is, the CAV responds to: **a**, **b** the HV only; **c**, **d** both the HV and the CHV

$$h_{02} = \frac{h_0 + h_1}{2} = \frac{s_2 - s_0 - l_0 - l_1}{2}\,, \tag{4.43}$$

between the CAV and CHV, whereas W is the speed policy introduced in (3.33). Notice that control law u_{01} is in fact the same as the one used in the ACC algorithm (3.30). We remark that acceleration feedback terms could also be included (cf. (4.6)), although now we omit them for the sake of a simpler discussion.

The behavior of the connected vehicle system in Fig. 4.7 is illustrated by numerical simulations of (4.37), (4.39) and (4.41) in Fig. 4.8. In all simulations, the human drivers have the same behavior. The CAV's controller is set such that it keeps the same equilibrium headway as the preceding HV does to avoid conservatism. The two rows of Fig. 4.8 show two different selection of control gains for the CAV corresponding to points J and K in Fig. 4.9b. In panels (a) and (b) of Fig. 4.8, the CAV responds to the HV immediately ahead but not to the CHV ($\alpha_{02} = \beta_{02} = 0$). The design of the CAV's controller is string stable and the CAV brakes less than the HV does, yet the amount of speed reduction is in the same order of magnitude for the three vehicles. In panels (c) and (d) the CAV leverages the benefits of connectivity by responding

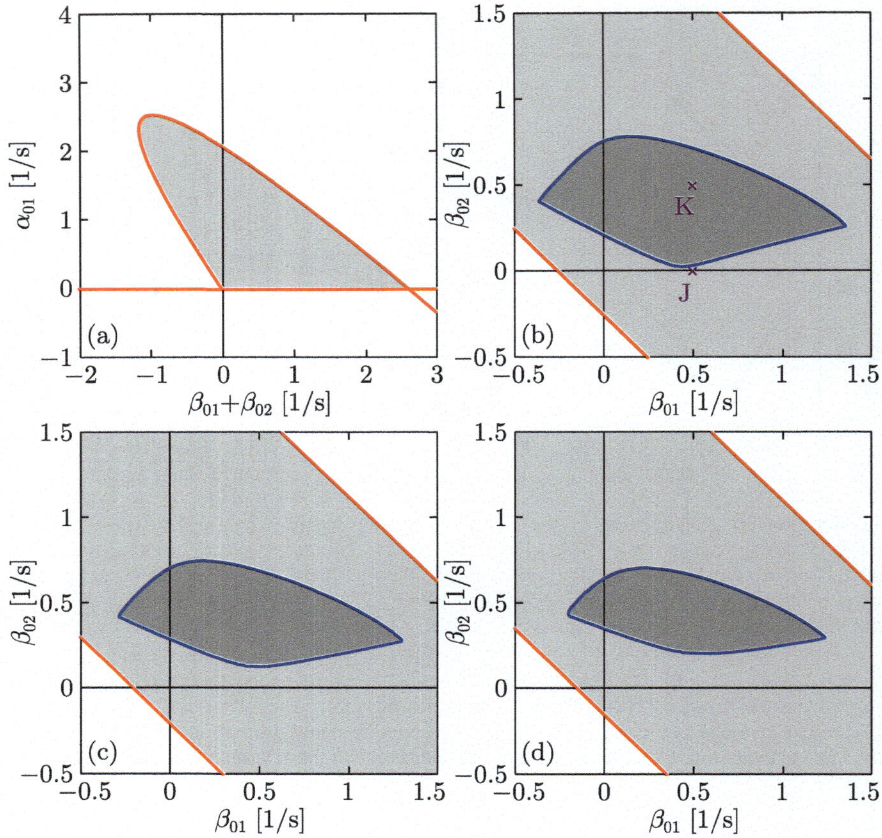

Fig. 4.9 Stability charts of system (4.48), (4.49) for human driver parameters $\alpha_{12} = 0.1\,[1/\mathrm{s}]$, $\beta_{12} = 0.6\,[1/\mathrm{s}]$, $\kappa_{12} = 0.6\,[1/\mathrm{s}]$ and $\sigma_{12} = 1\,[\mathrm{s}]$, and CAV control parameters $\alpha_{02} = 0$ and $\sigma_{01} = \sigma_{02} = 0.6\,[\mathrm{s}]$. Panel **a** shows the plant stability chart in the $(\beta_{01} + \beta_{02}, \alpha_{01})$ plane for $\kappa_{01} = 0.6\,[1/\mathrm{s}]$, whereas panels **b–d** show the plant and string stability charts in the (β_{01}, β_{02}) plane for $\alpha_{01} = 0.4\,[1/\mathrm{s}]$ and **b** $\kappa_{01} = 0.6\,[1/\mathrm{s}]$, **c** $\kappa_{01} = 0.8\,[1/\mathrm{s}]$, **d** $\kappa_{01} = 1\,[1/\mathrm{s}]$. The plant and head-to-tail string stable regions are shaded by light and dark gray, respectively. Points J(0.5,0) and K(0.5,0.5) correspond to the numerical simulations in Fig. 4.8 and frequency response plots in Fig. 4.10

to both the HV and the CHV ahead ($\beta_{02} \neq 0$). The response to the CHV results in a more desired behavior: the CAV starts to reduce its as speed early as the HV does in Fig. 4.8d. This way, despite the string unstable behavior of the HV, the CAV manages to drive significantly smoother (with less speed reduction) than even the CHV. Similar to Sect. 3.4, we call this desired behavior as *head-to-tail string stable*.

In what follows, we analyze the stability of the control design, while other performance measures will be addressed in Sects. 4.6 and 4.7. To this end, we linearize around the equilibrium solution (2.11) that can also be expressed in the form

$$
\begin{aligned}
s_i(t) &= v^* t + \bar{s}_i \,, \\
v_i(t) &\equiv v^*
\end{aligned}
\tag{4.44}
$$

with constants $\bar{s}_i$ that satisfy $h_i^* = \bar{s}_{i+1} - \bar{s}_i - l_i$. These position constants and the corresponding equilibrium headway values can be obtained by substituting (4.44) into (4.37), (4.39) and (4.41). For $v^* \le v_{\max}$, i.e., $W(v^*) = v^*$ one obtains

$$
\begin{aligned}
&\alpha_{12}\left(V_1(h_1^*) - v^*\right) = 0 \,, \\
&\alpha_{01}\left(V_0(h_0^*) - v^*\right) + \alpha_{02}\left(V_0\left((h_0^* + h_1^*)/2\right) - v^*\right) = 0 \,,
\end{aligned}
\tag{4.45}
$$

which can be solved for h_0^* and h_1^*.

Note that the equilibrium headway h_0^* ahead of the CAV and the equilibrium headway h_1^* ahead of the HV are not necessarily equal to each other (unless $V_0(h) = V_1(h)$), thus the CAV may keep a shorter or a longer headway than human drivers, depending on its design V_0. Furthermore, although $V_1(h_1^*) = v^*$ holds, $V_0(h_0^*) = v^*$ is not necessarily true, unless $\alpha_{02} = 0$ or $V_0(h) = V_1(h)$. If $\alpha_{02} \neq 0$ and $V_0(h) \neq V_1(h)$, the equilibrium headway h_0^* of the CAV depends on the equilibrium headway h_1^* of the HV, which is typically undesired from control design point of view. Since setting $V_0(h) = V_1(h)$ would require the CAV to know the range policy of the HV, one may rather choose $\alpha_{02} = 0$ to ensure that the equilibrium headway h_0^* of the CAV is independent from the HV's behavior. Still, the α_{02} term may be useful to improve performance in certain cases, for example, if vehicle 1 happens to be automated (a CAV instead of a HV) and it cooperates with CAV 0, then they can choose the same range policy.

Now let us consider the perturbations

$$
\begin{aligned}
\tilde{s}_i(t) &= s_i(t) - v^* t - \bar{s}_i \,, \\
\tilde{v}_i(t) &= v_i(t) - v^* ,
\end{aligned}
\tag{4.46}
$$

around the solution (4.44), (cf. (2.22)), and define the state and output for each vehicle as

$$
\mathbf{x}_i(t) = \begin{bmatrix} \tilde{s}_i(t) \\ \tilde{v}_i(t) \end{bmatrix}, \quad y_i(t) = \tilde{v}_i(t) \,,
\tag{4.47}
$$

$i = 0, 1, 2$. Then, the linearized dynamics can be written in the form

$$
\begin{aligned}
\dot{\mathbf{x}}_1(t) &= \mathbf{a}_1 \mathbf{x}_1(t) + \mathbf{a}_{12} \mathbf{x}_1(t - \sigma_{12}) + \mathbf{b}_{12} \mathbf{x}_2(t - \sigma_{12}) + \text{h.o.t.} \,, \\
y_1(t) &= \mathbf{c}_1 \mathbf{x}_1(t) \,, \\
\dot{\mathbf{x}}_0(t) &= \mathbf{a}_0 \mathbf{x}_0(t) + \mathbf{a}_{01} \mathbf{x}_0(t - \sigma_{01}) + \mathbf{b}_{01} \mathbf{x}_1(t - \sigma_{01}) \\
&\qquad + \mathbf{a}_{02} \mathbf{x}_0(t - \sigma_{02}) + \mathbf{b}_{02} \mathbf{x}_2(t - \sigma_{02}) + \text{h.o.t.} \,, \\
y_0(t) &= \mathbf{c}_0 \mathbf{x}_0(t) \,,
\end{aligned}
\tag{4.48}
$$

with coefficient matrices

$$\mathbf{a}_i = \begin{bmatrix} 0 & 1 \\ 0 & 0 \end{bmatrix}, \quad \mathbf{a}_{ij} = \begin{bmatrix} 0 & 0 \\ -\frac{\alpha_{ij}\kappa_{ij}}{j-i} & -(\alpha_{ij}+\beta_{ij}) \end{bmatrix}, \quad \mathbf{b}_{ij} = \begin{bmatrix} 0 & 0 \\ \frac{\alpha_{ij}\kappa_{ij}}{j-i} & \beta_{ij} \end{bmatrix}, \quad \mathbf{c}_i = \begin{bmatrix} 0 & 1 \end{bmatrix}, \tag{4.49}$$

where $\kappa_{ij} = V_i'(h_{ij}^*)$ and h_{ij} is defined by (4.42), (4.43).

To analyze plant and string stability, the linearized dynamics (4.48) can be transformed to the Laplace domain. Let us denote the Laplace transforms of $y_i(t)$ and $\mathbf{x}_i(t)$ by $Y_i(s)$ and $\mathbf{X}_i(s)$, which are related by

$$\mathbf{X}_i(s) = \begin{bmatrix} \frac{1}{s} \\ 1 \end{bmatrix} Y_i(s) , \tag{4.50}$$

according to (4.47). Since HV 1 responds to the motion of CHV 2 while CAV 0 responds to both HV 1 and CHV 2, the dynamics in Laplace domain become

$$\begin{aligned} Y_1(s) &= T_{12}(s)Y_2(s) , \\ Y_0(s) &= T_{01}(s)Y_1(s) + T_{02}(s)Y_2(s) , \end{aligned} \tag{4.51}$$

where zero initial conditions are considered for the linearized system. Similar to Sect. 3.4, the transfer functions T_{12}, T_{01} and T_{02} are called *link transfer functions* [251], since they characterize the response associated with each individual information link between vehicles.

One can also describe the response of the whole three-vehicle chain, from the head vehicle (CHV 2) to the tail vehicle (CAV 0), by combining the two lines of (4.51) into

$$Y_0(s) = G_{02}(s)Y_2(s) , \tag{4.52}$$

where G_{02} is the *head-to-tail transfer function* [251] and it is given by

$$G_{02}(s) = T_{01}(s)T_{12}(s) + T_{02}(s) , \tag{4.53}$$

cf. (3.104) for $N = 2$ where the connectivity-enabled term $T_{02}(s)$ is missing.

Taking the Laplace transform of (4.48) and substituting (4.50) lead to the link transfer functions:

$$\begin{aligned} T_{12}(s) &= \mathbf{c}_1 \left(s\mathbf{I} - \mathbf{a}_1 - \mathbf{a}_{12}\mathrm{e}^{-s\sigma_{12}} \right)^{-1} \mathbf{b}_{12}\mathrm{e}^{-s\sigma_{12}} \begin{bmatrix} \frac{1}{s} \\ 1 \end{bmatrix}, \\ T_{01}(s) &= \mathbf{c}_0 \left(s\mathbf{I} - \mathbf{a}_0 - \mathbf{a}_{01}\mathrm{e}^{-s\sigma_{01}} - \mathbf{a}_{02}\mathrm{e}^{-s\sigma_{02}} \right)^{-1} \mathbf{b}_{01}\mathrm{e}^{-s\sigma_{01}} \begin{bmatrix} \frac{1}{s} \\ 1 \end{bmatrix}, \\ T_{02}(s) &= \mathbf{c}_0 \left(s\mathbf{I} - \mathbf{a}_0 - \mathbf{a}_{01}\mathrm{e}^{-s\sigma_{01}} - \mathbf{a}_{02}\mathrm{e}^{-s\sigma_{02}} \right)^{-1} \mathbf{b}_{02}\mathrm{e}^{-s\sigma_{02}} \begin{bmatrix} \frac{1}{s} \\ 1 \end{bmatrix}, \end{aligned} \tag{4.54}$$

which, after substituting the coefficient matrices from (4.49), take the form

$$
\begin{aligned}
T_{12}(s) &= \frac{(\beta_{12}s + \alpha_{12}\kappa_{12})\mathrm{e}^{-s\sigma_{12}}}{s^2 + ((\alpha_{12} + \beta_{12})s + \alpha_{12}\kappa_{12})\,\mathrm{e}^{-s\sigma_{12}}}\,,\\
T_{01}(s) &= \frac{(\beta_{01}s + \alpha_{01}\kappa_{01})\mathrm{e}^{-s\sigma_{01}}}{s^2 + ((\alpha_{01} + \beta_{01})s + \alpha_{01}\kappa_{01})\,\mathrm{e}^{-s\sigma_{01}} + ((\alpha_{02} + \beta_{02})s + \alpha_{02}\kappa_{02}/2)\,\mathrm{e}^{-s\sigma_{02}}}\,,\\
T_{02}(s) &= \frac{(\beta_{02}s + \alpha_{02}\kappa_{02}/2)\mathrm{e}^{-s\sigma_{02}}}{s^2 + ((\alpha_{01} + \beta_{01})s + \alpha_{01}\kappa_{01})\,\mathrm{e}^{-s\sigma_{01}} + ((\alpha_{02} + \beta_{02})s + \alpha_{02}\kappa_{02}/2)\,\mathrm{e}^{-s\sigma_{02}}}\,.
\end{aligned}
\tag{4.55}
$$

For example, below we will choose $\alpha_{02} = 0$ and the identical delays $\sigma_{01} = \sigma_{02} = \sigma$, which yield the simplified formulas

$$
\begin{aligned}
T_{12}(s) &= \frac{\beta_{12}s + \alpha_{12}\kappa_{12}}{s^2\mathrm{e}^{s\sigma_{12}} + (\alpha_{12} + \beta_{12})s + \alpha_{12}\kappa_{12}}\,,\\
T_{01}(s) &= \frac{\beta_{01}s + \alpha_{01}\kappa_{01}}{s^2\mathrm{e}^{s\sigma} + (\alpha_{01} + \beta_{01} + \beta_{02})s + \alpha_{01}\kappa_{01}}\,,\\
T_{02}(s) &= \frac{\beta_{02}s}{s^2\mathrm{e}^{s\sigma} + (\alpha_{01} + \beta_{01} + \beta_{02})s + \alpha_{01}\kappa_{01}}\,,
\end{aligned}
\tag{4.56}
$$

cf. (3.105) where $\alpha_{12} = \alpha_{\mathrm{h}}$, $\beta_{12} = \beta_{\mathrm{h}}$, $\kappa_{12} = \kappa_{\mathrm{h}}$, and $\sigma_{12} = \tau$.

Having derived the link and head-to-tail transfer functions, one can analyze plant and string stability. Plant stability requires that the roots of the characteristic equation

$$
\mathrm{Den}(G_{02}(s)) = 0\,, \tag{4.57}
$$

are located in the left half of the complex plane; see Theorem 2.4. The characteristic equation is given by the denominator of the head-to-tail transfer function, which is the product of the denominators of the link transfer functions. Therefore, the characteristic equation can be decomposed into

$$
\begin{aligned}
&\mathrm{Den}(T_{12}(s)) = 0\,,\\
&\mathrm{Den}(T_{01}(s)) = \mathrm{Den}(T_{02}(s)) = 0\,,
\end{aligned}
\tag{4.58}
$$

and the denominator of each individual link transfer function must fulfill the plant stability conditions. This can be analyzed by repeating the steps presented in Chap. 2 and in Sect. 4.2. Note that human drivers are typically plant stable and only the second line of (4.58) is relevant from point of view of the CAV's control design.

For example, if one chooses $\alpha_{02} = 0$ and the delays are identical $\sigma_{01} = \sigma_{02} = \sigma$, then the plant stability boundaries reduce to (4.13), (4.14) such that $\alpha = \alpha_{01}$ and $\beta = \beta_{01} + \beta_{02}$. Therefore, the plant stability chart in the $(\beta_{01} + \beta_{02}, \alpha_{01})$ plane, as illustrated in Fig. 4.9a, matches the one in Fig. 4.4a. In the (β_{01}, β_{02}) plane, on the other hand, the plant stability boundaries are straight lines; see Fig. 4.9b–d for a selected value of α_{01}. As in all previous figures, the plant stable region is light gray and the plant stability boundaries are red in Fig. 4.9.

If every vehicle were automated, one could require all of them to attenuate the speed fluctuations relative to their immediate predecessors, that is, require $i+1$-to-i string stability for $i=0, 1$; see (2.70) in Theorem 2.2. This, indeed, would guarantee head-to-tail string stability; see (2.72) and (2.74) in Theorem 2.3 and (3.103) and (3.104) in Sect. 3.4. However, human drivers are often string unstable and the 2-to-1 string stability (from CHV 2 to HV 1) may not be ensured in this three-vehicle example. That is, $|T_{12}(\mathrm{j}\omega)| > 1$ may occur for some $\omega > 0$, depending on human driver behavior. Still, one can ensure head-to-tail string stability for the whole three-vehicle chain (from CHV 2 to CAV 0) by designing the CCC that satisfies

$$|G_{02}(\mathrm{j}\omega)| < 1\,, \quad \forall \omega > 0\,. \tag{4.59}$$

This guarantees that the speed fluctuations of the head vehicle (CHV 2) are attenuated by those of the tail vehicle (CAV 0), even if a member in the middle (HV 1) behaves string unstable.

Based on condition (4.59), head-to-tail string stability can be analyzed analogously to the string stability calculations in Chaps. 2 and 3. The underlying formulas, however, get more complicated as the head-to-tail transfer function contains the expressions of multiple link transfer functions. Although analytical results could be obtained (for example via the method in Sect. 2.3.3), we refrain from presenting them due to their algebraic complexity. Instead, we do a brute force numerical evaluation of (4.59) via calculating $|G_{02}(\mathrm{j}\omega)|$ for various values of control parameters and angular frequency ω.

The resulting string stability chart is plotted in Fig. 4.9b–d for three different values of the parameter κ_{01}.The head-to-tail string stable region is shown by the dark gray domain bounded by blue head-to-tail string stability curves. Notice that head-to-tail string stability can only be achieved when setting $\beta_{02} > 0$. That is, utilizing information from the distant vehicle (that can only be obtained by V2X connectivity) is necessary to ensure head-to-tail string stability. As κ_{01} is increased, i.e., the time headway of the CAV is decreased, the string stable domain shrinks, but selecting the right β_{02} still allows one to maintain head-to-tail string stability.

The frequency responses corresponding to points J and K in Fig. 4.9b are depicted in Fig. 4.10 where the three link transfer functions and the head-to-tail transfer function are both shown. In both panels, the HV shows string unstable behavior ($|T_{12}(\mathrm{j}\omega)| > 1$ for some $\omega > 0$). Panel (a) shows the case where the information shared by the CHV is not used by the CAV, i.e., $\beta_{02} = 0$ (that implies $T_{02}(\mathrm{j}\omega) \equiv 0$). Although the CAV responds in a string stable manner to the HV ($|T_{01}(\mathrm{j}\omega)| < 1$, $\forall \omega > 0$), this stable response is not able to compensate the string instability of the HV and head-to-tail string stability is not achieved ($|G_{02}(\mathrm{j}\omega)| > 1$ for some $\omega > 0$). In panel (b), on the other hand, the CAV responds to the CHV with $\beta_{02} \neq 0$ (that implies $T_{02}(\mathrm{j}\omega) \not\equiv 0$) and this ensures head-to-tail string stability ($|G_{02}(\mathrm{j}\omega)| < 1$, $\forall \omega > 0$).

This example illustrates that information obtained from multiple vehicles ahead via connectivity can improve control performance. In what follows, we show how

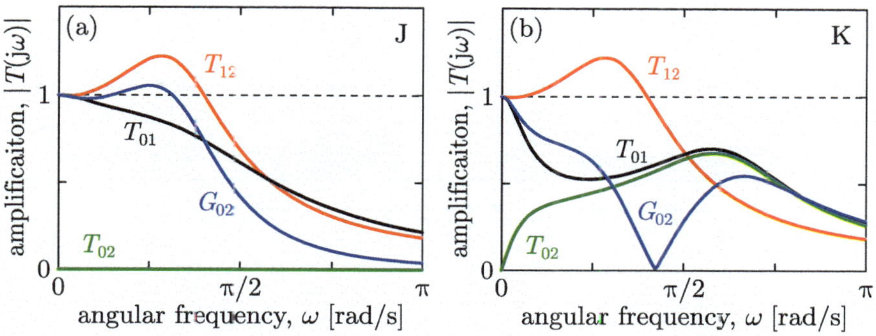

Fig. 4.10 Frequency response plots corresponding to the link transfer functions (4.55) and the head-to-tail transfer function (4.53) with human driver parameters $\alpha_{12} = 0.1\,[1/\text{s}]$, $\beta_{12} = 0.6\,[1/\text{s}]$, $\kappa_{12} = 0.6\,[1/\text{s}]$ and $\sigma_{12} = 1\,[\text{s}]$, and CAV control parameters $\alpha_{01} = 0.4\,[1/\text{s}]$, $\alpha_{02} = 0$, $\kappa_{01} = \kappa_{02} = 0.6\,[1/\text{s}]$ and $\sigma_{01} = \sigma_{02} = 0.6\,[\text{s}]$. Panels **a**, **b** are associated with point J(0.5,0) and point K(0.5,0.5) in Fig. 4.9b, respectively

the control design and these benefits scale up for general connectivity structures with arbitrary number and topology of connections between vehicles.

4.3.2 Vehicle Networks with General Connectivity Structure

In this section we study vehicle networks with arbitrary size and general connectivity structure. We consider a scenario where a connected automated vehicle (CAV) follows a string of vehicles consisting of non-connected human-driven vehicles (HVs) and connected human-driven vehicles (CHVs). We do not require these preceding vehicles to be automated although they can potentially possess various levels of automation. The main assumption we make is that each vehicle responds to vehicles ahead only (and not to vehicles behind) that implies no cooperation between vehicles. Nevertheless, this assumption, with some care, can be relaxed by following the approach and analyses presented in this section.

Figure 4.11 presents two motivating examples of such connected vehicle networks, both containing four vehicles. In Fig. 4.11a a CAV is preceded by two HVs and a CHV, i.e., the penetration of connectivity is 50% while the penetration of automation is 25%. Although HV 1 is not connected, the CAV can detect it and respond to it by the help of on-board sensors. The information transmitted by CHV 3 can improve the CAV's response by making the CAV more clairvoyant about upcoming events. Yet, the CAV may be unaware of the state (or even the existence) of HV 2. In contrast, Fig. 4.11b shows a case where a CAV is preceded by three CHVs, i.e., the penetration of connectivity is 100% and the penetration of automation is 25%. We remark that the arrows indicate the direction of information flow only for the information that is utilized for motion control. As a matter of fact, V2X connections are bidirectional

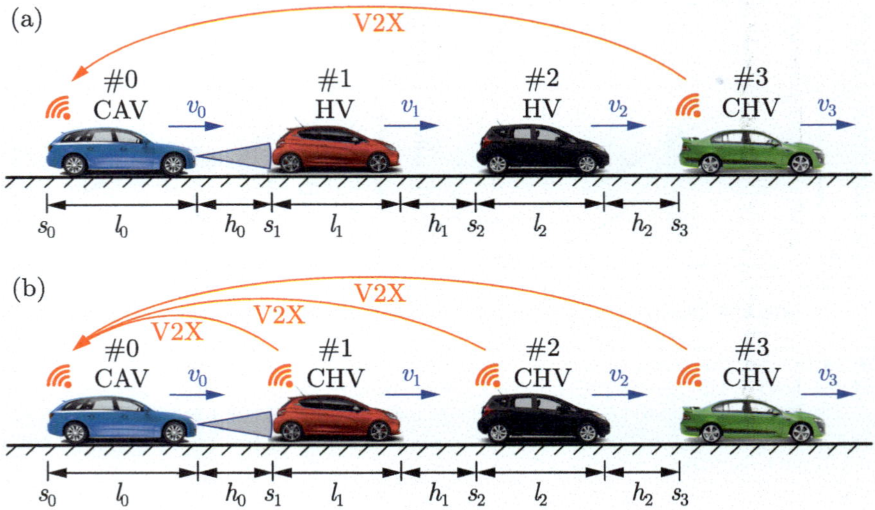

Fig. 4.11 Illustration of connected vehicle networks where a connected automated vehicle (CAV) responds to connected human-driven vehicles (CHVs) ahead via connected cruise control (CCC). Panel **a** shows a scenario with lean penetration of connectivity where some of the vehicles are non-connected human-driven vehicles (HVs). Panel **b** shows a case with full penetration of connectivity. Note that the penetration of automation is the same in both cases

and all vehicles are able to exchange information with each other, even the CHVs, although it is not indicated in the figure.

Our goal is to analyze the dynamics of connected vehicle networks without restricting the number of vehicles or connections. As a first step, one needs to systematically describe the connection topology or, more precisely, the topology of information flow describing which vehicle responds to which other vehicle. Consider a connected vehicle network consisting of $n+1$ vehicles (i.e., a CAV and n predecessors, $n \geq 1$), as depicted in Fig. 4.12. These vehicles can be considered as the nodes of a graph, whose edges represent that a vehicle is responding to another. In graph theory, topology is often characterized by a so-called *adjacency matrix*. The adjacency matrix proved to be a useful tool in many fields of science where network topology is concerned, with applications ranging from electric grids to social networks. Now we use this tool to describe connected vehicle networks.

We define the adjacency matrix $\boldsymbol{\eta} \in \mathbb{R}^{(n+1)\times(n+1)}$ such that its element in the i-th row and j-th column is

$$\eta_{ij} = \begin{cases} 1 & \text{if vehicle } i \text{ responds to vehicle } j\,, \\ 0 & \text{otherwise}\,, \end{cases} \tag{4.60}$$

$i, j = 0, 1, \ldots, n$. Note that although V2X communication is bidirectional, the response of vehicles is usually not, since they respond to what happens ahead rather

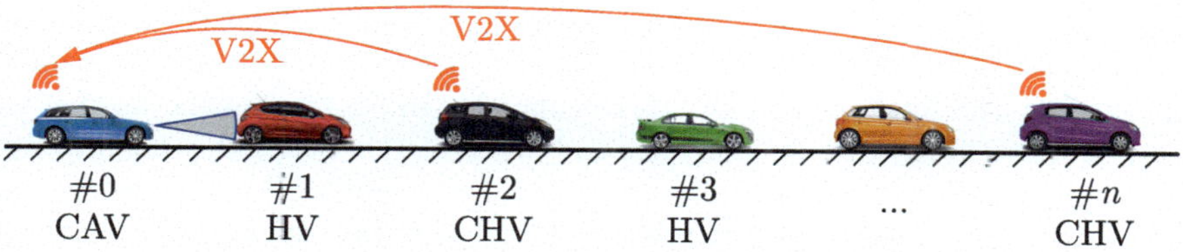

Fig. 4.12 Illustration of a general connected vehicle network consisting of $n+1$ vehicles where a connected automated vehicle (CAV) uses connected cruise control (CCC) to follow a string of n vehicles including human-driven vehicles (HVs) and connected human-driven vehicles (CHVs)

than behind. Therefore, $\boldsymbol{\eta}$ is not symmetric. In fact, if each vehicle responds to vehicles traveling ahead only, then $\eta_{ij} = 0$ for $j \le i$ that makes $\boldsymbol{\eta}$ an upper-triangular matrix. For example, the adjacency matrix corresponding to the connected vehicle network in Fig. 4.11a is

$$\boldsymbol{\eta} = \begin{bmatrix} 0 & 1 & 0 & 1 \\ 0 & 0 & 1 & 0 \\ 0 & 0 & 0 & 1 \\ 0 & 0 & 0 & 0 \end{bmatrix}, \tag{4.61}$$

while the one in Fig. 4.11b differs so that the second zero in the first row becomes a one. In what follows, we restrict ourselves to upper-triangular adjacency matrices.

Having established the adjacency matrix $\boldsymbol{\eta}$ that describes the *direct* links between vehicles, one can also determine the *indirect* ways vehicle j affects vehicle i. For instance, in Fig. 4.11a the CAV is affected by CHV 3 in a direct and an indirect way: the CAV responds to CHV 3 directly and it also responds HV 1 who responds to HV 2 that finally leads to CHV 3. In this case, we say that there are two paths leading from CHV 3 to the CAV. As a matter of fact, graph theory provides an explicit formula to calculate the number N_{ij} of paths from vehicle j to vehicle i if the adjacency matrix $\boldsymbol{\eta}$ is upper-triangular [184], as given by the following theorem.

Theorem 4.1 *In a directed graph whose adjacency matrix $\boldsymbol{\eta} \in \mathbb{R}^{(n+1)\times(n+1)}$ is upper-triangular, the number of paths from node j to node i is given by*

$$N_{ij} = (\boldsymbol{\eta})_{ij} + (\boldsymbol{\eta}^2)_{ij} + \ldots + (\boldsymbol{\eta}^n)_{ij} = \sum_{q=1}^{n} (\boldsymbol{\eta}^q)_{ij}, \tag{4.62}$$

$i, j = 0, \ldots, n$, $j > i$, where the q-th power indicates the number of paths of length q by taking the i-th row and j-th column of $\boldsymbol{\eta}^q$, $q = 1, \ldots, n$.

For example, the number of paths between CHV 3 and CAV 0 in Fig. 4.11a is $N_{03} = 1 + 0 + 1 = 2$.

The adjacency matrix and the number of paths characterizes well the topology of connections, however, the dynamic response associated with each connection is also important to describe the overall dynamics of connected vehicle networks.

Therefore, we derive the equations of motion for each vehicle and the corresponding transfer functions in order to quantify the dynamic characteristics of each link within the network. The final outcome of this calculation will be a systematic method to construct head-to-tail transfer functions and analyze plant and string stability for connected vehicle networks.

The dynamics of human-driven vehicles (HVs and CHVs) are given by

$$\begin{aligned} \dot{s}_i(t) &= v_i(t) \,, \\ \dot{v}_i(t) &= \mathrm{sat}\big(u_{i,i+1}(t-\sigma_{i,i+1})\big) \,, \end{aligned} \tag{4.63}$$

where the desired acceleration $u_{i,i+1}$ saturates according to the acceleration capabilities of the vehicle and $\sigma_{i,i+1}$ is the time delay that models driver reaction time and powertrain dynamics. The desired acceleration can be captured by human driver models of form

$$u_{i,i+1} = f_{i,i+1}(s_i, s_{i+1}, v_i, v_{i+1}) \,, \tag{4.64}$$

such as the optimal velocity model (OVM)

$$u_{i,i+1} = \alpha_{i,i+1}\big(V_i(s_{i+1}-s_i-l_i)-v_i\big) + \beta_{i,i+1}(v_{i+1}-v_i) \,, \tag{4.65}$$

see the detailed description in Chap. 2. Note that the model contains the expression of the headway $h_i = s_{i+1} - s_i - l_i$; cf. (2.1) and (4.1). We also remark that the human drivers may have different range policies V_i and driver parameters $\alpha_{i,i+1}$, $\beta_{i,i+1}$ and $\sigma_{i,i+1}$.

Similarly, we describe the dynamics of CAVs by the double integrator model with time delay and saturation:

$$\begin{aligned} \dot{s}_i(t) &= v_i(t) \,, \\ \dot{v}_i(t) &= \mathrm{sat}\left(\sum_{j=i+1}^{i+p} u_{ij}(t-\sigma_{ij})\right) . \end{aligned} \tag{4.66}$$

We assume that a CAV may respond to at most p preceding vehicles, where each response is associated with an individual desired acceleration and delay, denoted by u_{ij} and σ_{ij}. In case CAV i is not responding to vehicle j, we take $u_{ij} = 0$, otherwise the individual control inputs can be prescribed by control laws of the form

$$u_{ij} = F_{ij}(s_i, s_j, v_i, v_j, \dot{v}_j) \,, \tag{4.67}$$

such as the controllers

$$u_{ij} = \alpha_{ij}\,\big(V_i(h_{ij}) - v_i\big) + \beta_{ij}(W(v_j) - v_i) \,, \tag{4.68}$$

cf. (4.41), or

$$u_{ij} = \alpha_{ij} \big(V_i(h_{ij}) - v_i\big) + \beta_{ij}(W(v_j) - v_i) + \gamma_{ij}\dot{v}_j \;, \tag{4.69}$$

cf. (4.6), where h_{ij} indicates the average headway between vehicles j and i:

$$h_{ij} = \frac{1}{j-i}\sum_{k=i}^{j-1} h_k = \frac{1}{j-i}\left(s_j - s_i - \sum_{k=i}^{j-1} l_k\right), \tag{4.70}$$

cf. the special cases (4.42) and (4.43). Indeed, the speed policy W and the saturation function sat could also be vehicle dependent and have an index i, but we omit this detail for the sake of a simpler discussion.

Throughout this section we will use (4.68) as control law for CAVs to demonstrate results but, similar to previous sections, all arguments generalize for (4.67). We remark that the number of vehicles between two connected ones may be unknown in practice, i.e., if vehicle i obtains information from another vehicle it may not know its index j (unless $j = i + 1$). Therefore, the average headway h_{ij} may not be known in practice, and we will often drop this term by setting $\alpha_{ij} = 0$ for $j > i + 1$. This also eliminates the dependence of the CAV's equilibrium headway on the behavior of preceding human drivers; see the discussion after (4.45). Yet, the α_{ij} and h_{ij} terms may be useful under stronger assumptions, e.g., assuming known vehicle indices when considering a platoon of CAVs such as in cooperative adaptive cruise control (CACC).

Now we analyze the dynamics and stability properties of connected vehicle networks. Consider the dynamics (4.63), (4.65) and (4.66), (4.68) for a network of $n + 1$ vehicles ($i = 0, \ldots, n$) shown in Fig. 4.12. There exists the equilibrium (4.44) with constant speed v^* and constant headways $h_i^* = \bar{s}_{i+1} - \bar{s}_i - l_i$. For human drivers, the equilibrium headway can be obtained from $v^* = V_i(h_i^*)$, cf. (2.13). Whereas for CAVs, if the saturation limits in sat and W are not exceeded, the equilibrium headway is given by

$$\sum_{j=i+1}^{j=i+p} \alpha_{ij} \big(V_i(h_{ij}^*) - v^*\big) = 0 \;, \tag{4.71}$$

that can be solved for h_{ij}^* to finally express h_i^*. If CAVs do not respond to the position of vehicles farther ahead, i.e., if $\alpha_{ij} = 0$ for $j > i + 1$, then $v^* = V_i(h_i^*)$ holds. If all vehicles have the same range policy $V_i = V$, then they have the same equilibrium headway h^* given by $v^* = V(h^*)$, regardless of α_{ij}.

The linearized dynamics around this equilibrium can be given by the state and output (4.47) constructed from the perturbations (4.46) in the form

$$\begin{aligned}
\dot{\mathbf{x}}_i(t) &= \mathbf{a}_i\mathbf{x}_i(t) + \sum_{j=i+1}^{i+p} \big(\mathbf{a}_{ij}\mathbf{x}_i(t - \sigma_{ij}) + \mathbf{b}_{ij}\mathbf{x}_j(t - \sigma_{ij})\big) + \text{h.o.t.}, \\
y_i(t) &= \mathbf{c}_i\mathbf{x}_i(t) \;.
\end{aligned} \tag{4.72}$$

$i = 0, \ldots, n$, where the coefficient matrices are listed in (4.49). Note that the linear dynamics of HVs, CHVs, and CAVs are given by the same equations after unifying the notation: we write α_{ij}, β_{ij}, κ_{ij} and σ_{ij} for HVs and CHVs as well while assuming $\alpha_{ij} = 0$, $\beta_{ij} = 0$ for $j > i + 1$. The linear dynamics transform to Laplace domain in the form

$$Y_i(s) = \sum_{j=i+1}^{i+p} T_{ij}(s) Y_j(s) , \tag{4.73}$$

where the link transfer functions are given by

$$T_{ij}(s) = \mathbf{c}_i \left(s\mathbf{I} - \mathbf{a}_i - \sum_{j=i+1}^{i+p} \mathbf{a}_{ij} \mathrm{e}^{-s\sigma_{ij}} \right)^{-1} \mathbf{b}_{ij} \mathrm{e}^{-s\sigma_{ij}} \begin{bmatrix} \frac{1}{s} \\ 1 \end{bmatrix}, \tag{4.74}$$

after utilizing the relation (4.50) between state and output. By substituting the coefficient matrices (4.49), the specific expressions of the link transfer functions become

$$T_{ij}(s) = \frac{\left(\beta_{ij} s + \frac{\alpha_{ij}\kappa_{ij}}{j-i} \right) \mathrm{e}^{-s\sigma_{ij}}}{s^2 + \sum_{j=i+1}^{i+p} \left((\alpha_{ij} + \beta_{ij}) s + \frac{\alpha_{ij}\kappa_{ij}}{j-i} \right) \mathrm{e}^{-s\sigma_{ij}}} , \tag{4.75}$$

with $\kappa_{ij} = V_i'(h_{ij}^*)$ where h_{ij} is defined by (4.70) for $i, j = 0, \ldots, n,\ j > i$.

The link transfer functions T_{ij} characterize the dynamic response along the links listed in the adjacency matrix $\boldsymbol{\eta}$. We combine the information encoded into T_{ij} and $\boldsymbol{\eta}$ by constructing the so-called *weighted adjacency matrix* [251]. The weighted adjacency matrix $\mathbf{T} \in \mathbb{R}^{(n+1)\times(n+1)}$ has the same structure as the adjacency matrix $\boldsymbol{\eta}$: the element in the i-th row and j-th column is 0 if $\eta_{ij} = 0$ and it is T_{ij} otherwise. For example, the weighted adjacency matrix associated with Fig. 4.11a and (4.61) reads

$$\mathbf{T} = \begin{bmatrix} 0 & T_{01} & 0 & T_{03} \\ 0 & 0 & T_{12} & 0 \\ 0 & 0 & 0 & T_{23} \\ 0 & 0 & 0 & 0 \end{bmatrix} , \tag{4.76}$$

while the one corresponding to Fig. 4.11b contains T_{02} in the first row (instead of the second zero).

The weighted adjacency matrix contains information about the overall response G_{ij} of vehicle i to vehicle j defined by

$$Y_i(s) = G_{ij}(s) Y_j(s) , \tag{4.77}$$

that involves both direct and indirect connections. According to [251], it can be calculated based on the following theorem.

Theorem 4.2 *In a connected vehicle network whose weighted adjacency matrix* $\mathbf{T} \in \mathbb{R}^{(n+1)\times(n+1)}$ *is upper-triangular, the* j*-to-*i *transfer function is given by*

$$G_{ij} = (\mathbf{T})_{ij} + (\mathbf{T}^2)_{ij} + \ldots + (\mathbf{T}^n)_{ij} = \sum_{q=1}^{n} (\mathbf{T}^q)_{ij} , \tag{4.78}$$

$i, j = 0, \ldots, n,\ j > i.$

Note that (4.78) is analogous to (4.62).

This construction leads to the *head-to-tail transfer function* as the special case $i = 0$ and $j = n$:

$$Y_0(s) = G_{0n}(s) Y_n(s) . \tag{4.79}$$

To evaluate the head-to-tail transfer function with less computations, [251] provides an elegant alternative method summarized in the following theorem.

Theorem 4.3 *In a connected vehicle network whose weighted adjacency matrix* $\mathbf{T} \in \mathbb{R}^{(n+1)\times(n+1)}$ *is upper-triangular, the head-to-tail transfer function can be calculated as*

$$G_{0n} = \det\left(\hat{\mathbf{T}} - \hat{\mathbf{I}}\right) , \tag{4.80}$$

where $\mathbf{I} \in \mathbb{R}^{(n+1)\times(n+1)}$ *is the identity matrix and hat denotes the top right* $n \times n$ *block of the matrices.*

For example, $\mathbf{T}$ in (4.76) leads to the head-to-tail transfer function

$$G_{03} = \det\left(\hat{\mathbf{T}} - \hat{\mathbf{I}}\right) = \det \begin{bmatrix} T_{01} & 0 & T_{03} \\ -1 & T_{12} & 0 \\ 0 & -1 & T_{23} \end{bmatrix} = T_{01}T_{12}T_{23} + T_{03} . \tag{4.81}$$

Note that (4.78) leads to the same result with more computations.

The transfer functions allow one to evaluate the stability of the connected vehicle network. To achieve plant stability, the denominator of the head-to-tail transfer function G_{0n} must have its roots in the left half of the complex plane according to the following theorem that is analogous to Theorem 2.6 (see also Theorem 2.4).

Theorem 4.4 *A connected vehicle network consisting of* $n+1$ *vehicles traveling in an open chain configuration is plant stable at the linear level if and only if the characteristic roots given by the solutions of the characteristic equation*

$$\mathrm{Den}\big(G_{0n}(s)\big) = 0 , \tag{4.82}$$

have negative real parts, that is, $\mathrm{Re}(s_k) < 0$, $k = 1, \ldots, \infty$.

These eigenvalues are the same as the union of those corresponding to the individual link transfer functions; cf. Theorem 2.4. From the point of view of the CAV 0, the denominators of T_{0j} must be analyzed and these are the same for all j, cf. (4.74). In the special case of $\alpha_{0j} = 0$, $j > 1$ and identical delays $\sigma_{0j} = \sigma$, $j \geq 1$, this leads to the plant stability boundaries given by (4.13), (4.14) with $\alpha = \alpha_{01}$ and $\beta = \sum_{j=1}^{p} \beta_{0j}$, while the plant stability chart becomes qualitatively the same as in Fig. 4.9.

Meanwhile, the transfer functions G_{ij} are related to the *j-to-i string stability* condition by

$$\big|G_{ij}(\mathrm{j}\omega)\big| < 1 \,, \quad \forall \omega > 0 \,, \tag{4.83}$$

that leads to the notion of *head-to-tail string stability* for $i = 0$ and $j = n$.

Theorem 4.5 *A connected vehicle network consisting of $n + 1$ vehicles traveling in an open chain configuration is head-to-tail string stable if and only if the head-to-tail transfer function G_{0n} satisfies*

$$|G_{0n}(\mathrm{j}\omega)| < 1 \,, \quad \forall \omega > 0 \,. \tag{4.84}$$

Notice the analogy between Theorems 2.3 and 4.5. Evaluating string stability analytically, however, becomes challenging for large connected vehicle networks. In some cases, analytical results may still be achievable, for example by taking $G_{0n}(\mathrm{j}\omega) = \exp(-\mathrm{j}K)$ according to the approach in Sect. 2.3.3 that may lead to a linear system of equations for the string stability boundaries in terms of α_{ij} and β_{ij}. Nonetheless, if analytical results are too complex or not available, one can evaluate the inequality in (4.84) numerically for various values of ω and other parameters.

Stability charts corresponding to the scenario in Fig. 4.11b are plotted in Figs. 4.13 and 4.14, where the plant stable region is light gray bounded by red and the head-to-tail string stable domain is dark gray encircled by blue. In both figures, panel (a) shows the plant stability chart in the $(\beta_{01} + \beta_{02} + \beta_{03}, \alpha_{01})$ plane, that matches the charts in Figs. 4.4a and 4.9a. Figure 4.13b–d depict the stability charts in the (β_{01}, β_{03}) plane for various values of β_{02}, while Fig. 4.14b–d show them in the (β_{02}, β_{03}) plane for various values of β_{01}. The scenario in Fig. 4.13b demonstrates that head-to-tail string stability cannot be achieved without response to multiple vehicles ahead (when $\beta_{02} = \beta_{03} = 0$). Meanwhile, although nonzero β_{02} and β_{03} gains are beneficial, one should not increase their values too much but should address stability carefully.

The performance of connected cruise control is illustrated in Fig. 4.15 by numerical simulations of the four-vehicle network in Fig. 4.11b. Four cases are considered in the four rows of Fig. 4.15 with the same CHV behavior but different control parameters for the CAV, which correspond to points J, K, L and M in Figs. 4.13b, c and 4.14c. In all four cases, the CAV keeps the same equilibrium headway as the CHVs, and it behaves in an $i + 1$-to-i string stable manner as opposed to CHVs. In case J shown in panels (a) and (b), the CAV does not respond to CHV 2 and CHV 3, it only responds to its immediate predecessor CHV 1. That is, the potential benefits of connectivity from response to multiple vehicles ahead are not exploited by the CAV. This leads to a head-to-tail string unstable case where the CAV brakes about as much as the lead vehicle. In case K shown in panels (c) and (d), the CAV responds to both CHV 1 and CHV 2 but not to CHV 3. Response to multiple vehicles allows the CAV to mitigate the head-to-tail string instability and drive smoother than the lead vehicle.

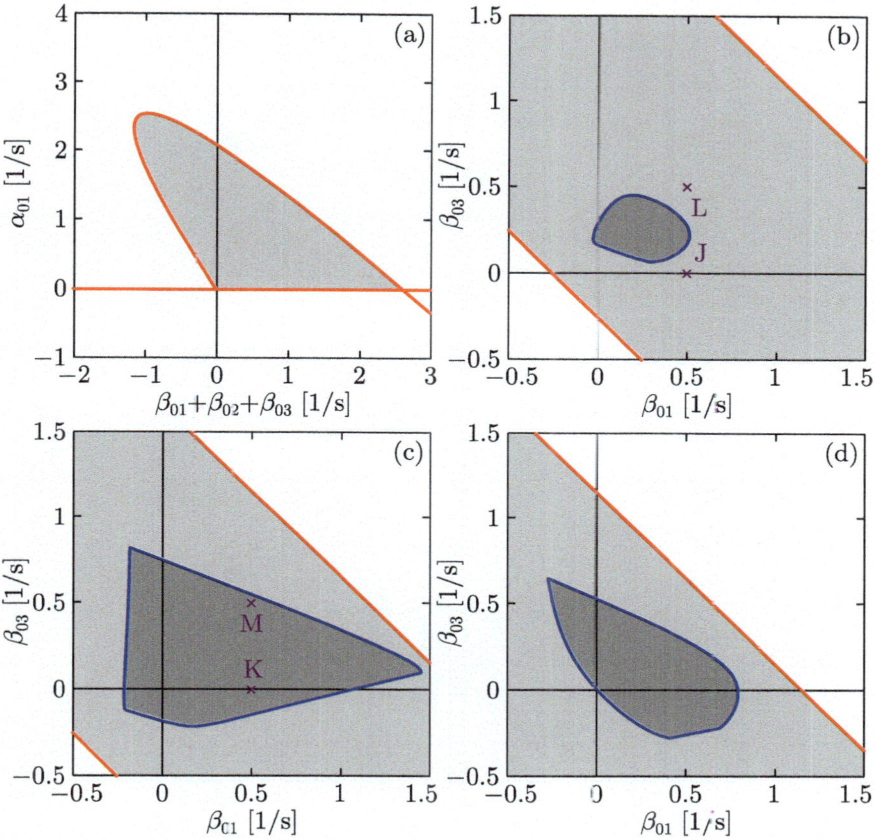

Fig. 4.13 Stability charts of the connected vehicle network illustrated in Fig. 4.11b given by system (4.72), (4.49) for human driver parameters $\alpha_{12} = \alpha_{23} = 0.1\,[1/\text{s}]$, $\beta_{12} = \beta_{23} = 0.6\,[1/\text{s}]$, $\kappa_{12} = \kappa_{23} = 0.6\,[1/\text{s}]$ and $\tau_1 = \tau_2 = 1\,[\text{s}]$, and CAV control parameters $\alpha_{02} = \alpha_{03} = 0$, $\kappa_{01} = 0.6\,[1/\text{s}]$ and $\sigma_{01} = \sigma_{02} = \sigma_{03} = 0.6\,[\text{s}]$. Panel **a** shows the plant stability chart in the $(\beta_{01} + \beta_{02} + \beta_{03}, \alpha_{01})$ plane, whereas panels **b–d** show the plant and string stability charts in the (β_{01}, β_{03}) plane for $\alpha_{01} = 0.4\,[1/\text{s}]$ and **b** $\beta_{02} = 0\,[1/\text{s}]$, **c** $\beta_{02} = 0.5\,[1/\text{s}]$, **d** $\beta_{02} = 1\,[1/\text{s}]$. The plant and head-to-tail string stable regions are shaded by light and dark gray, respectively

In case L shown in panels (e) and (f), the CAV responds to CHV 1 and CHV 3, which makes the CAV more clairvoyant compared to previous case: it starts to increase its headway early in panel (e) that allows a milder speed reduction in panel (f). Finally, the CAV responds to all three preceding vehicles in case M shown in panels (g) and (h), which leads to the smoothest driving behavior.

While this example already illustrates the significant benefits that connectivity brings to control design, we show further benefits considering various performance measures in the next sections.

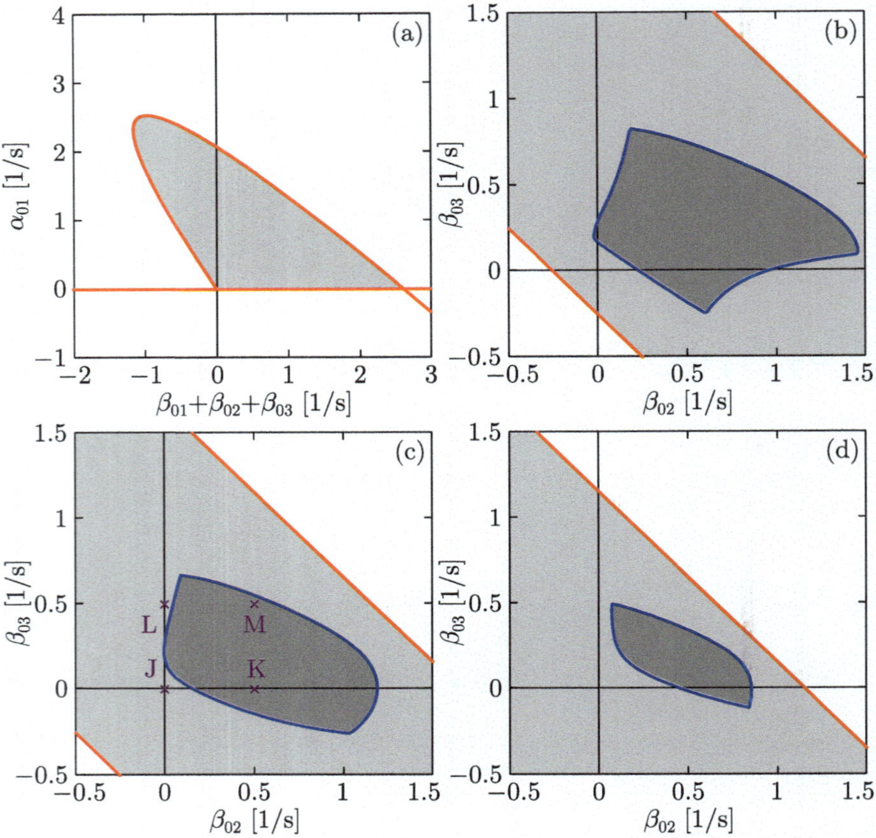

Fig. 4.14 Stability charts of the connected vehicle network illustrated in Fig. 4.11b given by system (4.72), (4.49) for human driver parameters $\alpha_{12} = \alpha_{23} = 0.1\,[1/\mathrm{s}]$, $\beta_{12} = \beta_{23} = 0.6\,[1/\mathrm{s}]$, $\kappa_{12} = \kappa_{23} = 0.6\,[1/\mathrm{s}]$ and $\tau_1 = \tau_2 = 1\,[\mathrm{s}]$, and CAV control parameters $\alpha_{02} = \alpha_{03} = 0$, $\kappa_{01} = 0.6\,[1/\mathrm{s}]$ and $\sigma_{01} = \sigma_{02} = \sigma_{03} = 0.6\,[\mathrm{s}]$. Panel **a** shows the plant stability chart in the $(\beta_{01} + \beta_{02} + \beta_{03}, \alpha_{01})$ plane, whereas panels **b–d** show the plant and string stability charts in the (β_{02}, β_{03}) plane for $\alpha_{01} = 0.4\,[1/\mathrm{s}]$ and **b** $\beta_{01} = 0\,[1/\mathrm{s}]$, **c** $\beta_{01} = 0.5\,[1/\mathrm{s}]$, **d** $\beta_{01} = 1\,[1/\mathrm{s}]$. The plant and head-to-tail string stable regions are shaded by light and dark gray, respectively. Points J, K, L and M correspond to the numerical simulations in Fig. 4.15

4.4 Mitigating Congestion with Connected Cruise Control in Mixed Traffic Systems

In Sect. 3.4 we investigated mixed traffic where automated vehicles (AVs), driven by ACC, were mixed into the flow of human-driven vehicles (HVs) in an open chain. We showed that for large enough AV penetration rates head-to-tail string stability can be achieved since the string stable AVs can compensate for the string

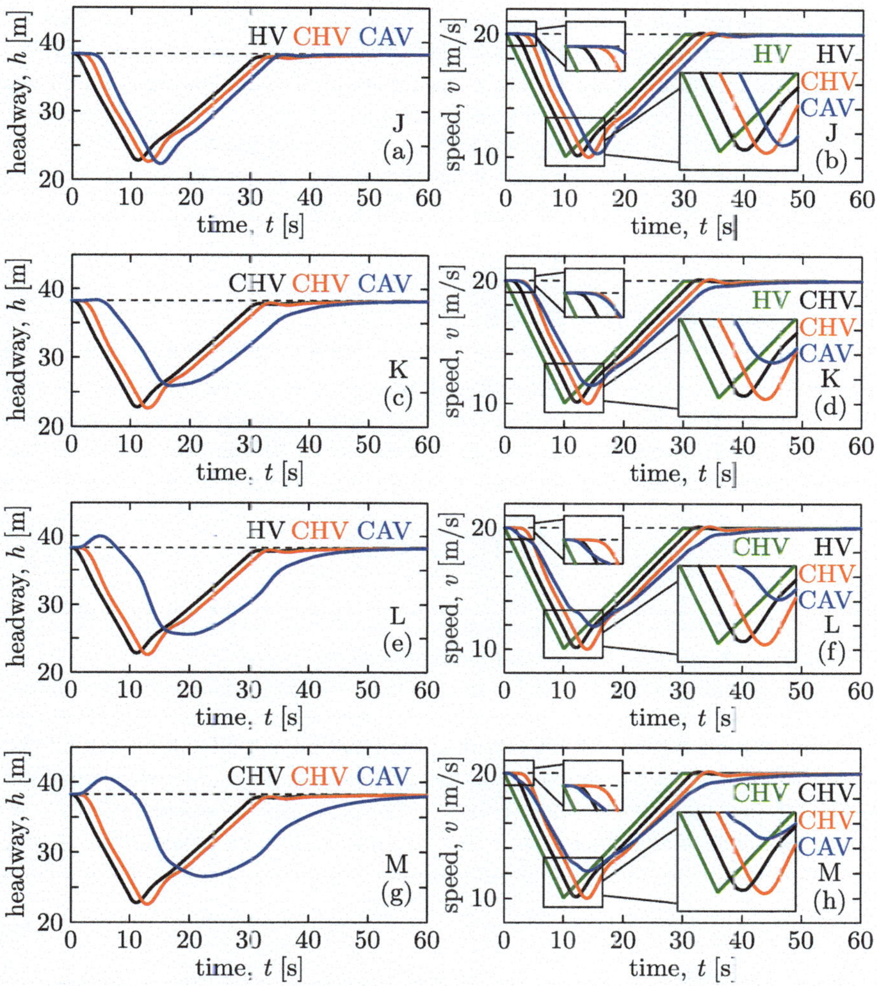

Fig. 4.15 Numerical simulation of the connected vehicle network in Fig. 4.11b where the CHVs and CAV are governed by (4.63), (4.65) and (4.66), (4.68). The lead vehicle uses constant -1 and $0.5\,[\mathrm{m/s^2}]$ acceleration to slow down from 20 to 10 [m/s] and to speed back up. The range policy is piecewise linear (2.5), (2.6) and (3.32) for each vehicle with $h_{\rm st} = 5$ [m], $h_{\rm go} = 55$ [m] and $v_{\max} = 30$ [m/s]. The initial conditions over $t \le 0$ are $v_i(t) = v^* = 20$ [m/s], $h_i(t) = V_i^{-1}(v^*) = 38.33$ [m], $i = 0, 1, 2$. The human driver parameters are $\alpha_{12} = \alpha_{23} = 0.1$ [1/s], $\beta_{12} = \beta_{23} = 0.6$ [1/s] and $\tau_1 = \tau_2 = 1$ [s]. The CAV's control parameters are $\alpha_{01} = 0.4$ [1/s], $\alpha_{02} = \alpha_{03} = 0$, $\beta_{01} = 0.5$ [1/s] and $\sigma_{01} = \sigma_{02} = \sigma_{03} = 0.6$ [s]. The rest of the parameters are $l = 5$ [m], $a_{\min} = 7\,[\mathrm{m/s^2}]$ and $a_{\max} = 3\,[\mathrm{m/s^2}]$. The four rows of the figure correspond to points J(0,0), K(0.5,0), L(0,0.5) and M(0.5,0.5) in the (β_{02}, β_{03}) plane in Fig. 4.14c, that is, the CAV responds to: **a**, **b** neither CHV 2 nor CHV 3; **c**, **d** CHV 2 but not CHV 3; **e**, **f** CHV 3 but not CHV 2; **g**, **h** both CHV 2 and CHV 3

unstable behavior of human drivers. In this section we repeat the same analysis by mixing connected automated vehicles (CAVs), driven by CCC, into the flow of HVs. We demonstrate the positive effects of connectivity by showing that head-tail-string stability can be achieved for smaller CAV penetration compared to the AV penetration rates needed. Finally, we also conduct stability analysis for mixed traffic systems in ring configuration by using the tools described in Sect. 2.3.3. For more details on stabilizing traffic with CAVs, please refer to [78, 133].

4.4.1 Open Chain of Vehicles

Consider an open chain of mixed traffic, illustrated in Fig. 4.16, with $N = N_{\rm HV} + N_{\rm CAV}$ follower vehicles, where $N_{\rm CAV}$ and $N_{\rm HV}$ denote the numbers of CAVs and HVs, respectively. Similar to the AV penetration rate (3.102) in Sect. 3.4 we define the CAV penetration rate as

$$p = \frac{N_{\rm CAV}}{N} = \frac{N_{\rm CAV}}{N_{\rm HV} + N_{\rm CAV}} \,. \tag{4.85}$$

For example, for the scenarios shown in Fig. 4.16 where every n-th vehicle is CAV we have $N = N_{\rm CAV} n$ and the penetration rate becomes $p = 1/n$.

First we present numerical simulations that demonstrate the behavior of the mixed vehicle chain with various penetrations of CAVs. We use the same model for HVs and CAVs as in Figs. 4.8 and 4.15. HVs are described by the car-following model (4.63), (4.65), which includes saturation at constant acceleration limits, time delay, and the OVM with piecewise linear range policy (2.5), (2.6). The parameters of HVs are chosen such that these vehicles are string unstable. CAVs are governed by (4.66),

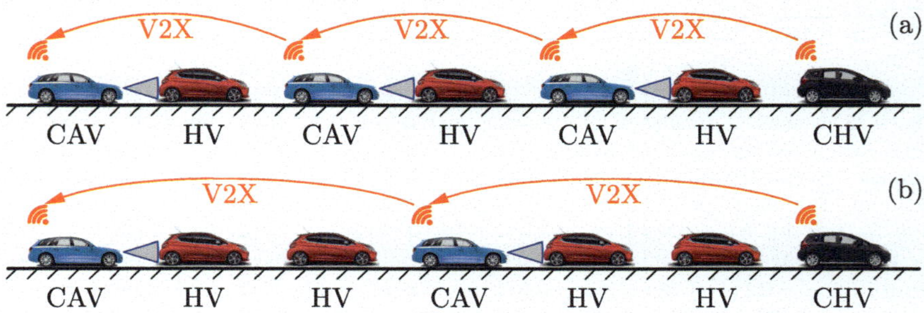

Fig. 4.16 Illustration of mixed traffic flow consisting of both connected automated vehicles (CAVs) that use connected cruise control (CCC), human-driven vehicles (HVs) and connected human-driven vehicles (CHVs). Panel **a** shows a scenario where every second vehicle behind the CHV is CAV, while panel **b** depicts a case where every third vehicle is CAV. Note that the penetrations of connectivity and automation are different in these two cases

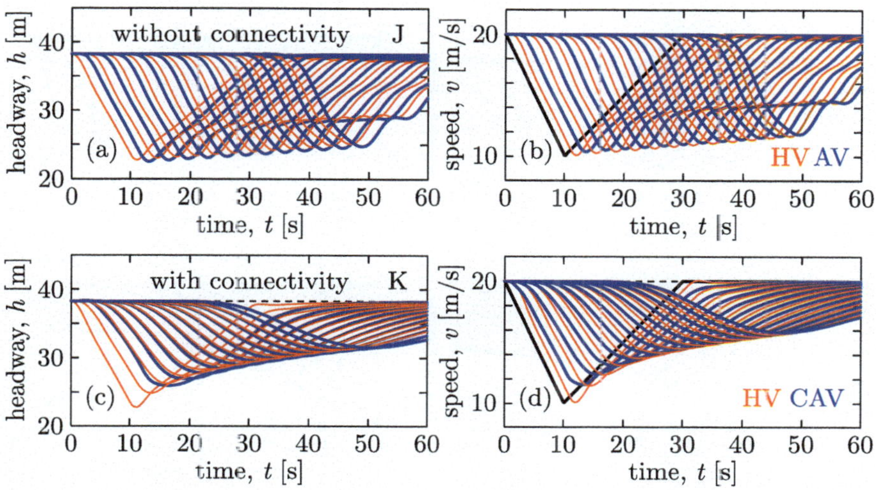

Fig. 4.17 Numerical simulations of mixed traffic illustrated in Fig. 4.16a for the case of 24 follower vehicles with CAV penetration rate $p = 50\%$ (i.e., every second vehicle is CAV). The same models and parameters are used as in Fig. 4.8. The two rows of the figure correspond to points J(0.5,0) and K(0.5,0.5) in the (β_{01}, β_{02}) plane in Fig. 4.9b. That is, CAVs do not respond to the information from connectivity and behave as AVs executing ACC in panels (**a**, **b**), while CAVs respond to the connected vehicle ahead of them using CCC in panels (**c**, **d**)

(4.68), that is, CCC with saturation and delay, the piecewise linear range policy (3.32) and the speed policy (3.33) shown in Fig. 3.7.

Figure 4.17 shows simulation results for 24 follower vehicles with $p = 50\%$ penetration of CAVs by considering every second vehicle to be connected and automated. This scenario is represented in Fig. 4.16a, and it can also be viewed as repeating the 3-vehicle scenario in Fig. 4.7 and concatenating it. The simulation parameters are chosen to match those of the 3-vehicle setup that was simulated previously in Fig. 4.8. Therefore, the simulated response of the first 3 vehicles of the chain in Fig. 4.17 is identical to the response of the 3-vehicle scenario in Fig. 4.8. The top and bottom rows of Fig. 4.17 correspond to the top and bottom rows of Fig. 4.8, respectively, and points J and K in Fig. 4.9b. That is, CAVs do not respond to the information from connectivity ($\beta_{02} = 0$) and behave as non-connected automated vehicles (AVs) executing ACC in Fig. 4.17a, b, while CAVs leverage connectivity ($\beta_{02} \neq 0$) in Fig. 4.17c, d.

In both cases of Fig. 4.17, head-to-tail string stability is observed for the vehicle chain in the sense that the tail vehicle reduces its speed less than the head vehicle. This is accomplished thanks to the high penetration of CAVs that drive smoothly and overcome the string unstable behavior of HVs. Recall that in Sect. 3.4 head-to-tail string stability was accomplished in a similar fashion with high penetration of ACC-driven AVs in mixed traffic; cf. Fig. 3.15. In comparison to ACC, connectivity and CCC have a clear advantage: the response of CAVs to the information from

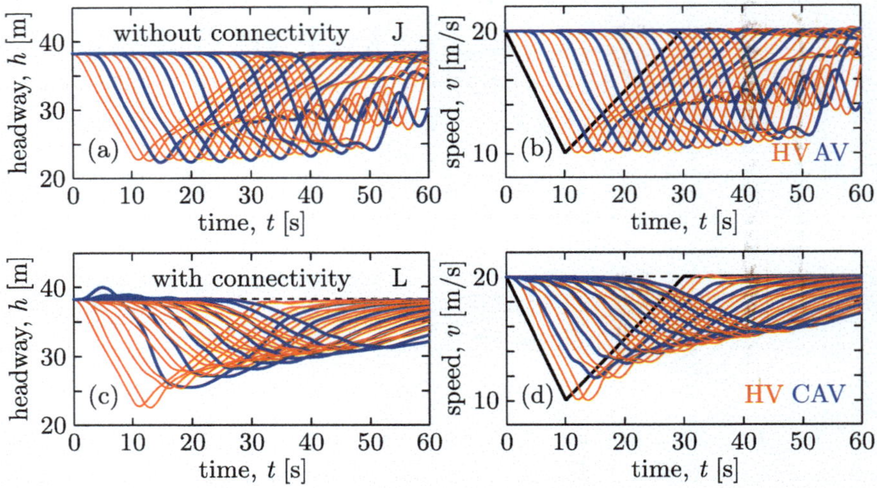

Fig. 4.18 Numerical simulations of mixed traffic illustrated in Fig. 4.16b for the case of 24 follower vehicles with CAV penetration rate $p = 33.33\%$ (i.e., every third vehicle is CAV). The same models and parameters are used as in the first and third rows of Fig. 4.15. The two rows of the figure correspond to points J(0.5,0) and L(0.5,0.5) in the (β_{01}, β_{03}) plane in Fig. 4.13b. That is, CAVs do not respond to the information from connectivity and behave as AVs executing ACC in panels (**a**, **b**), while CAVs respond to the connected vehicle ahead of them using CCC in panels (**c**, **d**)

connectivity significantly reduces speed perturbations along the vehicle chain in Fig. 4.17d and makes the traffic flow smoother compared to the non-connected case in Fig. 4.17b.

The simulations are repeated for $p = 33.33\%$ penetration of CAVs in Fig. 4.18, with every third vehicle being connected and automated. This car-following scenario is shown in Fig. 4.16b, and it is the concatenation of the 4-vehicle CCC setup in Fig. 4.11a. The same parameters are selected as previously for the 4-vehicle setup, hence the simulated responses of the first 4 vehicles in Fig. 4.18 match those in the first and third rows of Fig. 4.15. The two rows of Fig. 4.18 correspond to cases without connectivity ($\beta_{03} = 0$) and with connectivity ($\beta_{03} \neq 0$); cf. points J and L in Fig. 4.13b. Note that points J and L are both head-to-tail string unstable in the stability charts (since HVs are highly string unstable), hence the underlying control gains could result in the growth of certain speed perturbations. Nonetheless, the selected control gains work well for the particular speed perturbation in Fig. 4.17, and the tail vehicle reduces its speed less than head vehicle in both cases. While the speed perturbation decreases only slightly for the case without connectivity in Fig. 4.18b, it decays significantly with connectivity in Fig. 4.18d.

The effect of CAV penetration on the behavior of mixed traffic is demonstrated by simulations in Fig. 4.19. The first row shows the case of human-driven traffic, while the other rows correspond to $p = 16.67\%$, $p = 33.33\%$ and $p = 50\%$ CAV penetration levels considering every 6th, 3rd and 2nd vehicles to be CAVs, respec-

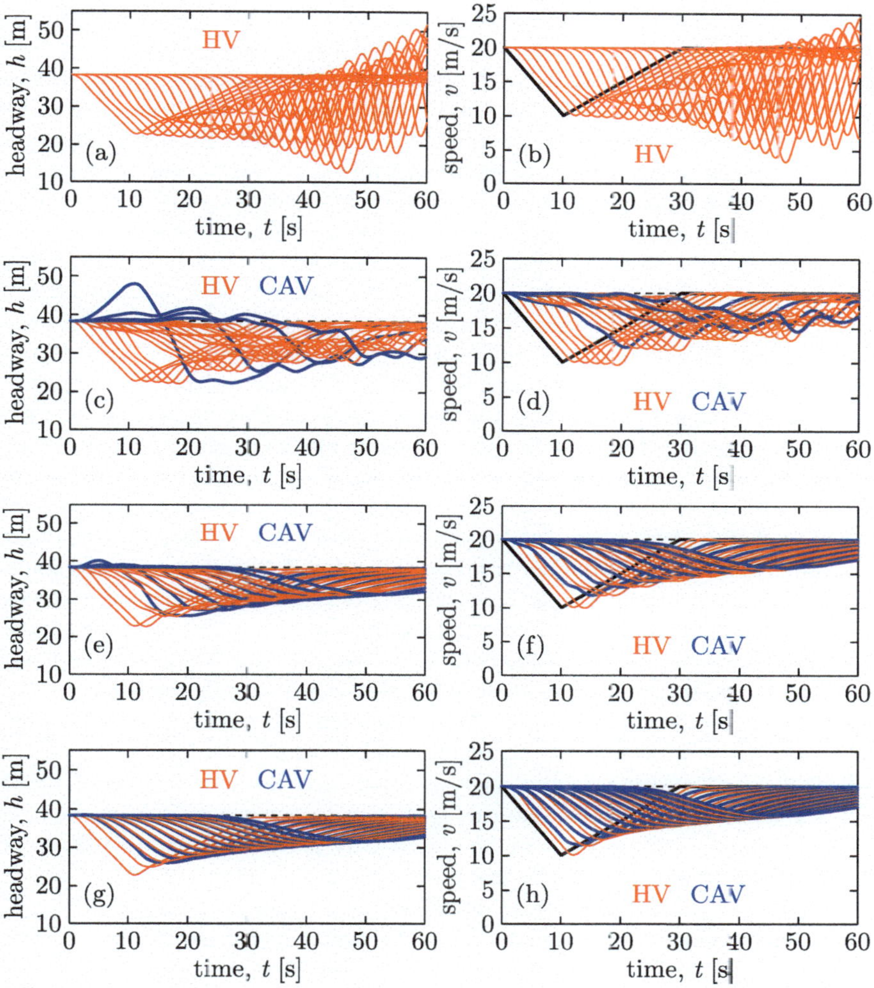

Fig. 4.19 Numerical simulations of mixed traffic for the case of 24 follower vehicles with CAV penetration rates **a**, **b** $p = 0\%$, **c**, **d** $p = 16.67\%$, **e**, **f** $p = 33.33\%$ and **g**, **h** $p = 50\%$. That is, the traffic is human driven in the top row while every n-th vehicle is CAV in the bottom three rows with $n = 6$, $n = 3$ and $n = 2$, respectively. The same models and parameters are used as in Figs. 4.17 and 4.18 while choosing $\beta_{01} = 0.5\,[1/\mathrm{s}]$ and $\beta_{0n} = 0.5\,[1/\mathrm{s}]$

tively. Each CAV leverages connectivity and responds to the closest CAV ahead (with the same $\beta_{0n} \neq 0$ gain for $n = 6$, $n = 3$ and $n = 2$, respectively). In the case of human-driven traffic in Fig. 4.19a, b, speed perturbations amplify as they propagate backward along the vehicle chain due the string instability of human drivers. For $p = 16.67\%$ penetration of CAVs in Fig. 4.19c, d, the perturbation level still grows along the human-driven sections of the chain while it decreases significantly at the CAVs. Overall, the speed perturbation of the tail vehicle becomes noticeably smaller than that of the head vehicle thanks to the smooth driving of CAVs. Similar trend is observed for $p = 33.33\%$ in Fig. 4.19e, f and for $p = 50\%$ in Fig. 4.19g, h: the perturbations decay along the vehicle chain which indicates head-to-tail string stable behavior. Note that these two cases are identical to Figs. 4.18c, d and 4.17c, d. Importantly, we can observe that head-to-tail string stability is not impacted significantly by the reduction of CAV penetration from 50 to 33.33% and to 16.67%.

These results highlight that CAVs with well-designed CCC algorithms hold the promise of achieving string stability and mitigating traffic congestion in mixed traffic flows for a wide range of penetration levels, including relatively low penetration. This was difficult to accomplish without connectivity using AVs executing ACC in Sect. 3.4; cf. Fig. 3.15 where head-to-tail string instability is observed for lower penetrations such as $p = 16.67\%$. We remark that Fig. 4.19 is the connected counterpart of Fig. 3.15, that is, AVs are replaced by CAVs (while the human-driven cases in Figs. 4.19a, b and 3.15a, b are identical). Comparing Figs. 4.19 and 3.15 clearly demonstrates the benefits of connectivity: CCC leads to a much smoother response with smaller speed perturbations than ACC. This is achieved by the behavior that CAVs start to decelerate earlier than AVs, which leads to milder speed reduction, and it may also result in a temporary increase in headway.

Next we conduct string stability analysis at the linear level for the case of mixed vehicle chains. We consider the setup in Fig. 4.16 and assume that: (i) the tail vehicle is CAV; (ii) each CAV responds to the closest CAV ahead of it; and (iii) the front CAV responds to the lead vehicle that is connected (i.e., either a CHV, as shown in the figure, or a CAV). We denote the indices of all vehicles by $i = 0, 1, \ldots, N$ and the indices of connected vehicles (i.e., CAVs and the lead vehicle) by i_k where $k = 0, 1, \ldots, N_{\mathrm{CAV}}$. That is, $i_0 = 0$ refers to the tail CAV and $i_{N_{\mathrm{CAV}}} = N$ marks the lead vehicle.

The head-to-tail string stability analysis follows that presented in Sect. 3.4 for mixed traffic with AVs. The expression of the head-to-tail transfer function, however, is modified according to Sect. 4.3 to take connectivity into account. We use Theorem 4.5 to evaluate head-to-tail string stability based on the condition

$$|G_{0N}(\mathrm{j}\omega)| < 1\,, \quad \forall \omega > 0\,, \tag{4.86}$$

cf. (3.103), where $G_{0N}(s)$ is the head-to-tail transfer function of the entire vehicle chain. This can be constructed as in [136]

$$G_{0N}(s) = \prod_{k=0}^{N_{\text{CAV}}-1} G_{i_k,i_{k+1}}(s)\,, \tag{4.87}$$

cf. (3.104), where $G_{i_k,i_{k+1}}(s)$ is the head-to-tail transfer function from CAV $k+1$ to CAV k given by

$$G_{i_k,i_{k+1}}(s) = \prod_{i=i_k}^{i_{k+1}-1} T_{i,i+1}(s) + T_{i_k,i_{k+1}}(s)\,, \tag{4.88}$$

see (4.53), (4.81). As before, the link transfer functions $T_{i,i+1}(s)$ describe the response of individual vehicles to the motion of their immediate predecessors, while the link transfer function $T_{i_k,i_{k+1}}(s)$ describes the response of CAV k to CAV $k+1$ by using connectivity.

Finally, to compare vehicle chains of different length, we take the N-th root and use scaled head-to-tail transfer function

$$G(s) = \sqrt[N]{G_{0N}(s)}\,, \tag{4.89}$$

cf. (3.109). Accordingly, the head-to-tail string stability condition becomes

$$|G(\mathrm{j}\omega)| < 1\,, \quad \forall\omega > 0\,, \tag{4.90}$$

cf. (3.110).

In order to simplify the expression of the head-to-tail transfer function $G(s)$ and relate it to the penetration rate p, we assume that every n-th vehicle is CAV and the total number of follower vehicles is $N = N_{\text{CAV}}n$. Then the penetration becomes $p = 1/n$. Furthermore, the indices of connected vehicles are $i_k = kn$, and we have

$$G_{0N}(s) = G_{0n}(s)\, G_{n,2n}(s) \cdots G_{N-n,N}(s)\,, \tag{4.91}$$

cf. (4.87), where

$$G_{kn,(k+1)n}(s) = T_{kn,kn+1}(s)\, T_{kn+1,kn+2}(s) \cdots T_{(k+1)n-1,(k+1)n}(s) + T_{kn,(k+1)n}(s)\,, \tag{4.92}$$

cf. (4.88). For example, for $k = 0$ we have

$$G_{0n}(s) = T_{01}(s)\, T_{12}(s) \cdots T_{n-1,n}(s) + T_{0n}(s)\,, \tag{4.93}$$

cf. (4.53) and (4.81).

In addition, assume that HVs are identical (described by link transfer function $T_{\text{h}}(s)$) while CAVs are also identical (described by $T_{01}(s)$ and $T_{0n}(s)$). Then the head-to-tail transfer functions further simplify to

$$G_{0N}(s) = G_{0n}^{N_{\text{CAV}}}(s)\,, \tag{4.94}$$

and

$$G_{0n}(s) = T_{01}(s) T_{\rm h}^{n-1}(s) + T_{0n}(s) \,, \tag{4.95}$$

while the scaled head-to-tail transfer function becomes

$$G(s) = \sqrt[n]{G_{0n}(s)} \,, \tag{4.96}$$

which directly depends on the penetration rate $p = 1/n$.

As for the link transfer functions, one may use

$$T_{\rm h}(s) = \frac{\beta_{\rm h} s + \alpha_{\rm h} \kappa_{\rm h}}{s^2 {\rm e}^{s\tau} + (\alpha_{\rm h} + \beta_{\rm h}) s + \alpha_{\rm h} \kappa_{\rm h}} \,, \tag{4.97}$$

for HVs with OVM parameterization, cf. (3.105), and

$$\begin{aligned} T_{01}(s) &= \frac{\beta_{01} s + \alpha \kappa}{s^2 {\rm e}^{s\sigma} + (\alpha + \beta_{01} + \beta_{0n}) s + \alpha \kappa} \,, \\ T_{0n}(s) &= \frac{\beta_{0n} s}{s^2 {\rm e}^{s\sigma} + (\alpha + \beta_{01} + \beta_{0n}) s + \alpha \kappa} \,, \end{aligned} \tag{4.98}$$

for CAVs, cf. (3.106) and (4.55), (4.56), (4.75), where we drop the subscripts of α, κ and σ for simplicity.

The corresponding frequency responses are plotted in Fig. 4.20 for different CAV penetration levels. In each panel of the figure, the red curve shows the response of HVs that are string unstable (i.e., $|T_{\rm h}({\rm j}\omega)| > 1$ for some angular frequency $\omega > 0$). The black curve depicts the link transfer function $T_{01}(s)$ of CAVs, which also represents the response of fully-automated non-connected traffic with AVs driven by ACC. The blue curves show the head-to-tail frequency response of mixed traffic. Each panel corresponds to different values of the CCC gain β_{0n} that is associated with the response to the information from connectivity.

Figure 4.20a represents the case without connectivity ($\beta_{0n} = 0$) where CAVs behave as AVs executing ACC. While the CAVs are string stable (i.e., $|T_{01}({\rm j}\omega)| < 1$ for all $\omega > 0$), head-to-tail string stability cannot be achieved (i.e., $|G({\rm j}\omega)| > 1$ for some $\omega > 0$) if their penetration is not high enough. The vehicle chain is head-to-tail string unstable for the penetration levels $p = 16.67\%$, $p = 33.33\%$ and $p = 50\%$ in this case. Recall that similar behavior was observed in Fig. 3.16 in Sect. 3.4. However, by leveraging connectivity (with $\beta_{0n} \neq 0$), head-to-tail string stability is achieved by CAVs at $p = 50\%$ penetration rate in Fig. 4.20b–d. (Note that the red, black, green and dark blue curves in Fig. 4.20a, d match those in Fig. 4.10a, b, except that $\sqrt[n]{G_{0n}(s)}$ is used for plotting instead of $G_{0n}(s)$.) For $p = 33.33\%$ penetration, the responses in Fig. 4.20b, d are head-to-tail string unstable while the string stability limit is reached in Fig. 4.20c. This also highlights that the choice of the gain β_{0n} affects head-to-tail string stability and CAV performance. The optimal selection of control gains with respect to certain performance criteria will be discussed in Sects. 4.6 and 4.7.

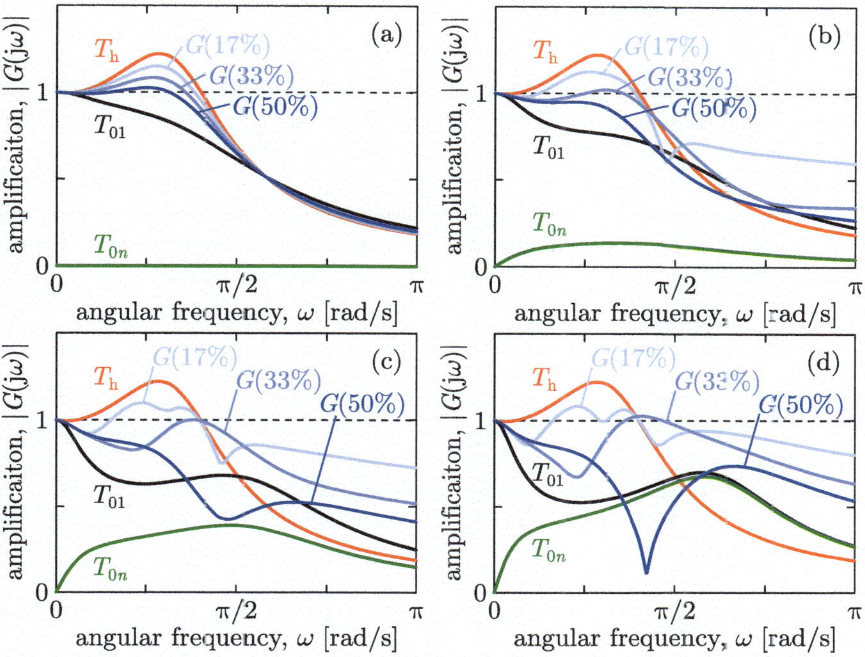

Fig. 4.20 Frequency responses of vehicle chains for mixed traffic scenarios based on the scaled head-to-tail transfer function (4.96) while using the link transfer functions (4.97) for HVs and (4.98) for CAVs. The frequency response curves are depicted for penetration rates $p = 0\%$ (red) and $p = 16.67, 33.33, 50\%$ associated with $n = 6, 3, 2$, respectively (blue). The HV parameters are $\alpha_h = 0.1\,[1/s]$, $\beta_h = 0.6\,[1/s]$, $\kappa_h = 0.6\,[1/s]$ and $\tau = 1\,[s]$. The CAV parameters are $\alpha = 0.4\,[1/s]$, $\beta_{01} = 0.5\,[1/s]$, $\kappa = 0.6\,[1/s]$, $\sigma = 0.6\,[s]$, and **a** $\beta_{0n} = 0\,[1/s]$ (corresponding to ACC), **b** $\beta_{0n} = 0.1\,[1/s]$, **c** $\beta_{0n} = 0.3\,[1/s]$ and **d** $\beta_{0n} = 0.5\,[1/s]$

Based on the frequency response plots in Fig. 4.20 head-to-tail string stability requires relatively high penetration of CAVs such as $p = 50\%$. Note, however, that the underlying string stability analysis considers perturbations with any angular frequency ω, which is a strict requirement. As opposed, the simulation results in Fig. 4.19 are restricted to a specific, representative perturbation signal. In such a case, the speed perturbation of the tail vehicle is significantly smaller than that of the head vehicle in simulation even for penetrations as low as $p = 16.67\%$. Nevertheless, for both analysis and simulation the benefits of connectivity over non-connected scenarios are clearly showcased. We also remark that in the analysis above a "the worst-case scenario" was considered for the human drivers, as they are not only string unstable but they are also identical. That is, they amplify signals at exactly the same angular frequency domains. The diversity of real human drivers makes it easier to compensate their string instability, and this can be achieved using much smaller CAV penetration rates. This will be illustrated using real-world experiments in Chap. 5.

4.4.2 Vehicles on a Ring

Next we analyze the stability of mixed traffic systems in ring configuration. We consider the dynamics at the linear level and derive stability charts based on the analysis outlined in Sects. 2.3.2 and 2.3.3.

Consider the scenario in Fig. 4.21, where $N = N_{\rm HV} + N_{\rm CAV}$ vehicles travel on a ring, with $N_{\rm CAV}$ number of CAVs and $N_{\rm HV}$ number of HVs. The linearized response of the N vehicles can be characterized by the head-to-tail transfer function $G_{0N}(s)$ introduced in (4.87). Since vehicle 0 is the same as vehicle N on the ring, we obtain the characteristic equation

$$G_{0N}(\lambda) = 1\,, \tag{4.99}$$

for the ring configuration [136]. This equation was formalized as (2.171) for a ring of N identical HVs, and it is also the linearized representation of the periodic boundary condition in (2.91).

As in the previous subsection, we assume that every n-th vehicle is a CAV (i.e., $N = N_{\rm CAV} n$), and both the HVs and CAVs are identical. Then, substituting (4.94) the characteristic equation becomes

$$G_{0n}^{N_{\rm CAV}}(\lambda) = 1\,, \tag{4.100}$$

where $G_{0n}(s)$ is given by (4.95). By taking the $N_{\rm CAV}$-th root and substituting (4.95), we obtain the characteristic equation of the ring configuration in the form

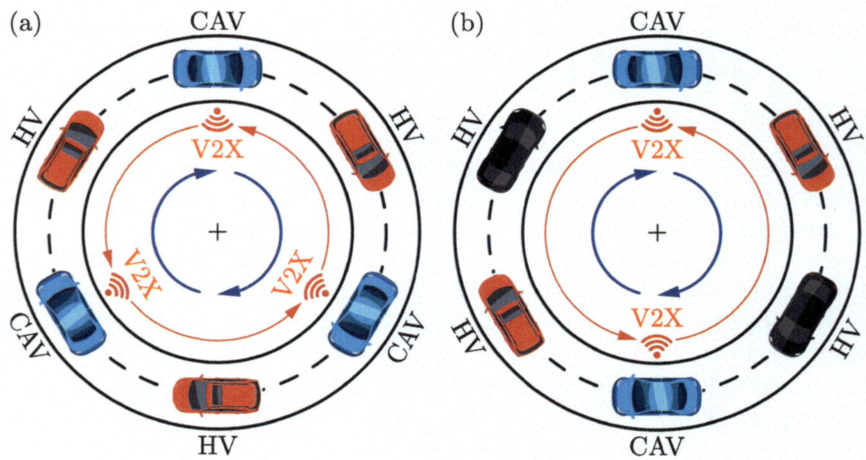

Fig. 4.21 Illustration of mixed traffic flow on a ring consisting of connected automated vehicles (CAVs) that use connected cruise control (CCC) and human-driven vehicles (HVs). Panel **a** shows a scenario where every second vehicle is a CAV, while panel **b** depicts a case where every third vehicle is a CAV

$$T_{01}(\lambda)\Gamma(\lambda) + T_{0n}(\lambda) = \mathrm{e}^{-\mathrm{j}\frac{2k\pi}{N_{\mathrm{CAV}}}}\,, \tag{4.101}$$

where

$$\Gamma(\lambda) = T_{\mathrm{h}}^{n-1}(\lambda)\,, \tag{4.102}$$

is the head-to-tail response of the $n-1$ HVs between two subsequent CAVs, while $k = 0, 1, \ldots, N_{\mathrm{CAV}} - 1$ is the wave number [136]. Notice the analogy with the characteristic equation (2.170) of human-driven traffic on a ring. By substituting the link transfer functions in (4.97) and (4.98) we obtain

$$\lambda^2 \mathrm{e}^{\lambda\sigma} + (\alpha + \beta_{01} + \beta_{0n})\lambda + \alpha\kappa - \big((\beta_{01}\lambda + \alpha\kappa)\Gamma(\lambda) + \beta_{0n}\lambda\big)\mathrm{e}^{\mathrm{j}\frac{2k\pi}{N_{\mathrm{CAV}}}} = 0\,. \tag{4.103}$$

Based on Theorem 2.7, the ring configuration is asymptotically stable if the roots of the characteristic equation (4.103) have negative real parts ($\mathrm{Re}(\lambda_l) < 0$, $l = 1, \ldots, \infty$). We derive the stability boundaries in the (β_{01}, β_{0n}) plane by considering $\lambda = 0$ and $\lambda = \mathrm{j}\omega$, $\omega > 0$. Substituting $\lambda = 0$ into (4.103) and recognizing that $\Gamma(0) = 1$ leads to

$$\alpha\kappa\left(1 - \mathrm{e}^{\mathrm{j}\frac{2k\pi}{N_{\mathrm{CAV}}}}\right) = 0\,. \tag{4.104}$$

Thus we obtain $\alpha = 0$ as stability boundary (which does not show up in the (β_{01}, β_{0n}) plane). Furthermore, notice that $\lambda = 0$ is a root of the characteristic equation also if $k = 0$ (regardless of the value of α). This root corresponds to the translational symmetry of the ring and can be disregarded.

To calculate the $\lambda = \mathrm{j}\omega$ stability boundaries, we treat the cases of $k = 0$ and $k > 0$ separately. For $k = 0$, the characteristic equation (4.103) simplifies to

$$\lambda^2 \mathrm{e}^{\lambda\sigma} + \alpha\lambda + (\beta_{01}\lambda + \alpha\kappa)(1 - \Gamma(\lambda)) = 0\,. \tag{4.105}$$

By substituting $\lambda = \mathrm{j}\omega$ and separating real and imaginary parts, we obtain

$$\begin{aligned} \text{Re:} &\; -\omega^2\cos(\omega\sigma) + \alpha\kappa(1 - \Gamma_{\mathrm{R}}) + \beta_{01}\omega\Gamma_{\mathrm{I}} = 0\,,\\ \text{Im:} &\; -\omega^2\sin(\omega\sigma) + \alpha\omega + \beta_{01}\omega(1 - \Gamma_{\mathrm{R}}) - \alpha\kappa\Gamma_{\mathrm{I}} = 0\,, \end{aligned} \tag{4.106}$$

where $\Gamma_{\mathrm{R}} = \mathrm{Re}\big(\Gamma(\mathrm{j}\omega)\big)$ and $\Gamma_{\mathrm{I}} = \mathrm{Im}\big(\Gamma(\mathrm{j}\omega)\big)$. This leads to the stability boundary

$$\beta_{01} = \frac{\hat{\omega}^2\cos(\hat{\omega}\sigma) - \alpha\kappa(1 - \hat{\Gamma}_{\mathrm{R}})}{\hat{\omega}\hat{\Gamma}_{\mathrm{I}}}\,, \tag{4.107}$$

where $\hat{\Gamma}_{\mathrm{R}} = \mathrm{Re}\big(\Gamma(\mathrm{j}\hat{\omega})\big)$, $\hat{\Gamma}_{\mathrm{I}} = \mathrm{Im}\big(\Gamma(\mathrm{j}\hat{\omega})\big)$, and $\hat{\omega}$ is given by the solution of

$$-\omega^2\big((1 - \Gamma_{\mathrm{R}})\cos(\omega\sigma) - \Gamma_{\mathrm{I}}\sin(\omega\sigma)\big) + \alpha\kappa\big((1 - \Gamma_{\mathrm{R}})^2 + \Gamma_{\mathrm{I}}^2\big) - \alpha\omega\Gamma_{\mathrm{I}} = 0\,. \tag{4.108}$$

This stability boundary appears as a vertical line in the (β_{01}, β_{0n}) plane.

For $k > 0$, substituting $\lambda = \mathrm{j}\omega$ into (4.103) provides

$$\begin{aligned} -\omega^2 \cos(\omega\sigma) + \alpha\kappa\,(1 - \Gamma_\mathrm{R}\cos K + \Gamma_\mathrm{I}\sin K) \\ +\beta_{01}\omega\,(\Gamma_\mathrm{R}\sin K + \Gamma_\mathrm{I}\cos K) + \beta_{0n}\omega\sin K = 0\,, \\ -\omega^2\sin(\omega\sigma) + (\alpha + \beta_{01} + \beta_{0n})\omega + \beta_{01}\omega\,(-\Gamma_\mathrm{R}\cos K + \Gamma_\mathrm{I}\sin K) \\ -\alpha\kappa\,(\Gamma_\mathrm{R}\sin K + \Gamma_\mathrm{I}\cos K) - \beta_{0n}\omega\cos K = 0\,, \end{aligned} \tag{4.109}$$

where $K = \frac{2k\pi}{N_\mathrm{CAV}}$. This can be rearranged to the stability boundaries

$$\beta_{01} = \frac{a}{c}\,, \qquad \beta_{0n} = \frac{b}{c}\,, \tag{4.110}$$
$$\begin{aligned} a &= \omega^2\big(\cos(\omega\sigma) - \cos(\omega\sigma - K)\big) + \alpha\omega\sin K - \alpha\kappa\big((\Gamma_\mathrm{R}+1)(1-\cos K) + \Gamma_\mathrm{I}\sin K\big)\,, \\ b &= -\omega^2\big(\cos(\omega\sigma) - \Gamma_\mathrm{R}\cos(\omega\sigma - K) - \Gamma_\mathrm{I}\sin(\omega\sigma - K)\big) - \alpha\omega\,(\Gamma_\mathrm{R}\sin K + \Gamma_\mathrm{I}\cos K) \\ &\quad + \alpha\kappa\big(1 - 2\Gamma_\mathrm{R}\cos K + 2\Gamma_\mathrm{I}\sin K + \Gamma_\mathrm{R}^2 + \Gamma_\mathrm{I}^2\big)\,, \\ c &= \omega\big((\Gamma_\mathrm{R} - 1)\sin K - \Gamma_\mathrm{I}\,(1 - \cos K)\big)\,. \end{aligned}$$

The stability charts of mixed traffic on a ring with HVs and CAVs executing CCC are depicted in Fig. 4.22. The stability boundaries (4.107) and (4.110) are colored based on the value of the wave number $k = 0, 1, \ldots, N_\mathrm{CAV} - 1$ (according to the color code used in Sect. 2.3.2), while the stable region is shaded gray. Figure 4.22a, b correspond to the scenario of Fig. 4.21a where every second vehicle is CAV (i.e., $n = 2$ and the penetration rate of CAVs is $p = 50\%$), with different number N of vehicles. Meanwhile, Fig. 4.22c, d correspond to the scenario of Fig. 4.21b where every third vehicle is CAV (i.e., $n = 3$ and $p = 33.33\%$). These scenarios can be viewed as repeating the corresponding open chain CCC setup along the ring: the 3-vehicle setup of Fig. 4.7 and the 4-vehicle setup of Fig. 4.11a, respectively.

Consequently, the stable regions of the mixed ring configuration in Fig. 4.22a, b closely approximate the head-to-tail string stable domain shown in Fig. 4.9b that was derived for the corresponding 3-vehicle setup in open chain. Similarly, the stable domains in Fig. 4.22c, d approximate that of the 4-vehicle case in Fig. 4.13b. This is also highlighted by the dashed blue curves in Fig. 4.22 that are copies of the head-to-tail string stability boundaries of the corresponding open chain in Figs. 4.9b and 4.13b. As the number N of vehicles (and also the number N_CAV of CAVs) is increased, the ring and open chain results match more closely.

The relationship between the open chain and ring configurations was established for human-driven traffic in Sect. 2.3.3. On the one hand, it was highlighted that the stable region of the ring approaches the string stable domain of the open chain as the number N of vehicles is increased to $N \to \infty$. On the other hand, it was also revealed that the stability boundaries of the ring are in fact samples of the string stability boundaries of the open chain. As a matter of fact, by considering a continuous wave number $K \in [0, 2\pi)$ (rather than $K = \frac{2k\pi}{N_\mathrm{CAV}}$) in (4.110), one obtains the head-to-tail

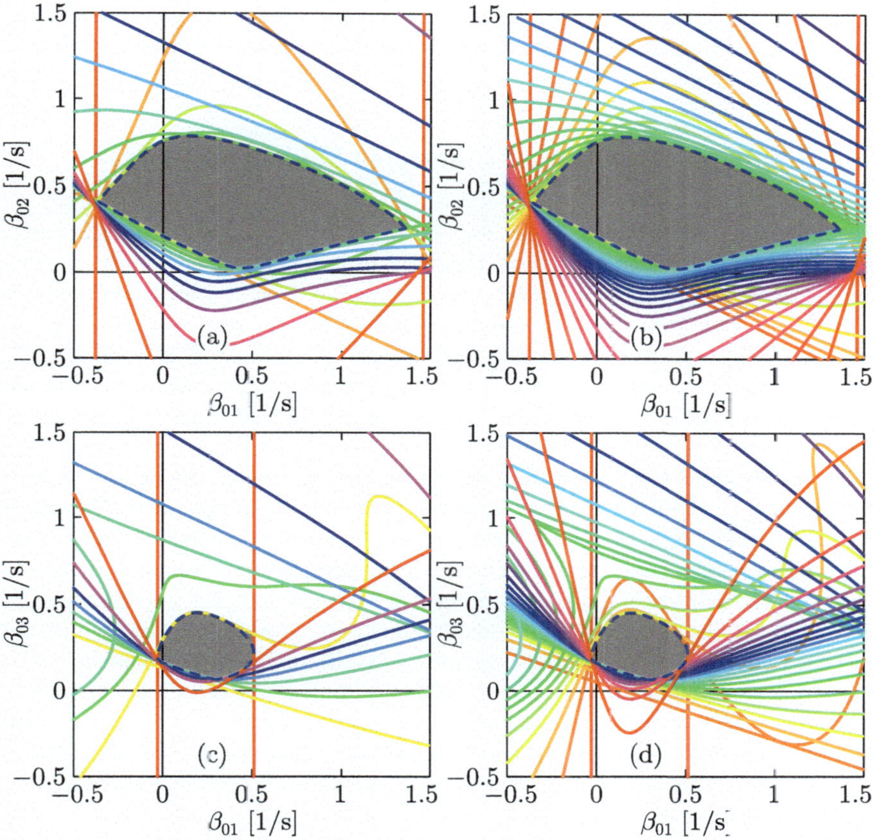

Fig. 4.22 Stability chart of mixed traffic on a ring based on the link transfer functions (4.97) for HVs and (4.98) for CAVs. The number of vehicles is **a**, **c** $N = 24$ and **b**, **d** $N = 60$, while every n-th vehicle is CAV with **a**, **b** $n = 2$ (corresponding to $p = 50\%$ penetration rate of CAVs) and **c**, **d** $n = 3$ (i.e., $p = 33.33\%$ penetration). The HV parameters are $\alpha_\mathrm{h} = 0.1\,[1/\mathrm{s}]$, $\beta_\mathrm{h} = 0.6\,[1/\mathrm{s}]$, $\kappa_\mathrm{h} = 0.6\,[1/\mathrm{s}]$ and $\tau = 1\,[\mathrm{s}]$. The CAV parameters are $\alpha = 0.4\,[1/\mathrm{s}]$, $\kappa = 0.6\,[1/\mathrm{s}]$ and $\sigma = 0.6\,[\mathrm{s}]$. Note that the parameters match those of Figs. 4.9b and 4.13b, and the head-to-tail string stability boundaries from those figures are shown here by dashed blue curves for reference. The wave numbers k of the stability boundaries are distinguished using different colors

string stability boundaries of the open chain CCC setup with $(n+1)$ vehicles (i.e., a connected head vehicle, $n-1$ HVs and a CAV).

Therefore, analyzing the head-to-tail transfer function $G_{0n}(s)$ of the $(n+1)$-vehicle open chain CCC scenario both provides guidance on designing the control gains and helps characterize the behavior of mixed traffic where the penetration of CAVs is around $p = 1/n$. The analysis shows that CAVs are able to achieve head-to-tail string stability and thereby mitigate congestion if their penetration in the mixed traffic flow is high enough. Note that the analysis above only holds at the linear level. The next section is dedicated to analyzing the nonlinear behavior of mixed traffic.

4.5 Nonlinear Dynamics of Connected Vehicle Systems

The majority of the previous discussions on string stability was based on linearizing the underlying dynamics. However, nonlinearities in connected vehicle systems may significantly impact the performance of CAVs. These nonlinearities include, for example, the saturation of accelerations in (4.63), (4.66), the range policy and speed policy in (4.65), (4.68) and the resistance terms discussed in Chap. 3; cf. Figs. 2.2, 3.2, and 3.7. It was demonstrated in Sect. 2.4 that nonlinearities can fundamentally shape the behavior of human-driven traffic and can potentially lead to so-called *phantom jams*.

In what follows, we investigate the nonlinear dynamics of CCC and the occurrence of phantom jams in mixed traffic systems that involve HVs, CHVs and CAVs, which was published in [110, 136]. This section is deeply rooted in the theory of nonlinear dynamical systems, hence readers who are less interested in this field may skip to the next section. Interested readers are suggested to read Sect. 2.4 as preliminary.

Section 2.4 discussed phantom jams in human-driven traffic. It was highlighted that, due to nonlinear dynamics, two qualitatively different behaviors may be possible for the same human drivers: smooth traffic flow or heavy congestion with largely fluctuating vehicle speeds. The congestion may arise unexpectedly—hence the name phantom jam—depending on perturbations like the braking of an individual driver. The underlying nonlinear dynamical system is called *bistable*, since it has two coexisting stable solutions: the equilibrium corresponding to smooth traffic and a periodic solution (specifically, a limit cycle) associated with oscillating motion.

Bistable behavior and phantom jams also occur in mixed traffic systems that involve CHVs, HVs and CAVs executing CCC, which we discuss next. Specifically, we investigate the nonlinear dynamics of traffic flows where HVs are described by (4.63), (4.65) with the piecewise cubic range policy (2.5), (2.9) whereas CAVs are governed by (4.66), (4.68) with the piecewise linear range policy (3.32) and the speed policy (3.33). That is, the sources of nonlinearities are: the saturation of accelerations, the range policies of HVs and CAVs and the speed policy of CAVs, as illustrated in Figs. 2.2, 3.2, and 3.7.

For simplicity, we consider scenarios where each HV is identical, and denote their parameters shortly as $\alpha_{i,i+1} = \alpha_{\mathrm{h}}$, $\beta_{i,i+1} = \beta_{\mathrm{h}}$, $\sigma_{i,i+1} = \tau$ and $V_i(h) = V_{\mathrm{h}}(h)$. Similarly, CAVs are identical with parameters $\alpha_{i,i+1} = \alpha$, $\alpha_{i,i+j} = 0$ for $j > 1$, $\beta_{i,i+j} = \beta_{0j}$, $\sigma_{ij} = \sigma$ and $V_i(h) = V(h)$. We consider the open chain and ring configurations in Figs. 4.16 and 4.21. That is, every n-th vehicle is a CAV, and each CAV responds to the vehicle immediately ahead of them and the CAV n vehicles ahead (i.e., all β gains are zero except β_{01} and β_{0n}).

Figure 4.23 demonstrates bistability and phantom jams in mixed traffic. The open chain of vehicles shown in Fig. 4.16 is simulated, where every second vehicle is CAV

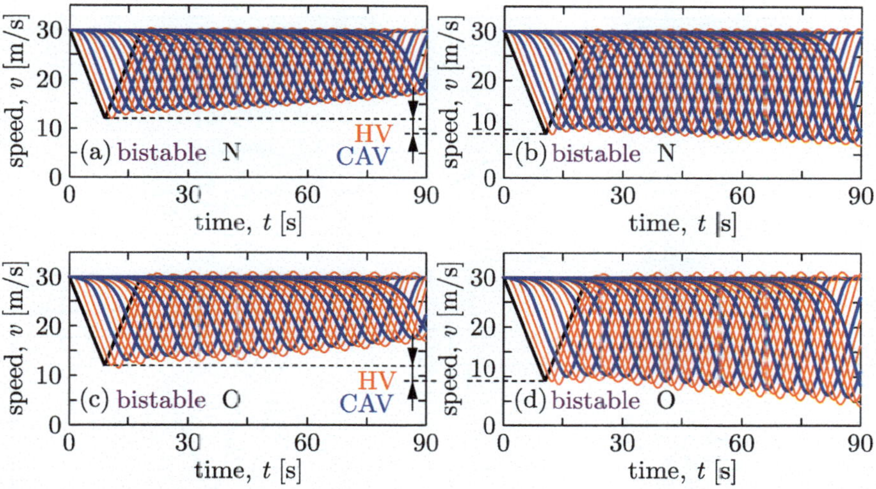

Fig. 4.23 Numerical simulation of the connected vehicle network in Fig. 4.15 where the HVs are governed by (4.63), (4.65) with the piecewise cubic range policy (2.5), (2.9), whereas the dynamics of CAVs are described by (4.66), (4.68) with the piecewise linear range policy (3.32), for the case of 48 follower vehicles with every n-th being a CAV. The speed of the lead vehicle is subject to a 18 [m/s] perturbation in panels **a** and **c**, and a 21 [m/s] perturbation in panels (**b**) and (**d**). The initial conditions over $t \leq 0$ are $v_i(t) = v_{\max}$ and $h_i(t) = h_{\mathrm{go}}$, $i = 0, 1, \ldots, 47$. The parameters are $a_{\min} = 7\,[\mathrm{m/s^2}]$, $a_{\max} = 3\,[\mathrm{m/s^2}]$, $h_{\mathrm{st}} = 5\,[\mathrm{m}]$, $h_{\mathrm{go}} = 55\,[\mathrm{m}]$ and $v_{\max} = 30\,[\mathrm{m/s}]$ for each vehicle; $\alpha_{\mathrm{h}} = 0.1\,[1/\mathrm{s}]$, $\beta_{\mathrm{h}} = 0.6\,[1/\mathrm{s}]$ and $\tau = 1\,[\mathrm{s}]$ for HVs; whereas $\alpha = 0.4\,[1/\mathrm{s}]$ and $\sigma = 0.6\,[\mathrm{s}]$ for CAVs. Panels **a, b** correspond to $n = 2$ shown in Fig. 4.16a and point N(0.25,0.1) in Fig. 4.25, while panels **c, d** are associated with $n = 3$ in Fig. 4.16b and point O(0.25,0.1) in Fig. 4.24

in panels (a) and (b) and every third vehicle is CAV in panels (c) and (d) of Fig. 4.23. The lead CHV reduces and then recovers its speed, while the vehicle chain responds to this perturbation. Panels (a) and (c) show that if the perturbation is small, CAVs are able to mitigate the onset of a congestion and speed perturbations get smaller and smaller along the vehicle chain. This implies convergence to smooth traffic flow. Panels (b) and (d), however, highlight that if the CHV's speed perturbation is large enough, perturbations tend to amplify along the vehicle chain and solutions become large-amplitude oscillating motions. This eventually leads to the formation of a phantom jam, which was triggered by the speed perturbation of the CHV.

To analyze this phenomenon in detail, we first consider the case of N vehicles driving in the ring configurations illustrated in Fig. 4.21. We use DDE-BIFTOOL [49, 194] to numerically investigate the dynamics that are coupled with the periodic boundary condition (2.91) and the headway constraint (2.92). We study the stability of the equilibrium (4.44) and the occurrence, amplitude and stability of limit cycles. That is, the analysis in Fig. 2.33 is repeated for mixed traffic.

Figure 4.24 shows the results for the case where every third vehicle is CAV; cf. Fig. 4.21b. By numerically computing equilibria and their stability for various (β_{01}, β_{03}) pairs, we first reproduced the stability chart of Fig. 4.22c in panel (a) of

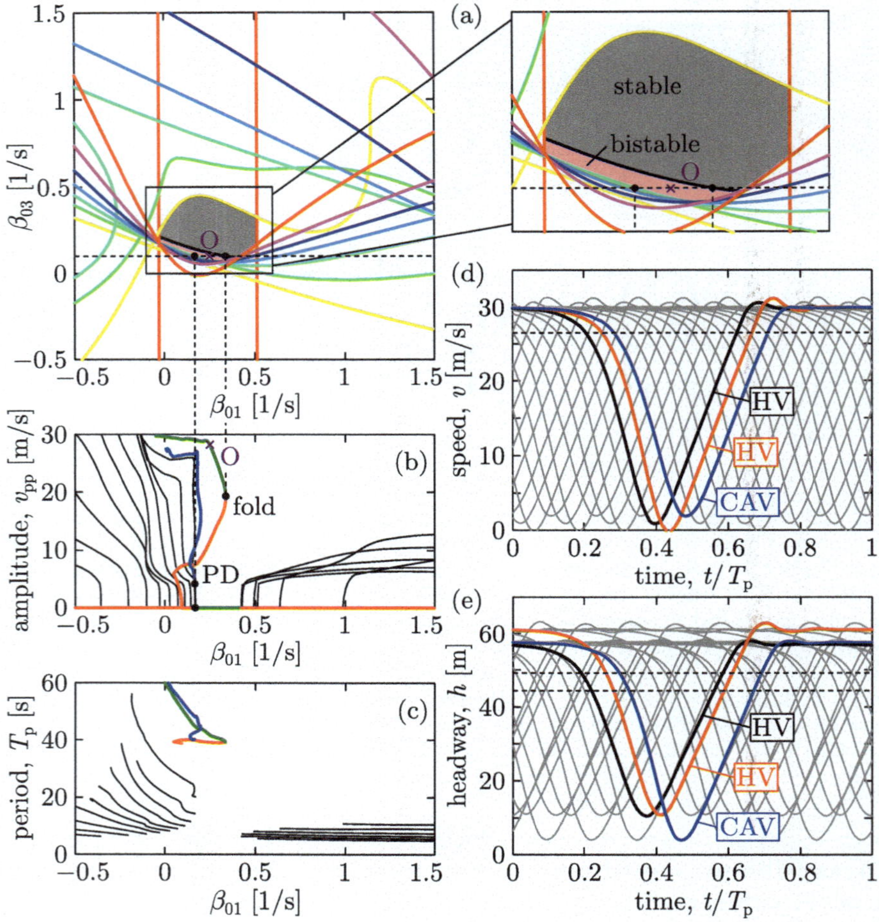

Fig. 4.24 Nonlinear dynamics of a connected vehicle network where $N = 24$ vehicles drive in ring configuration and every third vehicle is CAV; cf. Fig. 4.21b. HVs are governed by (4.63), (4.65) with the piecewise cubic range policy (2.5), (2.9), whereas the dynamics of CAVs are described by (4.66), (4.68) with the piecewise linear range policy (3.32). We consider the parameters $a_{\min} = 7\,[\mathrm{m/s^2}]$, $a_{\max} = 3\,[\mathrm{m/s^2}]$, $h_{\mathrm{st}} = 5\,[\mathrm{m}]$, $h_{\mathrm{go}} = 55\,[\mathrm{m}]$ and $v_{\max} = 30\,[\mathrm{m/s}]$ for each vehicle; $\alpha_{\mathrm{h}} = 0.1\,[1/\mathrm{s}]$, $\beta_{\mathrm{h}} = 0.6\,[1/\mathrm{s}]$ and $\tau = 1\,[\mathrm{s}]$ for HVs; whereas $\alpha = 0.4\,[1/\mathrm{s}]$ and $\sigma = 0.6\,[\mathrm{s}]$ for CAVs. The net ring length L is chosen such that $\kappa_{\mathrm{h}} = V_{\mathrm{h}}'(h_{\mathrm{h}}^*) = 0.6\,[1/\mathrm{s}]$, i.e., $h_{\mathrm{h}}^* = 44.43\,[\mathrm{m}]$ (which defines v^*, h^* for CAVs and L). Panels **b**, **c** are shown for $\beta_{03} = 0.1\,[1/\mathrm{s}]$, while panels **d**, **e** correspond to point O(0.25,0.1) marked in panels (**a**), (**b**)

Fig. 4.24. Along the colorful curves, we detected Hopf bifurcations [75]. These give rise to limit cycles which we computed numerically. By fixing β_{03} while varying β_{01} (see the dashed line in panel (a)), we continued a branch of limit cycles from each Hopf bifurcation point. The amplitude of these limit cycles (i.e., the peak-to-peak amplitude v_{pp} of the CAVs' speed fluctuations) and their stability are shown in panel (b), with the most important branch highlighted by thick line. Along this branch, the limit cycles are first unstable and their amplitude increases as β_{01} is varied. Then the branch folds, the limit cycles become stable, and their amplitude increases up to v_{max} as β_{01} is decreased. The time period of the limit cycles is plotted in panel (c). The limit cycle associated with point O is depicted in panels (d) and (e), with the motions of three vehicles highlighted by thick lines. These motions showcase large speed and headway fluctuations as part of a stop-and-go phantom jam. Note that the speed fluctuations of HVs are larger than those of their preceding vehicle, while CAVs tend to reduce speed fluctuations.

Overall, a similar qualitative picture is observed in Fig. 4.24b for mixed traffic as in Fig. 2.33b for human-driven traffic. The stable equilibrium coexists with a stable limit cycle for a certain range of β_{01} parameters, which is called bistability. In the bistable region, the traffic flow may either showcase smooth driving (i.e., convergence to the equilibrium) or a phantom jam (i.e., convergence to the stable limit cycle). Note that Fig. 4.24b reveals an additional phenomenon: the occurrence of period doubling (PD) bifurcation that gives rise to a stable limit cycle with doubled time period (see blue curve). The additional period-2 limit cycles were also considered when analyzing bistability. By repeating the calculations in panel (b) for various values of β_{03}, we identified the red shaded bistable domain in panel (a) of Fig. 4.24.

While Fig. 4.24 shows that bistability and phantom jams still exist in mixed traffic, a significant part of the linearly stable region is *not* bistable in panel (a). Thus, by using these charts, one can select controller parameters from the gray shaded globally stable region. This helps to avoid phantom jams associated with bistability, and ultimately leads to smooth traffic flow—even under nonlinear behavior. As such, CAVs can effectively eliminate congestions in mixed traffic.

Figure 4.25 shows the effect of increasing the penetration of CAVs in the traffic flow. As opposed to Fig. 4.24 where every third vehicle was CAV (corresponding to 33% penetration), Fig. 4.25 shows the scenario in which every second vehicle is CAV (i.e., 50% penetration). While the qualitative picture of the dynamics are similar, the increased number of CAVs clearly enlarges the size of the linearly stable region and reduces the relative size of the bistable region. This facilitates smooth, congestion-free traffic: phantom jams can be eliminated while keeping a large freedom to select the controller parameters of CAVs. This highlights the promise of CAVs in mitigating traffic congestions, provided that these vehicles appear in the traffic flow with large enough penetration.

Up until now we established the mathematical tools that allow one to design connected cruise control algorithms to regulate the longitudinal dynamics of connected automated vehicles while responding to the motion of connected human-driven vehicles in mixed traffic. The results were summarized using stability charts depicting the set of parameters that ensure plant stability and string stability. In the forthcom-

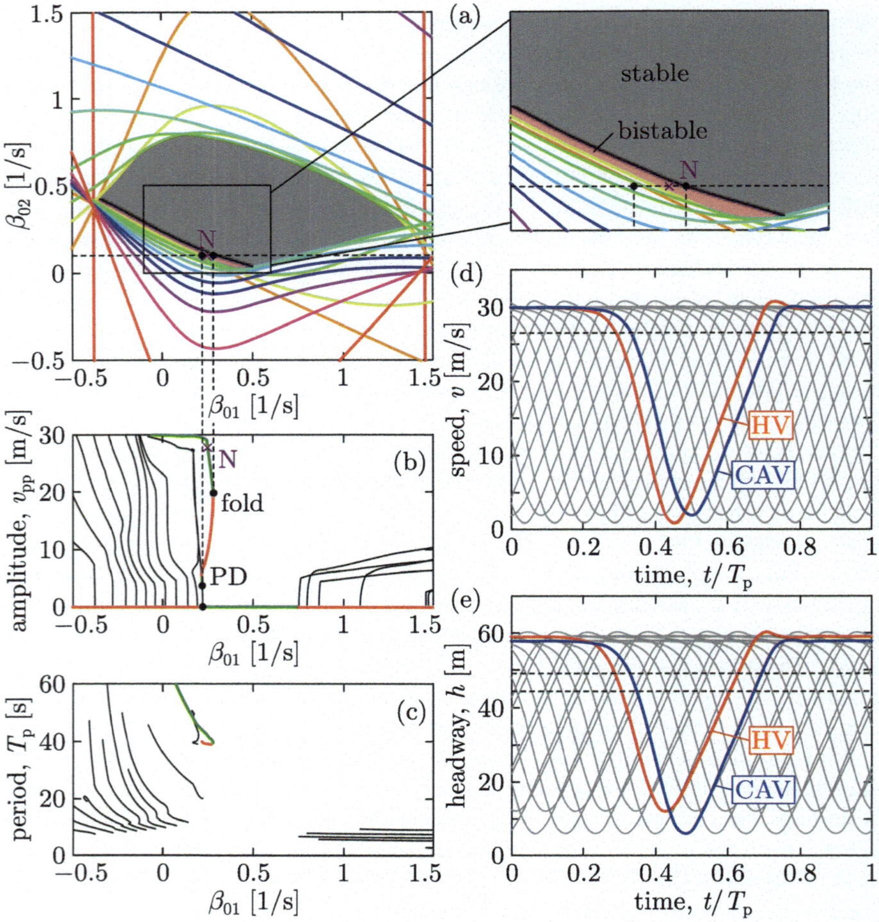

Fig. 4.25 Nonlinear dynamics of a connected vehicle network where $N = 24$ vehicles drive in ring configuration and every second vehicle is CAV; cf. Fig. 4.21a. The parameters are the same as in Fig. 4.24. Panels **b**, **c** are shown for $\beta_{02} = 0.1$ [1/s], while panels **d**, **e** correspond to point N(0.25,0.1) marked in panels (**a**), (**b**)

ing two sections we evaluate the performance of these controllers in two different aspects: energy efficiency and robustness against uncertainties in human behavior. These may allow one to select the particular parameters to be used from the plant and string stable set and achieve control designs that minimize energy consumption and are robust against human uncertainties.

4.6 Energy Efficiency in Connected Vehicle Systems

Here we establish a way to evaluate the energy consumption of longitudinal motion and discuss ways in which connected cruise controllers can be tuned to maximize energy efficiency based on [85, 191]. We evaluate wheel-to-distance energy consumption by calculating the energy per unit mass $w_i(t)$ consumed by vehicle i up to time t:

$$w_i(t) = \int_0^t v_i(\xi) g\big(\dot{v}_i(\xi) + p(v_i(\xi))\big) \mathrm{d}\xi \ , \tag{4.111}$$

for $v_i(\xi) \geq 0$. The integrand represents the power consumption per unit mass. This power measure relies on the product of the speed v_i and the acceleration command $\dot{v}_i + p(v_i)$, which is the sum of the acceleration $\dot{v}_i$ and the resistance $p(v_i)$ the vehicle must overcome; cf. (3.4) in Chap. 3.

The function g encodes how acceleration and braking commands relate to the energy consumption. For example, if one assumes that no energy is consumed or regenerated during brake commands, while energy is consumed during acceleration commands, then g becomes

$$g(x) = \max\{0, x\} \ . \tag{4.112}$$

Note that this function is often referred to as *rectified linear unit* or *ReLU function*, and it is illustrated in Fig. 4.26. The appropriate selection of g also allows modeling regenerative braking and various types of powertrains, i.e., vehicles with internal combustion engine, hybrid electric vehicles and battery electric vehicles. Below we use the simple choice in (4.112) to demonstrate our results.

To derive approximate analytical results for energy-optimal control parameters, we use the simplified measure

$$\varepsilon_i(t) = \int_0^t v_i(\xi) g\big(\dot{v}_i(\xi)\big) \mathrm{d}\xi \ , \tag{4.113}$$

as the approximation of (4.111) that neglects the resistance terms. Moreover, we consider a fixed time interval $t \in [0, T]$. We will often denote the final energy consumption $w_i(T)$ and $\varepsilon_i(T)$ shortly by w_i and ε_i.

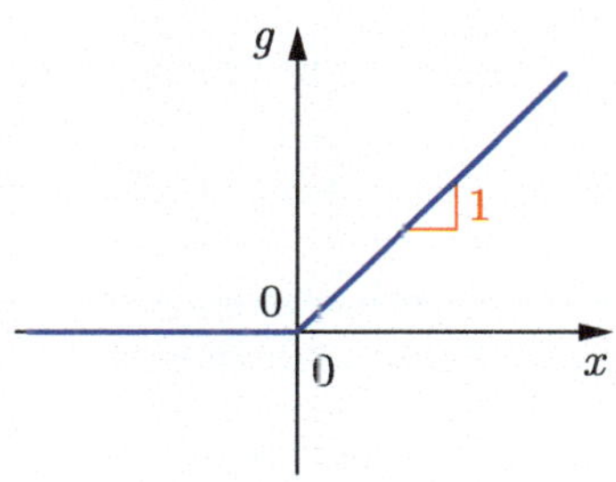

Fig. 4.26 The rectified linear unit (ReLU) function $g(x) = \max\{0, x\}$

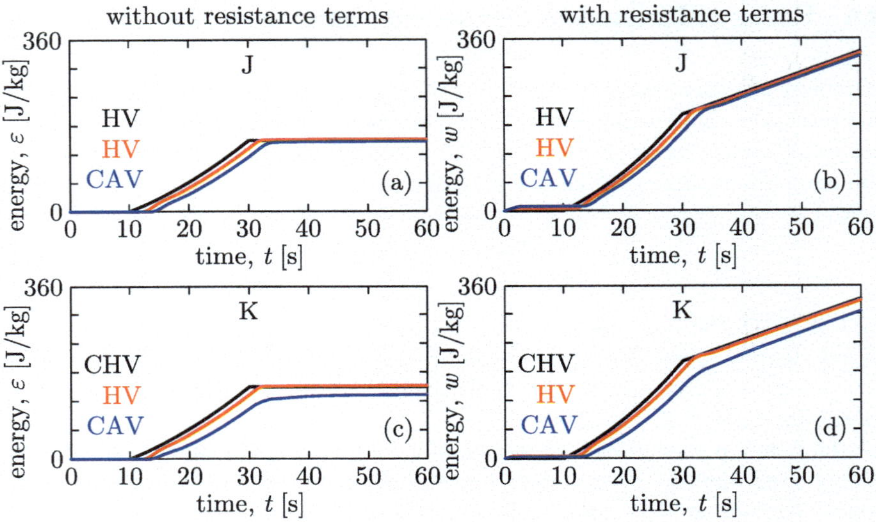

Fig. 4.27 Energy consumptions (4.111) and (4.113) corresponding to the numerical simulations of Fig. 4.8 for the connected vehicle network in Fig. 4.7. The two rows of the figure correspond to points J(0.5,0) and K(0.5,0.5) in the (β_{01}, β_{02}) plane in Fig. 4.9b, that is, the CAV responds to: **a**, **b** the HV only; **c**, **d** both the HV and the CHV. The left panels neglect the resistance terms while the right panels take them into account with parameters $k = 0.45$ [kg/m], $m_{\text{eff}} = m = 1500$ [kg], $\gamma = 0.01$, $g = 9.81$ [m/s^2] and $\phi = 0$

Figure 4.27 shows the energy consumptions (4.111) and (4.113) of the three-vehicle network illustrated in Fig. 4.7 based on the simulation results in Fig. 4.8. The two columns of the figure compare the energy consumption without and with taking resistance terms into account (given by ε_i and w_i). The plots highlight that the energy consumption is constant initially when the vehicles brake, and then it increases when they accelerate. Finally, during constant speed cruising, the energy consumption $\varepsilon_i(t)$ without resistance terms is constant, whereas $w_i(t)$ with resistance terms is linearly increasing. In these simulations the resistance terms were considered only in the energy measure and they were omitted from the right-hand side of the governing differential equations (4.37), (4.39), (4.41). In practice, this can be achieved by applying low-level controllers to cancel the effect of resistance forces on the dynamics [2].

Figure 4.27 also compares the energy efficiency of different vehicles and different control strategies for the CAV. The energy consumption is related to the smoothness of the driving behavior: if a vehicle brakes more than necessary, it typically results in higher energy consumption when accelerating back to the equilibrium speed, while smooth driving is usually more energy efficient. In Fig. 4.27, the HV has approximately the same energy consumption as the CHV. The CAV, on the other hand, can achieve better energy efficiency via smooth driving. The two rows of Fig. 4.27 compare the cases where the CAV responds to only one or both vehicles

Table 4.1 Total energy consumption corresponding to Figs. 4.27, 4.28 and 4.30. The CAV responds to the HV only in point J and to both the HV and the CHV in point K. The optimum points O_ε, O_w and O_J show the maximum achievable energy benefits

Point	Vehicle	ε [J/kg]	$\Delta\varepsilon$ [%]	w [J/kg]	Δw [%]
	CHV	150.0	0	332.8	0
	HV	152.7	+1.8	329.8	−0.9
J	CAV	148.3	−1.2	324.9	−2.4
K	CAV	133.7	−10.9	306.7	−7.8
O_ε	CAV	129.1	−13.9	303.2	−8.9
O_w	CAV	129.9	−13.4	302.4	−9.1
O_J	CAV	129.4	−13.7	303.3	−8.9

ahead (corresponding to the two rows of Fig. 4.8 and to points J and K in Fig. 4.9b). In panels (a) and (b), where the CAV responds to the HV only, the energy consumption of the CAV is slightly smaller than both the HV and the CHV. Such benefits are achievable even without utilizing connectivity. By exploiting connectivity, however, the CAV is able to outperform human drivers more significantly. Panels (c) and (d) indicate that the CAV's energy consumption reduces substantially when it responds to both the HV and the CHV.

These benefits are quantified by Table 4.1, which summarizes the total energy consumption of the three-vehicle network after 60 s of driving. The table shows the energy consumptions ε_i and w_i together with the relative improvements $\Delta\varepsilon_i = (\varepsilon_i - \varepsilon_n)/\varepsilon_n$ and $\Delta w_i = (w_i - w_n)/w_n$ compared to the leading CHV (vehicle $n = 2$). While the HV's energy performance is close to the CHV's (it is slightly higher without and slightly lower with considering resistance terms), the CAV outperforms both human drivers. In this example, the CAV saves 1–2% energy without utilizing information about the CHV, while the energy savings increase to 8–11% when relying on connectivity and responding to the CHV.

To optimize the CAV's energy efficiency, we evaluate the final energy consumption for various control parameters and select the optimal ones. This is illustrated in Fig. 4.28 where the (β_{01}, β_{02}) space of the control gains is explored. To obtain this figure, the simulations in Fig. 4.27 were repeated for a large number of (β_{01}, β_{02}) pairs in a "brute force" manner and the CAV's energy consumptions (4.111) and (4.113) were evaluated. The contours in Fig. 4.28 represent (β_{01}, β_{02}) pairs associated with the same energy consumption for the CAV, where the energy level is indicated by the color of the contours. The figure reveals that via connectivity, i.e., with the β_{02} term, we can descend to significantly lower energy levels compared to the case without connectivity, i.e., $\beta_{02} = 0$. The energy-optimal points that minimize ε_0 and w_0, denoted by O_ε and O_w, are indicated by black crosses. The energy consumptions in these optima are listed in Table 4.1: the optimal gains lead to 14 and 9% energy savings without and with considering resistance terms, respectively. Notice that the points O_ε and O_w are close to each other, implying that one may obtain

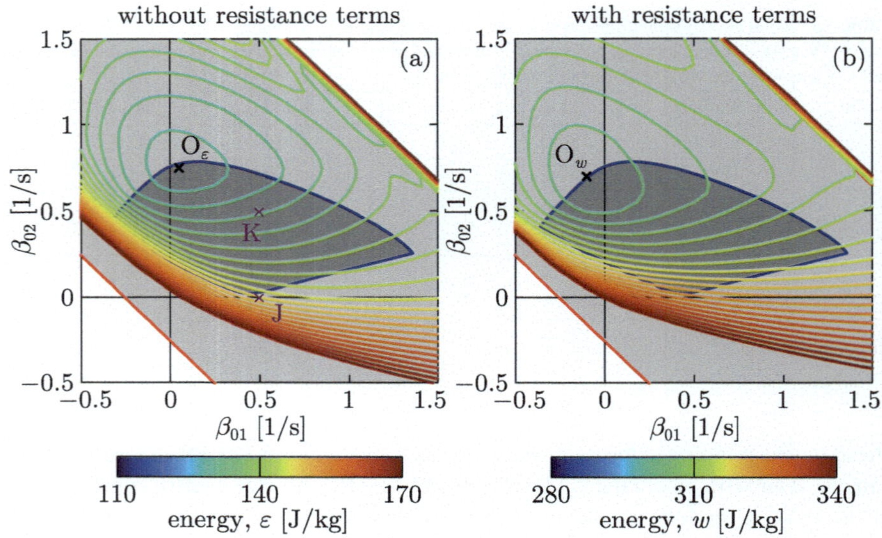

Fig. 4.28 Contour plot of the energy consumptions (4.111) and (4.113) obtained by numerical simulations. The contours are shown in the (β_{01}, β_{02}) plane, while all other parameters are the same as in Fig. 4.27. The contours overlay the stability chart of Fig. 4.9b

significant energy benefits in terms of w_0 even when control gains are optimized for ε_0 by neglecting resistance.

Note, however, that simulating the connected vehicle network for a large number of parameter combinations is computationally expensive, hence cannot be carried out in real time to tune the CAV's controller on board for energy efficiency. Meanwhile, the simulation results are specific to the selected speed profile of the lead vehicle, the human driver model, and the CAV's controller. To get more insight into how these affect the energy consumption, we conduct an approximate analysis with the ultimate goal of finding energy-optimal parameters under short computation time. In particular, we utilize a frequency domain approach to approximate the CAV's energy consumption and minimize it in a computationally efficient manner. The approach relies on the frequency content of the CAV's speed signal, which is obtained using the Fourier series of the preceding vehicles' speed signals and the CAV's linear response given by the link and head-to-tail transfer functions.

Consider the scenario discussed in Sect. 4.3.2: an open chain of $n + 1$ vehicles, where a CAV is following n HVs or CHVs, like in Fig. 4.12, with dynamics given by (4.66), (4.68) and (4.63), (4.65). Consider that the speed $v_i(t)$ of preceding vehicle i fluctuates around the equilibrium speed v^*. Assume that the speed fluctuations $\tilde{v}_i(t)$ can be decomposed into m harmonic components of angular frequency $\omega_k = k\Delta\omega = k2\pi/T$, amplitude ρ_{ik} and phase angle φ_{ik} based on Fourier series:

$$v_i(t) = v^* + \tilde{v}_i(t) = v^* + \sum_{k=1}^{m} \rho_{ik} \sin(\omega_k t + \varphi_{ik}) , \tag{4.114}$$

$i = 1, \ldots, n$ and $k = 1, \ldots, m$. Assuming linearized dynamics, the CAV responds to these speed fluctuations with harmonic speed in steady state, that consists of components of amplitude D_k and phase angle θ_k at angular frequency ω_k:

$$v_0(t) = v^* + \tilde{v}_0(t) = v^* + \sum_{k=1}^{m} D_k \sin(\omega_k t + \theta_k) . \tag{4.115}$$

These components can be calculated from the transfer functions describing the linearized dynamics.

If the head-to-tail transfer function G_{0n} is available, one can relate the speed fluctuations of the CAV to those of the leading vehicle n by

$$\begin{aligned} D_k &= \rho_{nk} \, |G_{0n}(\mathrm{j}\omega_k; \mathbf{p})| , \\ \theta_k &= \varphi_{nk} + \angle G_{0n}(\mathrm{j}\omega_k; \mathbf{p}) . \end{aligned} \tag{4.116}$$

Here we emphasize that the head-to-tail transfer function depends on design parameters denoted by $\mathbf{p}$, such as $\mathbf{p} = \begin{bmatrix} \beta_{01} & \beta_{02} & \ldots & \beta_{0n} \end{bmatrix}$, which shall be optimized to minimize energy consumption for the CAV. Note that (4.116) involves Fourier components related to the lead vehicle n only, it is not required to know the speeds of the other preceding vehicles $i = 1, \ldots, n-1$ and the frequency content thereof. However, calculating the head-to-tail transfer function G_{0n} requires knowing all elements of the weighted adjacency matrix $\mathbf{T}$, i.e., knowing how each vehicle responds to its predecessors in the open chain of $n+1$ vehicles, including the human-driven ones. The human driver models of Chap. 2 may help formulating the head-to-tail transfer function G_{0n}.

Alternatively, the need for human driver models can be eliminated if one calculates the CAV's response based on the data from all the preceding vehicles it responds to and the associated link transfer functions T_{0i}. As opposed to human driver behavior, these link transfer functions are known in the CAV's control design. In steady state, the response of the CAV can be given by

$$v_0(t) = v^* + \tilde{v}_0(t) = v^* + \sum_{i=1}^{n} \sum_{k=1}^{m} D_{ik} \sin(\omega_k t + \theta_{ik}) , \tag{4.117}$$

where D_{ik} and θ_{ik} represent the amplitudes and phase angles associated with the response to vehicle i. These can be calculated from the link transfer functions T_{0i} as

$$\begin{aligned} D_{ik} &= \rho_{ik} \, |T_{0i}(\mathrm{j}\omega_k; \mathbf{p})| , \\ \theta_{ik} &= \varphi_{ik} + \angle T_{0i}(\mathrm{j}\omega_k; \mathbf{p}) , \end{aligned} \tag{4.118}$$

where the link transfer functions depend on the CAV's control parameters $\mathbf{p}$.

In (4.117) the trigonometric terms of angular frequency ω_k can be combined into one, which leads to (4.115) with

$$\begin{aligned} D_k &= \sqrt{\left(\sum_{i=1}^{n} D_{ik}\cos\theta_{ik}\right)^2 + \left(\sum_{i=1}^{n} D_{ik}\sin\theta_{ik}\right)^2}\,, \\ \tan\theta_k &= \frac{\sum_{i=1}^{n} D_{ik}\sin\theta_{ik}}{\sum_{i=1}^{n} D_{ik}\cos\theta_{ik}}\,. \end{aligned} \tag{4.119}$$

While this approach does not require the knowledge of human driver models, it relies on data from multiple preceding vehicles (all vehicles the CAV responds to). This implies a more data-focused and less model-based approach.

Given the frequency content (4.115) of the CAV's speed, one can calculate the energy consumption. For example, for the simplified energy measure (4.113) that neglects the resistance terms we get

$$\begin{aligned} \varepsilon_0 &= \int_0^T v_0(t) g\big(\dot{v}_0(t)\big)\mathrm{d}t \\ &= \int_0^T \left(v^* + \sum_{k=1}^{m} D_k \sin(\omega_k t + \theta_k)\right) g\left(\sum_{k=1}^{m} D_k\omega_k \cos(\omega_k t + \theta_k)\right) \mathrm{d}t\,. \end{aligned} \tag{4.120}$$

This integral can be calculated analytically for simple, special cases such as the following.

Assume that the speed of the lead vehicle n consist of a single harmonic component of angular frequency ω, amplitude ρ and phase angle φ where the latter can be considered to be zero without loss of generality, that is,

$$v_n(t) = v^* + \rho\sin(\omega t)\,. \tag{4.121}$$

Then, the CAV's response involves a single harmonic only, associated with the same angular frequency ω, amplitude D and phase angle θ:

$$v_0(t) = v^* + D\sin(\omega t + \theta)\,, \tag{4.122}$$

where

$$\begin{aligned} D &= \rho\,|G_{0n}(\mathrm{j}\omega;\, \mathbf{p})|\,, \\ \theta &= \angle G_{0n}(\mathrm{j}\omega;\, \mathbf{p})\,. \end{aligned} \tag{4.123}$$

Considering braking losses according to (4.112), for the harmonic signal (4.123) the energy consumption (4.120) simplifies to

$$
\begin{aligned}
\varepsilon_0 &= \int_0^T v_0(t) g(\dot{v}_0(t)) \mathrm{d}t \\
&= \int_0^T \left(v^* + D\sin(\omega t + \theta)\right)\frac{1}{2}\left(D\omega\cos(\omega t + \theta) + \left|D\omega\cos(\omega t + \theta)\right|\right)\mathrm{d}t \\
&= 2v^* D \\
&= 2v^* \rho \left|G_{0n}(\mathrm{j}\omega; \mathbf{p})\right| ,
\end{aligned}
\tag{4.124}
$$

where we used the identity $g(x) = \max\{0, x\} = (x + |x|)/2$. Also, we exploited that $\int_0^T \cos(\omega t + \theta)\,\mathrm{d}t = 0$ and $\int_0^T \sin(\omega t + \theta)\cos(\omega t + \theta)\,\mathrm{d}t = 0$ for $\omega = 2\pi/T$. This indicates that the energy consumption is proportional to the magnitude of the head-to-tail transfer function, and is therefore closely related to head-to-tail string stability. While strictly speaking this result holds for speed signals with a single harmonic only, we can say, as a rule of thumb, that selecting string stable control gains is beneficial for energy efficiency.

For multiple harmonics, the energy consumption ε_0 in (4.120) cannot be calculated in a simple form, but it can be approximated in practically tractable way. In particular, ε_0 in (4.120) can be lower and upper bounded by class-$\mathcal{K}$ functions of a certain cost function established by the following theorem.

Theorem 4.6 *Consider the energy consumption ε_0 given by (4.120). If $v_0(t) \geq 0$ for all $t \in [0, T]$, $v_0(0) = 0$, and $v^* > m\bar{D}$ with $\bar{D} = \max_{k\in\{1,\ldots,m\}} D_k$, then there exist class-$\mathcal{K}$ functions α_1 and α_2 such that*

$$\alpha_1(J) \leq \varepsilon_0 \leq \alpha_2(J) , \tag{4.125}$$

where the quantity J is defined by

$$J = \sum_{k=1}^{m} D_k^2 \omega_k^2 , \tag{4.126}$$

with D_k expressed in (4.116) using the head-to-tail transfer function and in (4.118), (4.119) using the link transfer functions.

Here we give a proof of this statement, by following [85]. We remark that the proof is quite detailed and can be skipped by less interested readers.

Proof First, we approximate the ReLU function $g(x) = \max\{x, 0\} = (x + |x|)/2$ by a smooth function $\hat{g}(x)$ with parameter $d > 0$:

$$\hat{g}(x) = \frac{1}{2}\left(x + \frac{x^2}{\sqrt{d + x^2}}\right), \tag{4.127}$$

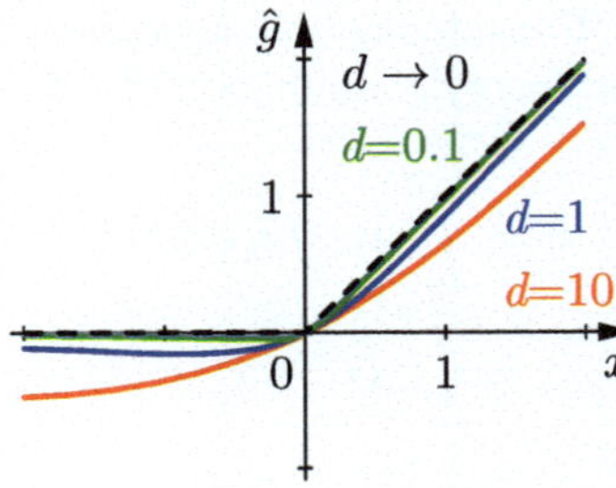

Fig. 4.29 The rectified linear unit (ReLU) function $g(x) = \max\{0, x\}$ and its smooth approximation $\hat{g}(x)$

where we have

$$\lim_{d \to 0} \hat{g}(x) = g(x) . \tag{4.128}$$

The smooth approximation of the ReLU is illustrated in Fig. 4.29 for different values of d.

Then, we take the expression of ε_0 in (4.120) and replace g with $\hat{g}$. We denote the resulting quantity by $\hat{\varepsilon}_0$. We derive lower and upper bounds for $\hat{\varepsilon}_0$, which yield bounds for ε_0 by taking the limit $d \to 0$. The expression of $\hat{\varepsilon}_0$ reads

$$\hat{\varepsilon}_0 = \int_0^T v_0(t)\hat{g}\left(\dot{v}_0(t)\right) \mathrm{d}t = \int_0^T \frac{1}{2}\big(v^* + \tilde{v}_0(t)\big) \left(\dot{v}_0(t) + \frac{\dot{v}_0(t)^2}{\sqrt{d + \dot{v}_0(t)^2}}\right) \mathrm{d}t , \tag{4.129}$$

where

$$\begin{aligned} \tilde{v}_0(t) &= \sum_{k=1}^{m} D_k \sin(\omega_k t + \theta_k) , \\ \dot{v}_0(t) &= \sum_{k=1}^{m} D_k \omega_k \cos(\omega_k t + \theta_k) , \end{aligned} \tag{4.130}$$

cf. (4.115).

Based on direct substitution of (4.130) and integration, the following terms drop:

$$\begin{aligned} \int_0^T \frac{1}{2} v^* \dot{v}_0(t) \mathrm{d}t &= 0 , \\ \int_0^T \frac{1}{2} \tilde{v}_0(t) \dot{v}_0(t) \mathrm{d}t &= 0 , \end{aligned} \tag{4.131}$$

and we obtain

$$\hat{\varepsilon}_0 = \int_0^T \frac{1}{2}\big(v^* + \tilde{v}_0(t)\big) \frac{\dot{v}_0(t)^2}{\sqrt{d + \dot{v}_0(t)^2}} \mathrm{d}t . \tag{4.132}$$

Note that the integrand is nonnegative for all $t \in [0, T]$ as we assume that the speed is nonnegative, $v_0(t) = v^* + \tilde{v}_0(t) \geq 0$. The remaining terms within the integral are

$$\begin{aligned}
\dot{v}_0(t)^2 &= \sum_{k=1}^{m} \sum_{l=1}^{m} D_k \omega_k D_l \omega_l \cos(\omega_k t + \theta_k) \cos(\omega_l t + \theta_l) \,, \\
\tilde{v}_0(t) \dot{v}_0(t)^2 &= \sum_{k=1}^{m} \sum_{l=1}^{m} \sum_{q=1}^{m} D_k D_l \omega_l D_q \omega_q \sin(\omega_k t + \theta_k) \cos(\omega_l t + \theta_l) \cos(\omega_q t + \theta_q) \,.
\end{aligned} \tag{4.133}$$

For $d + \dot{v}_0(t)^2$ in the denominator in (4.132), we have the following bounds if $\dot{v}_0(t) \not\equiv 0$

$$d \leq d + \dot{v}_0(t)^2 \leq d + \left(\sum_{k=1}^{m} D_k \omega_k \right)^2 \leq d + m \sum_{k=1}^{m} D_k^2 \omega_k^2 \leq M \sum_{k=1}^{m} D_k^2 \omega_k^2 \,, \tag{4.134}$$

$\forall t \in [0, T]$ with some $M > 0$. Here the third inequality is based on the Cauchy-Schwartz inequality, whereas the last inequality uses that for any d, D_k, ω_k and m there exists such $M > 0$ that the inequality holds (and $M \to m$ as $d \to 0$).

Combining (4.132) with $v^* + \tilde{v}_0(t) \geq 0$ and (4.134) gives the following bounds on the energy consumption $\hat{\varepsilon}_0$:

$$\frac{\int_0^T \left(v^* \dot{v}_0(t)^2 + \tilde{v}_0(t) \dot{v}_0(t)^2 \right) \mathrm{d}t}{2\sqrt{M \sum_{k=1}^{m} D_k^2 \omega_k^2}} \leq \hat{\varepsilon}_0 \leq \frac{\int_0^T \left(v^* \dot{v}_0(t)^2 + \tilde{v}_0(t) \dot{v}_0(t)^2 \right) \mathrm{d}t}{2\sqrt{d}} \,. \tag{4.135}$$

By substituting (4.133) and integrating, we get the following properties for the remaining integrals

$$\begin{aligned}
\int_0^T v^* \dot{v}_0(t)^2 \mathrm{d}t &= \frac{v^*}{2} T \sum_{k=1}^{m} D_k^2 \omega_k^2 \,, \\
-\frac{m\bar{D}}{2} T \sum_{k=1}^{m} D_k^2 \omega_k^2 \leq \int_0^T \tilde{v}_0(t) \dot{v}_0(t)^2 \mathrm{d}t &\leq \frac{m\bar{D}}{2} T \sum_{k=1}^{m} D_k^2 \omega_k^2 \,,
\end{aligned} \tag{4.136}$$

with $\bar{D} = \max_{k \in \{1,\ldots,m\}} D_k$. This finally leads to

$$\underline{c} \sqrt{\sum_{k=1}^{m} D_k^2 \omega_k^2} \leq \hat{\varepsilon}_0 \leq \bar{c} \sum_{k=1}^{m} D_k^2 \omega_k^2 \,, \tag{4.137}$$

with

$$\underline{c} = \frac{v^* - m\bar{D}}{4\sqrt{M}} T \,, \quad \bar{c} = \frac{v^* + m\bar{D}}{4\sqrt{d}} T \,. \tag{4.138}$$

If $v^* > m\bar{D}$ holds, then the lower bound is positive. Then, both the lower and upper bounds on $\hat{\varepsilon}_0$ are class-$\mathcal{K}$ functions of J given in (4.126), with $\alpha_1(J) = \underline{c}\sqrt{J}$ and $\alpha_2(J) = \bar{c}J$. Taking $d \to 0$ gives this property for ε_0 as well.

Note, however, that the upper bound on $\hat{\varepsilon}_0$ goes to infinity ($\bar{c} \to \infty$) when $d \to 0$. To resolve this issue one may take the derivative of (4.115) and show that

$$J \propto \int_0^T \dot{v}_0^2(t)\mathrm{d}t\,, \tag{4.139}$$

see [85, 191]. Then using Opial's inequality [25, 132] one may show that the energy consumption ε_0 given by (4.120) can be upper-bounded as

$$\varepsilon_0 = \int_0^T v_0(t)\max\{0, \dot{v}_0(t)\}\mathrm{d}t \le \int_0^T v_0(t)|\dot{v}_0(t)|\mathrm{d}t \le \frac{T}{2}\int_0^T \dot{v}_0^2(t)\mathrm{d}t\,, \tag{4.140}$$

if $v_0(t) \ge 0$, $\forall t \in [0, T]$, v_0 is absolutely continuous and $v_0(0) = 0$ (where the last inequality is from Corollary 1 in [25]). The upper bound in (4.140) is proportional to J defined in (4.139), that is, it is a class-$\mathcal{K}$ of J, which completes the proof. □

In the rest of this section, quantity J is termed *cost function*. In fact (4.139) is often used in the literature as cost function when optimizing for energy consumption. Since it determines both the lower and upper bounds of the energy consumption ε_0, minimizing the cost function J leads to control parameters $\mathbf{p}$ that are close to (but not necessary equal to) the energy optimum in terms of ε_0. Furthermore, recall that in the example of Fig. 4.28, the optimum with respect to ε_0 was close to optimal even in terms of the energy measure w_0 that involves the effect of resistance terms.

Introducing J in (4.126) gives a significant practical advantage: it can be calculated in a computationally efficient way, allowing control parameters to be optimized on board of the CAV. Namely, one can calculate the fast Fourier transform (FFT) of the incoming speed data on board to obtain the frequency content given by ω_k, ρ_{ik} and φ_{ik}. Then, one can compute the amplitudes D_k by evaluating either (4.116) using the head-to-tail transfer function G_{0n} or by evaluating (4.118), (4.119) using the link transfer functions T_{0i}, to finally obtain the cost function J by (4.126). This is a computationally tractable process and is much faster than evaluating the energy measures ε_0 and w_0 directly through numerical simulations.

The "brute force" evaluation of the cost function J for various (β_{01}, β_{02}) pairs in the three vehicle scenario of Fig. 4.7 is illustrated in Fig. 4.30. Panel (a) shows the contours of the energy consumption ε_0 for reference, that are the same as in Fig. 4.28a. Panel (b) shows the contours of the cost function J, which were obtained by the fast Fourier transform of the leading vehicle's speed signal in Fig. 4.8 and using the head-to-tail transfer function given by (4.53), (4.55). The contours of J follow the same structure as the contours of ε_0. The optimum point O_J minimizing the cost function J is located close to the optimum point O_ε minimizing the energy consumption ε_0. This allows one to achieve close to optimal performance in practice, since O_J can be obtained in a computationally efficient manner (while O_ε may be

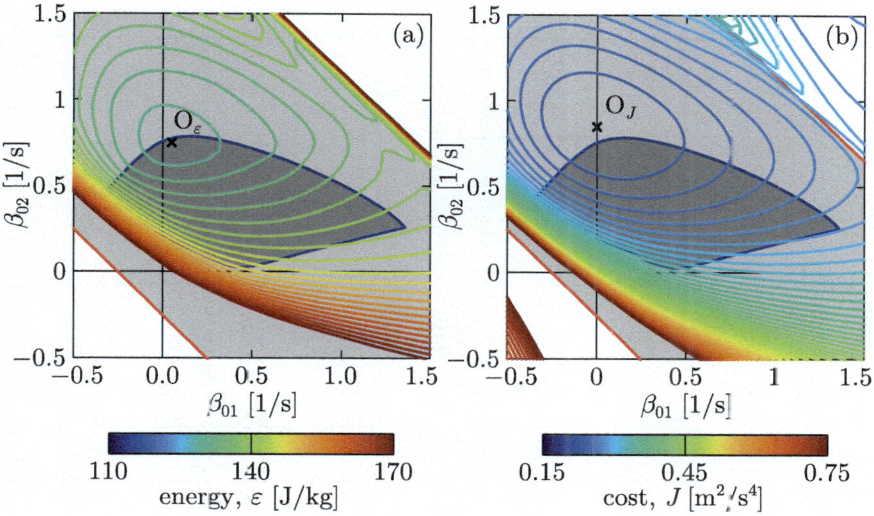

Fig. 4.30 Contour plot of the energy consumption (4.113) obtained by numerical simulations and the cost function (4.126) evaluated by frequency domain analysis. The contours are shown in the (β_{01}, β_{02}) plane, while all other parameters are the same as in Fig. 4.27. The contours overlay the stability chart of Fig. 4.9b

hard to find). The energy savings by selecting O_J are listed in the last row of Table 4.1. This point represents the practically achievable benefits and shows similar energy performance as points O_ε and O_w.

We further illustrate the energy performance of connected cruise control by analyzing the four-vehicle network shown in Fig. 4.11b. First, we show the energy consumption associated with the numerical simulation results of Fig. 4.15. The energy consumption as a function of time is illustrated in Fig. 4.31. In the four rows of the figure, the CAV responds to the following vehicles, respectively: neither CHV 2 nor CHV 3, CHV 2 but not CHV 3, CHV 3 but not CHV 2, both CHV 2 and CHV 3. That is, the CAV utilizes more information from connectivity and becomes more clairvoyant as we go from the top to the bottom of the figure. Accordingly, the energy consumption of the CAV reduces significantly both without (left) and with (right) considering the resistance terms.

The energy consumption of the three CHVs and the four different CAV controllers in Fig. 4.31 is listed in Table 4.2 (see points J, K, L and M). The subsequent CHVs consume more and more energy in terms of ε due to their string instability and excessive braking. Therefore, in point J, although the CAV consumes less energy than CHV 1 and CHV 2, its energy consumption ε_0 is still more than that of the leading CHV 3. However, utilizing the potential of connectivity by responding to multiple vehicles leads to energy savings. In point M, the energy benefits of the CAV reach 16 and 12% with respect to the leading CHV 3 in terms of ε and w, respectively.

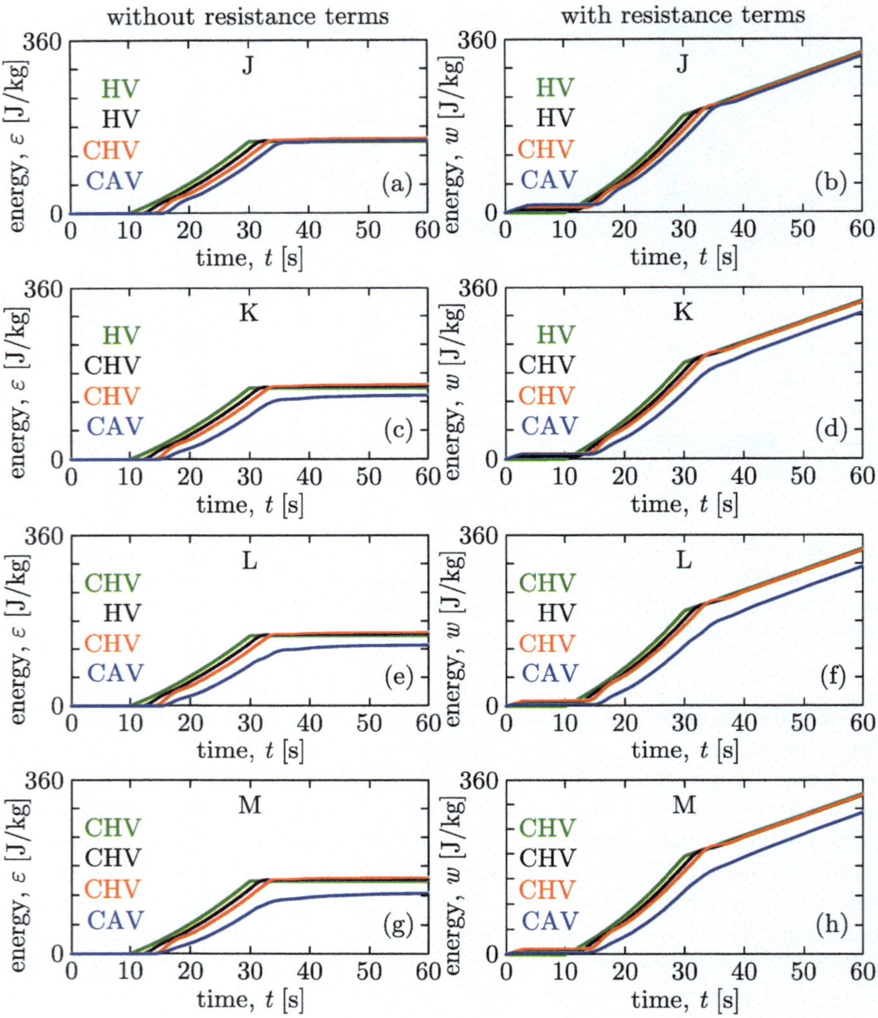

Fig. 4.31 Energy consumptions (4.111) and (4.113) corresponding to the numerical simulations of Fig. 4.15 for the connected vehicle network in Fig. 4.11b. The four rows of the figure correspond to points J(0,0), K(0.5,0), L(0,0.5) and M(0.5,0.5) in the (β_{02}, β_{03}) plane in Fig. 4.14c, that is, the CAV responds to: **a**, **b** neither CHV 2 nor CHV 3; **c**, **d** CHV 2 but not CHV 3; **e**, **f** CHV 3 but not CHV 2; **g**, **h** both CHV 2 and CHV 3. The left panels neglect the resistance terms while the right panels take them into account with parameters $k = 0.45$ [kg/m], $m_{\text{eff}} = m = 1500$ [kg], $\gamma = 0.01$, $g = 9.81$ [m/s^2] and $\phi = 0$

Table 4.2 Total energy consumption corresponding to Figs. 4.31 and 4.32. The CAV responds to neither CHV 2 nor CHV 3 in point J, to CHV 2 but not CHV 3 in point K, to CHV 3 but not CHV 2 in point L and to both CHV 2 and CHV 3 in point M. The optimum points O_ε, O_w and O_J show the maximum achievable energy benefits

Point	Vehicle	ε [J/kg]	$\Delta\varepsilon$ [%]	w [J/kg]	Δw [%]
	CHV 3	150.0	0	332.8	0
	CHV 2	152.7	+1.8	329.8	−0.9
	CHV 1	156.9	+4.6	330.2	−0.8
J	CAV 0	152.0	+1.4	325.8	−2.1
K	CAV 0	134.5	−10.4	307.9	−7.5
L	CAV 0	129.8	−13.4	295.2	−11.3
M	CAV 0	125.5	−16.3	293.8	−11.7
O_ε	CAV 0	123.6	−17.6	293.4	−11.8
O_w	CAV 0	124.5	−17.0	292.5	−12.1
O_J	CAV 0	123.9	−17.4	292.6	−12.1

The contours of energy consumption ε_0 and w_0 based on simulations and those of the cost function J based on frequency domain analysis are plotted in Fig. 4.32. The space of the (β_{02}, β_{03}) control gains is explored as a counterpart to the stability chart in Fig. 4.14c. The optimum points O_ε, O_w, and O_J with respect to the three measures are indicated by black crosses. Note that point O_ε in panel (a) is a local optimum, while the global optimum (marked by a thin black cross) is located close to the plant stability boundary and is therefore disregarded. The points O_ε, O_w, and O_J are located close to each other, near the corner of the head-to-tail string stable domain. The energy performance of these three points is listed at the bottom of Table 4.2. Selecting point O_J (which is practically achievable) leads to 17 and 12% energy savings compared to the leading CHV 3 in terms of ε and w, respectively.

4.7 Robustness Against Human Driver Uncertainties

Here we discuss the sensitivity of CCC design to the uncertainty of human behavior. Since CCC allows one to incorporate human-driven vehicles in the vehicle chains, the model parameters used for these vehicles influence the control design of connected automated vehicles. Above we used nominal human driver parameters that were identified experimentally using data from multiple drivers. However, the parameters typically differ for different drivers and may even change for a single driver over time.

Therefore, we consider the possible range of uncertain parameters rather than their nominal values. We assume that the parameters lie in a known parameter range while their exact value is unknown, and they are constant in time (i.e., their uncertainty is

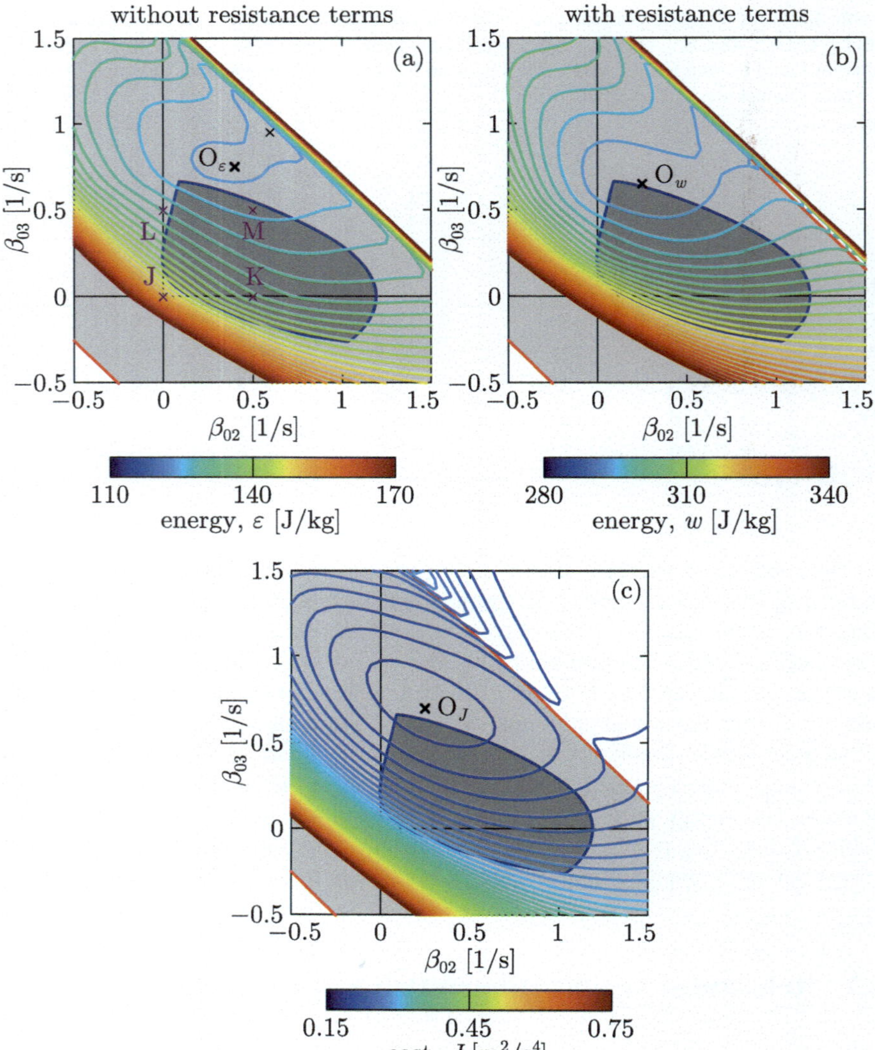

Fig. 4.32 Contour plot of the energy consumptions (4.111) and (4.113) obtained by numerical simulations and the cost function (4.126) evaluated by frequency domain analysis. The contours are shown in the (β_{02}, β_{03}) plane for $\beta_{01} = 0.5$ [1/s], while all other parameters are the same as in Fig. 4.31. The contours overlay the stability chart of Fig. 4.14c

also constant). The uncertain parameter range around the nominal parameters $\kappa_{\rm h}$, $\alpha_{\rm h}$, $\beta_{\rm h}$, and $\tau_{\rm h}$ are denoted and quantified as follows:

$$\begin{aligned} \kappa_{\rm h} + \rho_\kappa \delta_\kappa &\in [0.4, 1.2]\,[1/{\rm s}]\,, \\ \alpha_{\rm h} + \rho_\alpha \delta_\alpha &\in [0.05, 0.25]\,[1/{\rm s}]\,, \\ \beta_{\rm h} + \rho_\beta \delta_\beta &\in [0.25, 0.75]\,[1/{\rm s}]\,, \\ \tau + \rho_\tau \delta_\tau &\in [0.5, 1.5]\,[{\rm s}]\,, \end{aligned} \tag{4.141}$$

where $\rho_\kappa, \rho_\alpha, \rho_\beta, \rho_\tau \geq 0$ quantify the size of the uncertain parameter range and

$$\delta_\kappa, \delta_\alpha, \delta_\beta, \delta_\tau \in [-1, 1]\,, \tag{4.142}$$

allow us to "scan" the entire range. For example, for the values above we have $\rho_\kappa = 0.4\,[1/{\rm s}]$, $\rho_\alpha = 0.1\,[1/{\rm s}]$, $\rho_\beta = 0.25\,[1/{\rm s}]$ and $\rho_\tau = 0.5\,[{\rm s}]$. Our goal is to design a connected cruise controller that achieves plant and string stability for any possible parameter combinations in this range.

One way to achieve this goal is via "brute force" computation. Namely, one may consider different parameter combinations from the uncertainty regions listed above and generate string stability charts for each combination. Then the robust string stability domain can be found as the intersection of these domains, as illustrated in the top panels of Fig. 4.33 for the simple three-car network in Fig. 4.7. Each orange curve corresponds to a parameter combination at the edge of the parameter uncertainty region (i.e., $2^4 = 16$ curves are drawn on each panel). The top left panel corresponds to 10% uncertainty in each parameter while the top right panel corresponds to 25% uncertainty. The bottom panels depict the frequency response plots for point K. All of these curves stay below 1 on the left (suggesting that point K is robust against 10% uncertainty in human parameters), while on the right some curves rise above 1 (showing that point K is not robust against 25% uncertainty).

This "brute force" method, however, has multiple limitations. The string stability boundaries in Fig. 4.33 belong to the vertices of the uncertain parameter region, but in general there is no guarantee that ensuring string stability for these particular human parameters is sufficient to ensure string stability for parameters inside the uncertainty region. Indeed, one may "scan" the parameter ranges by selecting m different values of each parameter and plot m^4 string stability boundaries, but this becomes computationally expensive as m increases. Moreover, if a vehicle chain includes n human-driven vehicles, one shall draw m^{4n} boundaries (assuming that each human driver can still be represented by four parameters), which becomes computationally infeasible for large n. In order to resolve this issue we adopt tools from robust control [90, 264] that enable us to generate robust string stability charts in a computationally efficient fashion. These are already illustrated in Fig. 4.33 with the black regions on the top panels and the black curves on the bottom panels. Below we provide the technical details to generate these results. In particular, we use the method described in [79]; and for similar studies see [57, 118, 261].

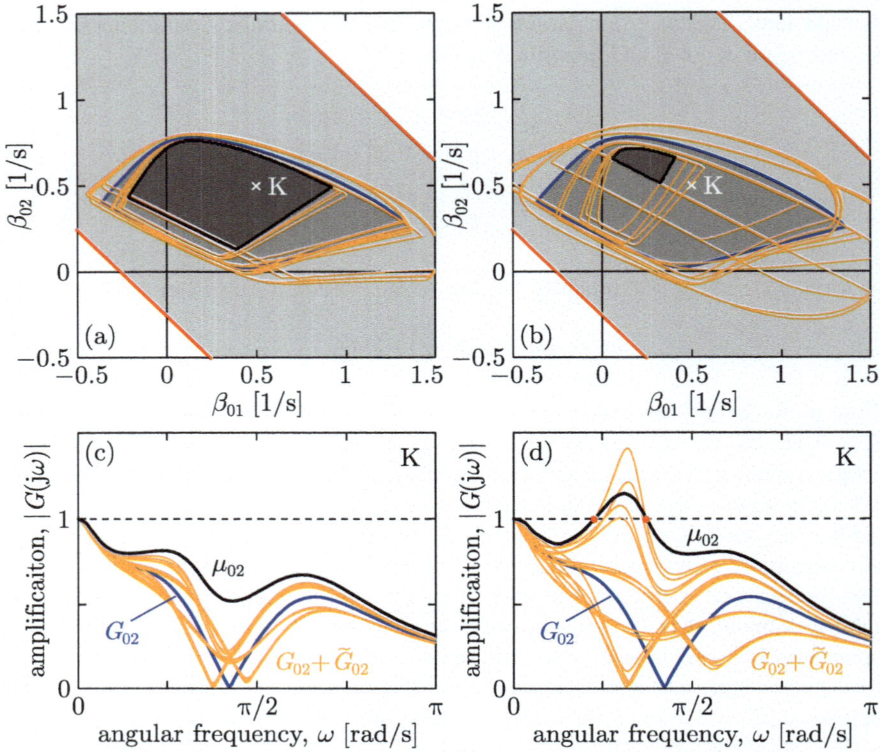

Fig. 4.33 Top: Stability chart under human parameter uncertainties; the nominal parameters are the same as those in Fig. 4.9b. The relative uncertainty of each human driver parameter is assumed to be the same, $\rho_\kappa/\kappa_h = \rho_\alpha/\alpha_h = \rho_\beta/\beta_h = \rho_\tau/\tau =: p$, where **a** $p = 10\%$ and **b** $p = 25\%$. Stability boundaries are plotted for all 16 combinations of $\alpha_h \pm \rho_\alpha$, $\beta_h \pm \rho_\beta$, $\kappa_h \pm \rho_\kappa$ and $\tau \pm \rho_\tau$. The black curves are obtained with robust stability analysis. Bottom: Frequency response plots; cf. Fig. 4.10b. Again the black curves are obtained via robust stability analysis. Notice that μ goes above 1 whenever any of the perturbed transfer functions goes above 1 (see red dots)

4.7.1 Robustness Without Time Delay

To illustrate the method, we first consider delay-free dynamics ($\tau = 0$, $\delta_\tau = 0$). For the remaining parameters, we define the uncertainty matrix

$$\boldsymbol{\delta} = \begin{bmatrix} \delta_\kappa & 0 & 0 \\ 0 & \delta_\alpha & 0 \\ 0 & 0 & \delta_\beta \end{bmatrix}, \tag{4.143}$$

and the following norm is used to quantify the largest parameter deviation:

$$\|\boldsymbol{\delta}\|_\infty = \max\{|\delta_\kappa|, |\delta_\alpha|, |\delta_\beta|\} . \tag{4.144}$$

The domain of $\boldsymbol{\delta}$ is denoted as $\boldsymbol{\delta} \in \mathbf{D}$, where $\mathbf{D}$ refers to diagonal matrices whose elements lie in $\mathbb{R}$. When $\|\boldsymbol{\delta}\|_\infty \leq 1$, the underlying parameters are in the admissible uncertain range, whereas $\|\boldsymbol{\delta}\|_\infty > 1$ is related to parameters lying outside the uncertain domain of interest.

Due to parameter uncertainty, the nominal transfer function $T(s)$ characterizing a human driver modifies to $T(s) + \tilde{T}(s; \boldsymbol{\delta})$, where we emphasize that the transfer function deviation $\tilde{T}$ depends on the parameter deviation $\boldsymbol{\delta}$ (and $\tilde{T}(s; \mathbf{0}) = 0$). For example, for the OVM we have

$$T(s) = \frac{\beta_\mathrm{h} s + \alpha_\mathrm{h} \kappa_\mathrm{h}}{s^2 + (\alpha_\mathrm{h} + \beta_\mathrm{h}) s + \alpha_\mathrm{h} \kappa_\mathrm{h}} , \tag{4.145}$$

and

$$T(s) + \tilde{T}(s; \boldsymbol{\delta}) = \frac{(\beta_\mathrm{h} + \rho_\beta \delta_\beta) s + (\alpha_\mathrm{h} + \rho_\alpha \delta_\alpha)(\kappa_\mathrm{h} + \rho_\kappa \delta_\kappa)}{s^2 + (\alpha_\mathrm{h} + \rho_\alpha \delta_\alpha + \beta_\mathrm{h} + \rho_\beta \delta_\beta) s + (\alpha_\mathrm{h} + \rho_\alpha \delta_\alpha)(\kappa_\mathrm{h} - \rho_\kappa \delta_\kappa)} . \tag{4.146}$$

Note that the explicit expression for $\tilde{T}(s; \boldsymbol{\delta})$ is nontrivial and its systematic calculation requires the mathematical tools introduced below. Rather than directly evaluating $\tilde{T}(s; \boldsymbol{\delta})$ for different values of $\boldsymbol{\delta}$, which is computationally very expensive, we will utilize μ-calculus, i.e., structured singular value analysis [46, 264], in order to evaluate plant and string stability under parameter uncertainties. In what follows, we derive the robust stability condition in a form that is suitable for μ-calculus.

For plant stability, the stability boundary must not be reached for any parameter deviation, i.e., we must achieve for all $\omega > 0$ that

$$\mathrm{Den}\Big(T(\mathrm{j}\omega) + \tilde{T}(\mathrm{j}\omega; \boldsymbol{\delta})\Big) \neq 0 , \quad \forall \boldsymbol{\delta} \in \mathbf{D} , \ \|\boldsymbol{\delta}\|_\infty \leq 1 . \tag{4.147}$$

This can also be stated as

$$\mathrm{Den}\Big(T(\mathrm{j}\omega) + \tilde{T}(\mathrm{j}\omega; \boldsymbol{\delta})\Big) = 0 \quad \Rightarrow \quad \|\boldsymbol{\delta}\|_\infty > 1 , \tag{4.148}$$

i.e., the plant stability boundary lies outside the admissible parameter range given by $\|\boldsymbol{\delta}\|_\infty \leq 1$. That is, amongst those $\boldsymbol{\delta}$ values that satisfy the equality even the smallest possible $\|\boldsymbol{\delta}\|_\infty$ is greater than 1. This can be written equivalently as

$$\min_{\boldsymbol{\delta} \in \mathbf{D}} \Big\{ \|\boldsymbol{\delta}\|_\infty : \mathrm{Den}\Big(T(\mathrm{j}\omega) + \tilde{T}(\mathrm{j}\omega; \boldsymbol{\delta})\Big) = 0 \Big\} > 1 . \tag{4.149}$$

For string stability, we require that the magnitude of the (perturbed) transfer function is less than 1 at any frequency, i.e., for all $\omega > 0$ we have

$$\left|T(\mathrm{j}\omega) + \tilde{T}(\mathrm{j}\omega; \boldsymbol{\delta})\right| < 1\,, \quad \forall \boldsymbol{\delta} \in \mathbf{D}\,, \ \|\boldsymbol{\delta}\|_\infty \le 1\,. \tag{4.150}$$

This can be rephrased by introducing an auxiliary variable $\delta_{\mathrm{ss}} \in \mathbb{C}$ as

$$\left(T(\mathrm{j}\omega) + \tilde{T}(\mathrm{j}\omega; \boldsymbol{\delta})\right)\delta_{\mathrm{ss}} \ne 1\,, \quad \forall \boldsymbol{\delta} \in \mathbf{D}\,, \ \|\boldsymbol{\delta}\|_\infty \le 1 \ \text{ and } \ \forall \delta_{\mathrm{ss}} \in \mathbb{C}\,, \ |\delta_{\mathrm{ss}}| \le 1\,. \tag{4.151}$$

Equivalently, this reads

$$1 - \left(T(\mathrm{j}\omega) + \tilde{T}(\mathrm{j}\omega; \boldsymbol{\delta})\right)\delta_{\mathrm{ss}} = 0 \quad \Rightarrow \quad \|\boldsymbol{\delta}\|_\infty > 1 \ \text{ or } \ |\delta_{\mathrm{ss}}| > 1\,, \tag{4.152}$$

i.e., string instability could happen outside the admissible parameter range $\|\boldsymbol{\delta}\|_\infty \le 1$ only. The condition $|\delta_{\mathrm{ss}}| > 1$ or $\|\boldsymbol{\delta}\|_\infty > 1$ is equivalent to $\max\{|\delta_{\mathrm{ss}}|, \|\boldsymbol{\delta}\|_\infty\} > 1$. Hence, analogously to (4.149), we can write

$$\min_{\boldsymbol{\delta}\in\mathbf{D}, \delta_{\mathrm{ss}}\in\mathbb{C}} \left\{\max\{\|\boldsymbol{\delta}\|_\infty, |\delta_{\mathrm{ss}}|\} : 1 - \left(T(\mathrm{j}\omega) + \tilde{T}(\mathrm{j}\omega; \boldsymbol{\delta})\right)\delta_{\mathrm{ss}} = 0\right\} > 1\,. \tag{4.153}$$

Since both (4.149) and (4.153) must hold, combining them leads to

$$\begin{aligned}\min_{\boldsymbol{\delta}\in\mathbf{D}, \delta_{\mathrm{ss}}\in\mathbb{C}} \Big\{ &\max\{\|\boldsymbol{\delta}\|_\infty, |\delta_{\mathrm{ss}}|\} : \\ &\mathrm{Den}(T(\mathrm{j}\omega) + \tilde{T}(\mathrm{j}\omega; \boldsymbol{\delta}))\left(1 - \left(T(\mathrm{j}\omega) + \tilde{T}(\mathrm{j}\omega; \boldsymbol{\delta})\right)\delta_{\mathrm{ss}}\right) = 0\Big\} > 1\,. \end{aligned} \tag{4.154}$$

We will use this as robust stability condition, which must hold for all $\omega > 0$. This particular form will be useful to apply the tools of μ-calculus, after a few more steps of organizing the formulas to a desired structure.

Since the auxiliary variable δ_{ss} plays a similar role in (4.154) as $\boldsymbol{\delta}$, let us collect them in a matrix

$$\hat{\boldsymbol{\delta}} = \mathrm{diag}\,(\boldsymbol{\delta}, \delta_{\mathrm{ss}}) = \begin{bmatrix} \boldsymbol{\delta} & \mathbf{0} \\ \mathbf{0} & \delta_{\mathrm{ss}} \end{bmatrix}, \tag{4.155}$$

with domain $\hat{\boldsymbol{\delta}} \in \hat{\mathbf{D}} = \mathbf{D} \times \mathbb{C}$. Then, notice that the singular values of $\hat{\boldsymbol{\delta}}$ are $\sigma(\hat{\boldsymbol{\delta}}) = \{|\delta_\kappa|, |\delta_\alpha|, |\delta_\beta|, |\delta_{\mathrm{ss}}|\}$, while the largest singular value $\bar{\sigma}(\hat{\boldsymbol{\delta}})$ is

$$\bar{\sigma}(\hat{\boldsymbol{\delta}}) = \max\{\|\boldsymbol{\delta}\|_\infty, |\delta_{\mathrm{ss}}|\}\,, \tag{4.156}$$

i.e., the expression appearing in (4.154).

Therefore, the robust stability condition (4.154) can also be stated as

$$\min_{\hat{\boldsymbol{\delta}}\in\hat{\mathbf{D}}} \left\{\bar{\sigma}(\hat{\boldsymbol{\delta}}) : \mathrm{Den}\left(T(\mathrm{j}\omega) + \tilde{T}(\mathrm{j}\omega; \boldsymbol{\delta})\right)\left(1 - \left(T(\mathrm{j}\omega) + \tilde{T}(\mathrm{j}\omega; \boldsymbol{\delta})\right)\delta_{\mathrm{ss}}\right) = 0\right\} > 1\,. \tag{4.157}$$

Finally, we can express the equality in this expression by the help of $\hat{\boldsymbol{\delta}}$ in the form

$$\text{Den}\left(T(\text{j}\omega)+\tilde{T}(\text{j}\omega;\boldsymbol{\delta})\right)\left(1-\left(T(\text{j}\omega)+\tilde{T}(\text{j}\omega;\boldsymbol{\delta})\right)\delta_{\text{ss}}\right)=\det\left(\mathbf{I}-\mathbf{m}(\text{j}\omega)\hat{\boldsymbol{\delta}}\right), \tag{4.158}$$

with an appropriate matrix $\mathbf{m}(\text{j}\omega)$ that we derive below.

Thus, the final, compact form of the robust stability condition becomes

$$\mu(\omega)<1\,,\quad \forall\omega>0\,, \tag{4.159}$$

with μ defined by

$$\mu(\omega)=\frac{1}{\min\limits_{\hat{\boldsymbol{\delta}}\in\hat{\mathbf{D}}}\left\{\bar{\sigma}(\hat{\boldsymbol{\delta}}):\det\left(\mathbf{I}-\mathbf{m}(\text{j}\omega)\hat{\boldsymbol{\delta}}\right)=0\right\}}\,, \tag{4.160}$$

see [79]. This expression is widely used in robust control, and μ is called the structured singular value [46, 264]. The robust stability condition (4.159) resembles the string stability condition $|T(\text{j}\omega)|<1, \forall\omega>0$. In fact, the graph of $\mu(\omega)$ and $|T(\text{j}\omega)|$ are identical for zero parameter uncertainty. Furthermore, there exist computational methods [51, 161, 247] and software packages [13] to estimate the value of μ via upper and lower bounds without having to analyze each possible human driver parameter combination in a "brute force" manner.

To obtain $\mu(\omega)$, the expression of $\mathbf{m}(\text{j}\omega)$ must be derived such that (4.158) holds. In the derivations below we drop the arguments $(\text{j}\omega)$ and $(\text{j}\omega;\boldsymbol{\delta})$ for compactness. We consider the four blocks of $\hat{\boldsymbol{\delta}}$ in (4.155) and separate $\mathbf{m}$ into four corresponding blocks:

$$\mathbf{m}=\begin{bmatrix}\mathbf{m}^{yu} & \mathbf{m}^{yw}\\ \mathbf{m}^{zu} & m^{zw}\end{bmatrix}, \tag{4.161}$$

where the superscripts will get their meaning below. Note that here $\mathbf{m}^{yu}$ represents a 3×3 matrix, m^{zw} is a scalar, while $\mathbf{m}^{zu}$ are $\mathbf{m}^{yw}$ are 1×3 and 3×1 matrices, respectively. We expand the determinant in (4.158) as

$$\begin{aligned}\det\left(\mathbf{I}-\mathbf{m}\,\hat{\boldsymbol{\delta}}\right)&=\det\begin{bmatrix}\mathbf{I}-\mathbf{m}^{yu}\boldsymbol{\delta} & -\mathbf{m}^{yw}\delta_{\text{ss}}\\ -\mathbf{m}^{zu}\boldsymbol{\delta} & 1-m^{zw}\delta_{\text{ss}}\end{bmatrix}\\ &=\det\left(\mathbf{I}-\mathbf{m}^{yu}\boldsymbol{\delta}\right)\left(1-\left(m^{zw}+\mathbf{m}^{zu}\boldsymbol{\delta}\left(\mathbf{I}-\mathbf{m}^{yu}\boldsymbol{\delta}\right)^{-1}\mathbf{m}^{yw}\right)\delta_{\text{ss}}\right),\end{aligned} \tag{4.162}$$

where we used the identity $\det\begin{bmatrix}\mathbf{a} & \mathbf{b}\\ \mathbf{c} & \mathbf{d}\end{bmatrix}=\det(\mathbf{a})\det\left(\mathbf{d}-\mathbf{c}\,\mathbf{a}^{-1}\mathbf{b}\right)$. This gives the desired equality (4.158) once we write $T+\tilde{T}$ in the form

$$T+\tilde{T}=m^{zw}+\mathbf{m}^{zu}\boldsymbol{\delta}\left(\mathbf{I}-\mathbf{m}^{yu}\boldsymbol{\delta}\right)^{-1}\mathbf{m}^{yw}\,. \tag{4.163}$$

This form has a special structure which can be organized as follows. Notice that $T+\tilde{T}$ establishes the relationship between an input w and an output z as

$$z = \left(T + \tilde{T}\right) w = m^{zw} w + \mathbf{m}^{zu} \boldsymbol{\delta} \left(\mathbf{I} - \mathbf{m}^{yu} \boldsymbol{\delta}\right)^{-1} \mathbf{m}^{yw} w \,, \tag{4.164}$$

where in our case $z = V_i$ and $w = V_{i+1}$ are the velocity perturbations of the human driver and its predecessor in Laplace domain. Note that $m^{zw} = T$, which can be obtained by taking $\boldsymbol{\delta} = \mathbf{0}$.

By introducing auxiliary variables $\mathbf{y}$ and $\mathbf{u}$ as

$$\mathbf{y} = \left(\mathbf{I} - \mathbf{m}^{yu} \boldsymbol{\delta}\right)^{-1} \mathbf{m}^{yw} w \,, \tag{4.165}$$

and

$$\mathbf{u} = \boldsymbol{\delta}\, \mathbf{y} \,, \tag{4.166}$$

we can write

$$\begin{bmatrix} \mathbf{y} \\ z \end{bmatrix} = \begin{bmatrix} \mathbf{m}^{yu} & \mathbf{m}^{yw} \\ \mathbf{m}^{zu} & m^{zw} \end{bmatrix} \begin{bmatrix} \mathbf{u} \\ w \end{bmatrix}, \tag{4.167}$$

or shortly

$$\begin{bmatrix} \mathbf{y} \\ z \end{bmatrix} = \mathbf{m} \begin{bmatrix} \mathbf{u} \\ w \end{bmatrix}, \tag{4.168}$$

where $\mathbf{m}$ is defined in (4.161). A graphical representation of this $\mathbf{m}$-$\boldsymbol{\delta}$ framework is shown in Fig. 4.34a using block diagrams. The black part of the diagram refers to the system with nominal parameters while the blue part represents the effects of parameter uncertainties.

Thus, what remains is to find the right auxiliary variables $\mathbf{y}$ and formulate the input-output relationship in the form (4.168), that yield matrix $\mathbf{m}$ required for μ in (4.160). The particular choice of $\mathbf{y}$ and expression of $\mathbf{m}$ for the OVM are as follows. We write the linearized dynamics of each human-driven vehicle (cf. (4.63), (4.65)) in Laplace domain (while considering zero delay) as

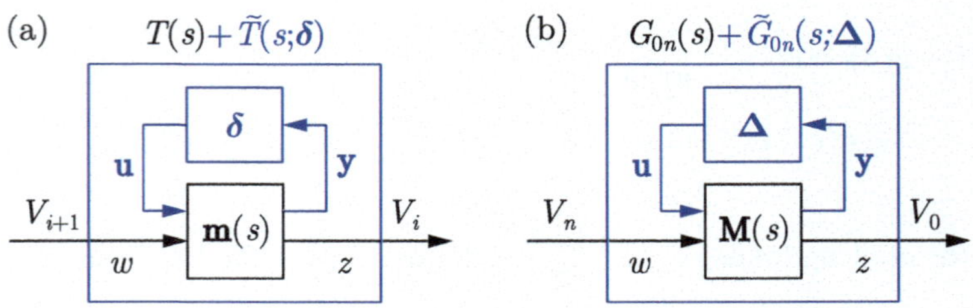

Fig. 4.34 **a** Block diagram representing the $\mathbf{m}$-$\boldsymbol{\delta}$ framework (4.166), (4.168). **b** Block diagram representing the $\mathbf{M}$-$\boldsymbol{\Delta}$ framework (4.198). The black parts of the diagrams refer to systems with nominal parameters while the blue parts represent the effects of parameter uncertainties

$$
\begin{aligned}
sH_i &= V_{i+1} - V_i \;, \\
sV_i &= U_i \;, \\
U_i &= (\alpha_\mathrm{h} + \rho_\alpha\delta_\alpha)\left((\kappa_\mathrm{h} + \rho_\kappa\delta_\kappa)H_i - V_i\right) + (\beta_\mathrm{h} + \rho_\beta\delta_\beta)(V_{i+1} - V_i) \;,
\end{aligned} \tag{4.169}
$$

and we choose **y** as the coefficients after δ_κ, δ_α and δ_β [79], to obtain

$$
\begin{aligned}
y_1 &= H_i \;, \\
y_2 &= (\kappa_\mathrm{h} + \rho_\kappa\delta_\kappa)H_i - V_i \;, \\
y_3 &= V_{i+1} - V_i \;, \\
z &= V_i \;.
\end{aligned} \tag{4.170}
$$

Hence the input variables become

$$
\begin{aligned}
u_1 &= \delta_\kappa y_1 \;, \\
u_2 &= \delta_\alpha y_2 \;, \\
u_3 &= \delta_\beta y_3 \;, \\
w &= V_{i+1} \;.
\end{aligned} \tag{4.171}
$$

Equations (4.169)–(4.171) imply

$$
\begin{aligned}
y_1 &= \frac{1}{s}(w - z) \;, \\
y_2 &= \frac{\kappa_\mathrm{h}}{s}(w - z) + \rho_\kappa u_1 - z \;, \\
y_3 &= w - z \;, \\
z &= \frac{1}{s}\left(\alpha_\mathrm{h} y_2 + \rho_\alpha u_2 + \beta_\mathrm{h} y_3 + \rho_\beta u_3\right) ,
\end{aligned} \tag{4.172}
$$

that has the form

$$
\begin{aligned}
\mathbf{y} &= \mathbf{m}_1 \begin{bmatrix} \mathbf{u} \\ w \end{bmatrix} + \mathbf{m}_2 z \;, \\
z &= \mathbf{m}_3 \begin{bmatrix} \mathbf{u} \\ w \end{bmatrix} + \mathbf{m}_4 \mathbf{y} \;,
\end{aligned} \tag{4.173}
$$

with

$$
\begin{aligned}
\mathbf{m}_1 &= \begin{bmatrix} 0 & 0 & 0 & \frac{1}{s} \\ \rho_\kappa & 0 & 0 & \frac{\kappa_\mathrm{h}}{s} \\ 0 & 0 & 0 & 1 \end{bmatrix} , \quad \mathbf{m}_2 = \begin{bmatrix} -\frac{1}{s} \\ -\frac{\kappa_\mathrm{h}}{s} - 1 \\ -1 \end{bmatrix} , \\
\mathbf{m}_3 &= \begin{bmatrix} 0 & \frac{\rho_\alpha}{s} & \frac{\rho_\beta}{s} & 0 \end{bmatrix} , \quad \mathbf{m}_4 = \begin{bmatrix} 0 & \frac{\alpha_\mathrm{h}}{s} & \frac{\beta_\mathrm{h}}{s} \end{bmatrix} .
\end{aligned} \tag{4.174}
$$

This can be reorganized to (4.168) using

$$\mathbf{m} = \begin{bmatrix} \mathbf{m}_1 + \mathbf{m}_2 \left(1 - \mathbf{m}_4\mathbf{m}_2\right)^{-1} \left(\mathbf{m}_3 + \mathbf{m}_4\mathbf{m}_1\right) \\ \left(1 - \mathbf{m}_4\mathbf{m}_2\right)^{-1} \left(\mathbf{m}_3 + \mathbf{m}_4\mathbf{m}_1\right) \end{bmatrix}. \tag{4.175}$$

This leads to the expression of matrix $\mathbf{m}$ in the form

$$\mathbf{m}(s) = \frac{1}{s^2+(\alpha_\mathrm{h}+\beta_\mathrm{h})s+\alpha_\mathrm{h}\kappa_\mathrm{h}} \begin{bmatrix} -\alpha_\mathrm{h} & -1 & -1 & s+\alpha_\mathrm{h} \\ s^2+\beta_\mathrm{h}s & -(\kappa_\mathrm{h}+s) & -(\kappa_\mathrm{h}+s) & s(\kappa_\mathrm{h}-\beta_\mathrm{h}) \\ -\alpha_\mathrm{h}s & -s & -s & s(s+\alpha_\mathrm{h}) \\ \alpha_\mathrm{h}s & s & s & \beta_\mathrm{h}s+\alpha_\mathrm{h}\kappa_\mathrm{h} \end{bmatrix} \begin{bmatrix} \rho_\kappa & 0 & 0 & 0 \\ 0 & \rho_\alpha & 0 & 0 \\ 0 & 0 & \rho_\beta & 0 \\ 0 & 0 & 0 & 1 \end{bmatrix}. \tag{4.176}$$

Note that $\mathbf{m}$ is a function of s, and it is evaluated in (4.160) only for $s = \mathrm{j}\omega$. Furthermore, we have the nominal transfer function $T(s) = m^{zw}(s)$ as the bottom right element of $\mathbf{m}(s)$ while the rest of the matrix can be used to calculate the perturbation $\tilde{T}(s) = \mathbf{m}^{zu}(s)\boldsymbol{\delta}\left(\mathbf{I} - \mathbf{m}^{yu}(s)\boldsymbol{\delta}\right)^{-1}\mathbf{m}^{yw}(s)$; cf. (4.163). Finally, the uncertainty sizes $\rho_\kappa, \rho_\alpha, \rho_\beta$ are collected in a separate matrix to increase the readability.

4.7.2 Robustness with Time Delay

Now we address uncertain delays based on the same framework presented above. In the presence of delays, the nominal transfer function becomes

$$T(s) = \frac{\beta_\mathrm{h}s + \alpha_\mathrm{h}\kappa_\mathrm{h}}{s^2\mathrm{e}^{s\tau} + (\alpha_\mathrm{h} + \beta_\mathrm{h})s + \alpha_\mathrm{h}\kappa_\mathrm{h}}, \tag{4.177}$$

whereas the perturbed transfer function is of the form

$$T(s) + \tilde{T}(s;\boldsymbol{\delta}) = \frac{(\beta_\mathrm{h} + \rho_\beta\delta_\beta)s + (\alpha_\mathrm{h} + \rho_\alpha\delta_\alpha)(\kappa_\mathrm{h} + \rho_\kappa\delta_\kappa)}{s^2\mathrm{e}^{s(\tau+\rho_\tau\delta_\tau)} + (\alpha_\mathrm{h}+\rho_\alpha\delta_\alpha + \beta_\mathrm{h}+\rho_\beta\delta_\beta)s + (\alpha_\mathrm{h}+\rho_\alpha\delta_\alpha)(\kappa_\mathrm{h}+\rho_\kappa\delta_\kappa)}. \tag{4.178}$$

Notice that this transfer function contains δ_τ in an exponential. This prevents the direct application of the $\mathbf{m}$-$\boldsymbol{\delta}$ framework where expressions are linear in terms of the parameters in $\boldsymbol{\delta}$.

The following technique allows us to overcome the nonlinearity introduced by the exponential. Instead of scanning all possible delays $\tau + \rho_\tau\delta_\tau$ with parameter $\delta_\tau \in [-1, 1]$, we introduce a complex parameter $\delta_\theta \in \mathbb{C}$, $|\delta_\theta| \leq 1$ and scan all possible values of $\mathrm{e}^{s(\tau+\rho_\tau\delta_\tau)}$. For δ_θ there is a corresponding function $\rho_\theta(s)$ that measures the magnitude of the uncertainty in $\mathrm{e}^{s(\tau+\rho_\tau\delta_\tau)}$. The collection of all uncertainties become

$$\boldsymbol{\delta} = \begin{bmatrix} \delta_\kappa & 0 & 0 & 0 \\ 0 & \delta_\alpha & 0 & 0 \\ 0 & 0 & \delta_\beta & 0 \\ 0 & 0 & 0 & \delta_\theta \end{bmatrix}. \tag{4.179}$$

Specifically, we define δ_θ and $\rho_\theta(s)$ via the Rekasius substitution [149] such that

$$\rho_\theta(s)\delta_\theta = \frac{1}{s}\frac{\mathrm{e}^{s\rho_\tau\delta_\tau}-1}{\mathrm{e}^{s\rho_\tau\delta_\tau}+1} = -\frac{\mathrm{j}}{s}\tan\left(\mathrm{j}\frac{s\rho_\tau\delta_\tau}{2}\right). \tag{4.180}$$

That is, instead of $\mathrm{e}^{s\rho_\tau\delta_\tau}$, we use a transformed expression thereof for algebraic convenience. This transformation allows us to handle the uncertain delay analogously to the delay-free case.

Accordingly, we define $\rho_\theta(s)$ as

$$\rho_\theta(s) = \frac{1}{s}\frac{\mathrm{e}^{s\rho_\tau}-1}{\mathrm{e}^{s\rho_\tau}+1} = -\frac{\mathrm{j}}{s}\tan\left(\mathrm{j}\frac{s\rho_\tau}{2}\right). \tag{4.181}$$

Dividing (4.180) with (4.181) relates δ_θ to δ_τ via

$$\delta_\theta = \frac{\tan\left(\mathrm{j}\frac{s\rho_\tau\delta_\tau}{2}\right)}{\tan\left(\mathrm{j}\frac{s\rho_\tau}{2}\right)}. \tag{4.182}$$

Note that we do not use parameter δ_τ in the **m-$\boldsymbol{\delta}$** framework, instead we vary δ_θ directly, without relying on (4.182). However, we must make sure that $\delta_\tau \in [-1, 1] \iff |\delta_\theta| \leq 1$ holds for $s = \mathrm{j}\omega$. We use (4.182) to verify this property. For $s = \mathrm{j}\omega$, δ_θ becomes real

$$\delta_\theta = \frac{\tan\left(\frac{\omega\rho_\tau\delta_\tau}{2}\right)}{\tan\left(\frac{\omega\rho_\tau}{2}\right)}. \tag{4.183}$$

Considering $\delta_\tau \in [-1, 1]$ we restrict the possible values of ω as $0 \leq \omega < \pi/\rho_\tau$. This ensures that the tangent function $\tan\left(\frac{\omega\rho_\tau\delta_\tau}{2}\right)$ monotonously increases with respect to $\delta_\tau \in [-1, 1]$, and hence we achieve $\delta_\theta \in [-1, 1]$. Therefore, when utilizing (4.181), robust stability analysis will hold only for frequencies $0 \leq \omega < \pi/\rho_\tau$. For zero delay uncertainty $\rho_\tau = 0$, this recovers the entire frequency range $\omega \geq 0$.

With these preliminaries, let us express $\mathrm{e}^{s\rho_\tau\delta_\tau}$ with $\rho_\theta(s)\delta_\theta$ by inverting (4.180):

$$\mathrm{e}^{s\rho_\tau\delta_\tau} = \frac{1+s\rho_\theta(s)\delta_\theta}{1-s\rho_\theta(s)\delta_\theta}. \tag{4.184}$$

Since δ_θ is not in an exponential and this equation can be rearranged to linear expressions of δ_θ, we can apply the **m-$\boldsymbol{\delta}$** calculation.

For our specific example with the optimal velocity model (4.63), (4.65), we can write the linearized dynamics with uncertain delay in Laplace domain as

$$
\begin{aligned}
sH_i &= V_{i+1} - V_i \,, \\
sV_i \mathrm{e}^{s(\tau+\rho_\tau\delta_\tau)} &= U_i \,, \\
U_i &= (\alpha_\mathrm{h} + \rho_\alpha\delta_\alpha)\big((\kappa_\mathrm{h} + \rho_\kappa\delta_\kappa)H_i - V_i\big) + (\beta_\mathrm{h} + \rho_\beta\delta_\beta)(V_{i+1} - V_i) \,,
\end{aligned}
\tag{4.185}
$$

cf. (4.169). That is, with the substitution of (4.184) we get

$$
\frac{sV_i}{1 - s\rho_\theta(s)\delta_\theta}\mathrm{e}^{s\tau} + \rho_\theta(s)\frac{s^2 V_i}{1 - s\rho_\theta(s)\delta_\theta}\delta_\theta \mathrm{e}^{s\tau} = U_i \,. \tag{4.186}
$$

Now, similarly to the delay-free scenario, we define the variables y_1, y_2, y_3, z and u_1, u_2, u_3, w as in (4.170) and (4.171). To address the uncertain delay, we pick the coefficient of δ_θ as variable y_4, define the corresponding input u_4, and express (4.186) with $z = V_i$ as

$$
\begin{aligned}
y_4 &= \frac{s^2 z}{1 - s\rho_\theta(s)\delta_\theta} \,, \\
u_4 &= \delta_\theta y_4 \,, \\
\big(sz &+ 2\rho_\theta(s)u_4\big)\mathrm{e}^{s\tau} = U_i \,.
\end{aligned}
\tag{4.187}
$$

Then, we can express y_1, y_2, y_3, y_4, z as

$$
\begin{aligned}
y_1 &= \frac{1}{s}(w - z) \,, \\
y_2 &= \frac{\kappa_\mathrm{h}}{s}(w - z) + \rho_\kappa u_1 - z \,, \\
y_3 &= w - z \,, \\
y_4 &= s\big(sz + \rho_\theta(s)u_4\big) \,, \\
z &= \frac{1}{s}\Big(\big(\alpha_\mathrm{h} y_2 + \rho_\alpha u_2 + \beta_\mathrm{h} y_3 + \rho_\beta u_3\big)\,\mathrm{e}^{-s\tau} - 2\rho_\theta(s)u_4\Big),
\end{aligned}
\tag{4.188}
$$

cf. (4.172). This leads to the form (4.173) with coefficient matrices

$$
\begin{aligned}
\mathbf{m}_1 &= \begin{bmatrix} 0 & 0 & 0 & 0 & \frac{1}{s} \\ \rho_\kappa & 0 & 0 & 0 & \frac{\kappa_\mathrm{h}}{s} \\ 0 & 0 & 0 & 0 & 1 \\ 0 & 0 & 0 & \rho_\theta(s)s & 0 \end{bmatrix}, \quad
\mathbf{m}_2 = \begin{bmatrix} -\frac{1}{s} \\ -\frac{\kappa_\mathrm{h}}{s} - 1 \\ -1 \\ s^2 \end{bmatrix}, \\
\mathbf{m}_3 &= \begin{bmatrix} 0 & \frac{\rho_\alpha}{s}\mathrm{e}^{-s\tau} & \frac{\rho_\beta}{s}\mathrm{e}^{-s\tau} & -\frac{2\rho_\theta(s)}{s} & 0 \end{bmatrix}, \quad
\mathbf{m}_4 = \begin{bmatrix} 0 & \frac{\alpha_\mathrm{h}}{s}\mathrm{e}^{-s\tau} & \frac{\beta_\mathrm{h}}{s}\mathrm{e}^{-s\tau} & 0 \end{bmatrix},
\end{aligned}
\tag{4.189}
$$

cf. (4.174). Based on (4.175), we finally obtain matrix $\mathbf{m}$ that appears in the expression (4.160) of μ:

$$\mathbf{m}(s) = \frac{1}{D(s)} \begin{bmatrix} -\alpha_h & -1 & -1 & 2e^{s\tau} & se^{s\tau} + \alpha_h \\ s^2e^{s\tau} + \beta_h s & -(\kappa_h + s) & -(\kappa_h + s) & 2(\kappa_h + s)e^{s\tau} & s(\kappa_h e^{s\tau} - \beta_h) \\ -\alpha_h s & -s & -s & 2se^{s\tau} & s(se^{s\tau} + \alpha_h) \\ \alpha_h s^3 & s^3 & s^3 & sq(s) & s^2(\beta_h s + \alpha_h \kappa_h) \\ \alpha_h s & s & s & -2se^{s\tau} & \beta_h s + \alpha_h \kappa_h \end{bmatrix} \times \begin{bmatrix} \rho_\kappa & 0 & 0 & 0 & 0 \\ 0 & \rho_\alpha & 0 & 0 & 0 \\ 0 & 0 & \rho_\beta & 0 & 0 \\ 0 & 0 & 0 & \rho_\theta(s) & 0 \\ 0 & 0 & 0 & 0 & 1 \end{bmatrix}, \tag{4.190}$$

cf. (4.176), where

$$q(s) = -s^2e^{s\tau} + (\alpha_h + \beta_h)s + \alpha_h\kappa_h \,, \tag{4.191}$$

and

$$D(s) = s^2e^{s\tau} + (\alpha_h + \beta_h)s + \alpha_h\kappa_h \,. \tag{4.192}$$

4.7.3 Robustness of Head-to-Tail Dynamics

Finally, we generalize the established framework to involve multiple vehicles and achieve robust connected cruise control (CCC). Consider the scenario in Sect. 4.3.2 where $n+1$ vehicles, indexed $i = 0, 1, \ldots, n$, follow each other in an open chain configuration; cf. Fig. 4.12. Vehicle 0 at the tail is assumed to be a CAV with known parameters without uncertainty. Vehicles $i = 1, \ldots, n-1$ are driven by humans (they are HVs or CHVs) and they have uncertain parameters while vehicle n is the lead vehicle whose motion acts as input for the vehicle chain. The human drivers $i = 1, \ldots, n-1$ are associated with parameter uncertainties given by the matrices $\boldsymbol{\delta}_{i,i+1}$, which can be organized together to quantify the uncertainty of the whole vehicle chain as

$$\boldsymbol{\Delta} = \begin{bmatrix} \boldsymbol{\delta}_{12} & \mathbf{0} & \cdots & \mathbf{0} \\ \mathbf{0} & \boldsymbol{\delta}_{23} & \cdots & \mathbf{0} \\ & & \ddots & \\ \mathbf{0} & \mathbf{0} & \cdots & \boldsymbol{\delta}_{n-1,n} \end{bmatrix}. \tag{4.193}$$

The parameter uncertainty affects the transfer functions $T_{i,i+1}(s) + \tilde{T}_{i,i+1}(s; \boldsymbol{\delta}_{i,i+1})$ of each human driver, and the weighted adjacency matrix $\mathbf{T}(s) + \tilde{\mathbf{T}}(s; \boldsymbol{\Delta})$ of the vehicle chain. The uncertainty is incorporated in the head-to-tail transfer function as

$$G_{0n}(s) + \tilde{G}_{0n}(s; \boldsymbol{\Delta}) = \det\left(\hat{\mathbf{T}}(s) + \tilde{\hat{\mathbf{T}}}(s; \boldsymbol{\Delta}) - \hat{\mathbf{I}}\right), \tag{4.194}$$

cf. Theorem 4.3. Ultimately, we intend to achieve plant and head-to-tail string stability robustly, for all possible values of the parameters ($\boldsymbol{\Delta} \in \mathbf{D}$, $\|\boldsymbol{\Delta}\|_\infty \leq 1$).

We conduct robust stability analysis based on μ-calculus, by analyzing the condition

$$\mu_{0n}(\omega) < 1\,, \quad \forall\omega > 0\,, \tag{4.195}$$

with head-to-tail μ function

$$\mu_{0n}(\omega) = \frac{1}{\min\limits_{\hat{\mathbf{\Delta}}\in\hat{\mathbf{D}}} \left\{\bar{\sigma}(\hat{\mathbf{\Delta}}) : \det\left(\mathbf{I} - \mathbf{M}(\mathrm{j}\omega)\hat{\mathbf{\Delta}}\right) = 0\right\}}, \tag{4.196}$$

where $\hat{\mathbf{\Delta}}$ contains the parameter uncertainties $\mathbf{\Delta}$ and the auxiliary parameter δ_{ss}

$$\hat{\mathbf{\Delta}} = \begin{bmatrix} \mathbf{\Delta} & \mathbf{0} \\ \mathbf{0} & \delta_{\mathrm{ss}} \end{bmatrix}, \tag{4.197}$$

while $\mathbf{M}$ characterizes the dynamics of the vehicle chain given in the form

$$\begin{aligned} \begin{bmatrix} \mathbf{y} \\ z \end{bmatrix} &= \mathbf{M}(s) \begin{bmatrix} \mathbf{u} \\ w \end{bmatrix}, \\ \mathbf{u} &= \mathbf{\Delta}\mathbf{y}\,, \end{aligned} \tag{4.198}$$

see [79]. Here, $\mathbf{y}$ and $\mathbf{u}$ collect output and input variables associated with each uncertain human driver:

$$\mathbf{y} = \begin{bmatrix} \mathbf{y}_1 \\ \mathbf{y}_2 \\ \vdots \\ \mathbf{y}_{n-1} \end{bmatrix}, \quad \mathbf{u} = \begin{bmatrix} \mathbf{u}_1 \\ \mathbf{u}_2 \\ \vdots \\ \mathbf{u}_{n-1} \end{bmatrix}. \tag{4.199}$$

The graphical representation of this **M**-**Δ** framework is shown by the block diagram in Fig. 4.34b. Again, the black part denotes the system with nominal parameters while the blue part represents the effects of parameter uncertainties.

To obtain $\mu_{0n}(\omega)$, we must formulate $\mathbf{M}(\mathrm{j}\omega)$. We achieve this based on the dynamics of each uncertain vehicle $i = 1, 2, \ldots, n-1$ given by $\mathbf{m}_{i,i+1}(s)$ and

$$\begin{aligned} \begin{bmatrix} \mathbf{y}_i \\ z_i \end{bmatrix} &= \mathbf{m}_{i,i+1}(s) \begin{bmatrix} \mathbf{u}_i \\ w_i \end{bmatrix}, \\ \mathbf{u}_i &= \boldsymbol{\delta}_{i,i+1}\mathbf{y}_i\,. \end{aligned} \tag{4.200}$$

Furthermore, we utilize that the input w_i of each vehicle is the output z_{i+1} of the preceding vehicle, and the lead vehicle's motion z_n is the input w to the whole system

$$\begin{aligned} w_i &= z_{i+1}\,, \\ w &= z_n\,. \end{aligned} \tag{4.201}$$

Finally, the overall output z of the vehicle chain is given by the response of the CAV, which is associated with link transfer functions $T_{0i}(s)$ according to

$$z = \sum_{i=1}^{n} T_{0i}(s) z_i \; . \tag{4.202}$$

Whenever the CAV is not responding to a preceding vehicle, we take $T_{0i}(s) = 0$. Equations (4.200)–(4.202) can be reorganized to the form (4.198) to finally obtain $\mathbf{M}(s)$. We show this calculation for two examples.

Consider the scenario in Fig. 4.7 with three vehicles: a leading CHV, a following HV with uncertain parameters, and a CAV responding to both the CHV and the HV via CCC. This scenario is associated with the nominal head-to-tail transfer function

$$G_{02}(s) = T_{01}(s) T_{12}(s) + T_{02}(s) \; , \tag{4.203}$$

and its uncertain counterpart

$$G_{02}(s) + \tilde{G}_{02}(s; \boldsymbol{\Delta}) = T_{01}(s)\big(T_{12}(s) + \tilde{T}_{12}(s; \boldsymbol{\delta}_{12})\big) + T_{02}(s) \; . \tag{4.204}$$

The equations describing the dynamics of the system are

$$\begin{aligned} \mathbf{y}_1 &= \mathbf{m}_{12}^{yu}(s)\mathbf{u}_1 + \mathbf{m}_{12}^{yw}(s) w_1 \; , \\ z_1 &= \mathbf{m}_{12}^{zu}(s)\mathbf{u}_1 + m_{12}^{zw}(s) w_1 \; , \\ z &= T_{01}(s) z_1 + T_{02}(s) z_2 \; , \end{aligned} \tag{4.205}$$

where $w_1 = z_2 = w$ and $m_{12}^{zw}(s) = T_{12}(s)$. By eliminating z_1 we can reach the form (4.198) with

$$\mathbf{M}(s) = \begin{bmatrix} \mathbf{m}_{12}^{yu}(s) & \mathbf{m}_{12}^{yw}(s) \\ \mathbf{m}_{12}^{zu}(s) T_{01}(s) & G_{02}(s) \end{bmatrix} . \tag{4.206}$$

Figure 4.35 represents how, in case of the scenario in Fig. 4.7, the $\mathbf{m}$-$\boldsymbol{\delta}$ framework (4.200) developed to describe the uncertain human-driven vehicle is embedded

Fig. 4.35 Block diagram of the system in Fig. 4.7 under human uncertainties

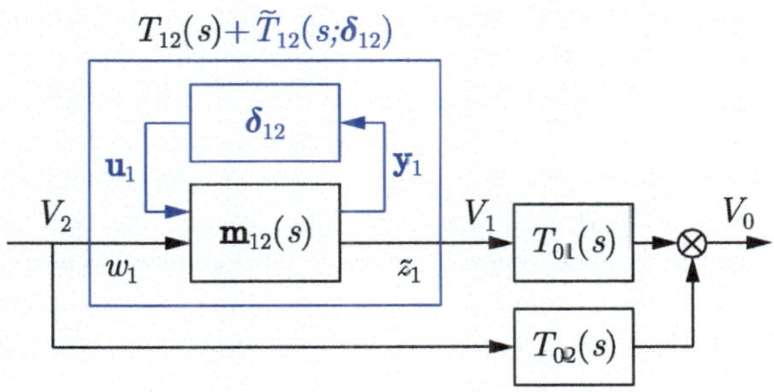

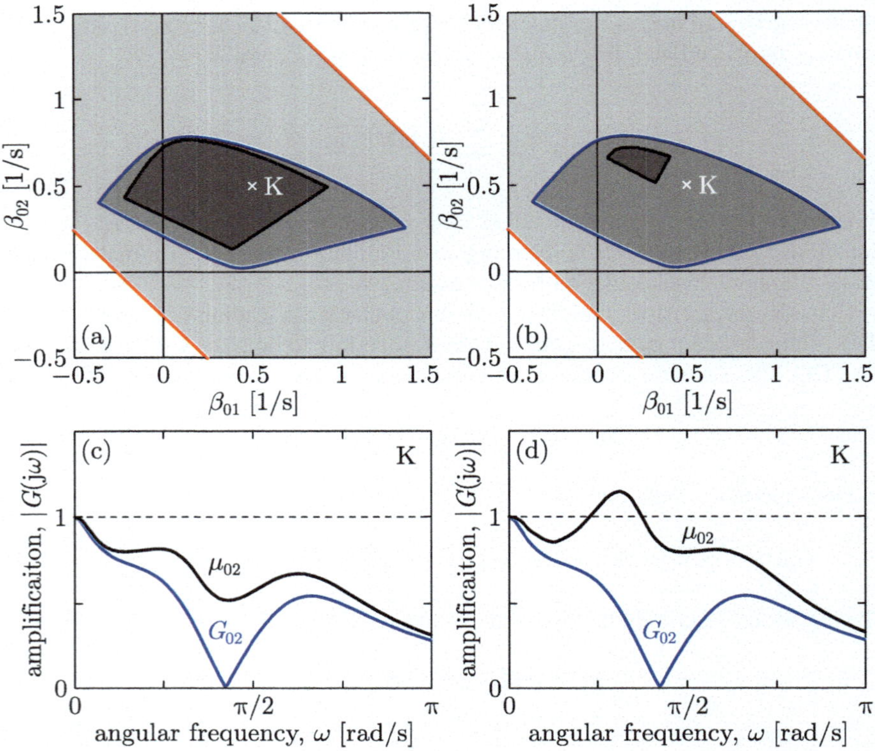

Fig. 4.36 Robust stability charts (top) and frequency response plots (bottom). The nominal parameters are the same as those in Fig. 4.9b, while the robust plots are the same as in Fig. 4.33

in the **M**-**Δ** framework (4.198) established for the whole system; cf. Fig. 4.34b. The structure of the nominal, black part resembles the connectivity in Fig. 4.7, while the blue part represents the uncertainty of the human driver. The resulting matrix **M** in (4.206) can be used to evaluate robust stability via (4.195), (4.196).

The corresponding robust stability charts are shown in Fig. 4.36 for 10% human parameter uncertainty (panel (a)) and 25% uncertainty (panel (b)), where black color indicates the robustly head-to-tail string stable domain over which (4.195) holds. The corresponding frequency response plots are shown in panels (c) and (d) for point K. The $\mu_{02}(\omega)$ curve representing the frequency response plot for robust head-to-tail string stability upper bounds the nominal frequency curve given by $G_{02}(j\omega)$. These μ values were estimated based on **M** in (4.206) using the `mussv` function of Matlab [13], and the multi-dimensional bisection method [11] was also used to speed up the computation of the robust stability boundaries.

The robust stability charts are depicted for different levels of uncertainty in Fig. 4.37. Notice that the black robustly head-to-tail string stable domain shrinks with increasing levels of uncertainty, and that the most robust (β_{01}, β_{02}) parameter

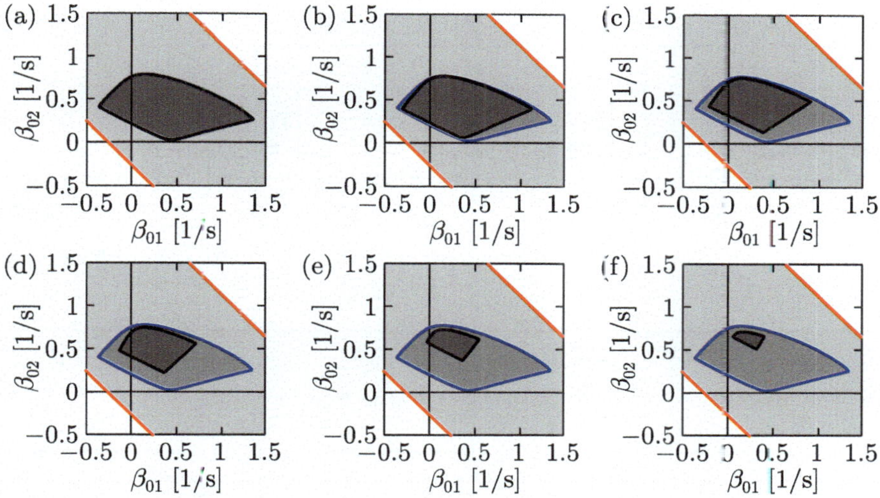

Fig. 4.37 Robust stability charts under different levels of uncertainty of human driver parameters $\rho_\kappa/\kappa_h = \rho_\alpha/\alpha_h = \rho_\beta/\beta_h = \rho_\tau/\tau =: p$, where **a** $p = 0\%$, **b** $p = 5\%$, **c** $p = 10\%$, **d** $p = 15\%$, **e** $p = 20\%$ and **f** $p = 25\%$

combination (where the robust domain shrinks to a point) is located at the top part of the dark gray nominal head-to-tail string stable domain.

Now consider the scenario in Fig. 4.11b with four vehicles: a leading CHV, two following CHVs with uncertain parameters, and a CAV responding to the three CHVs via CCC. This scenario is characterized by the nominal head-to-tail transfer function

$$G_{03}(s) = T_{01}(s)T_{12}(s)T_{23}(s) + T_{02}(s)T_{23}(s) + T_{03}(s) , \tag{4.207}$$

and the uncertain counterpart

$$\begin{aligned} G_{03}(s) + \tilde{G}_{03}(s; \boldsymbol{\Delta}) = {} & T_{01}(s)\Big(T_{12}(s) + \tilde{T}_{12}(s; \boldsymbol{\delta}_{12}))\Big)\Big(T_{23}(s) + \tilde{T}_{23}(s; \boldsymbol{\delta}_{23}))\Big) \\ & + T_{02}(s)\Big(T_{23}(s) + \tilde{T}_{23}(s; \boldsymbol{\delta}_{23}))\Big) + T_{03}(s). \end{aligned} \tag{4.208}$$

The structured equations of the dynamics are given by

$$\begin{aligned} \mathbf{y}_1 &= \mathbf{m}_{12}^{yu}(s)\mathbf{u}_1 + \mathbf{m}_{12}^{yw}(s)w_1 , \\ z_1 &= \mathbf{m}_{12}^{zu}(s)\mathbf{u}_1 + m_{12}^{zw}(s)w_1 , \\ \mathbf{y}_2 &= \mathbf{m}_{23}^{yu}(s)\mathbf{u}_2 + \mathbf{m}_{23}^{yw}(s)w_2 , \\ z_2 &= \mathbf{m}_{23}^{zu}(s)\mathbf{u}_2 + m_{23}^{zw}(s)w_2 , \\ z &= T_{01}(s)z_1 + T_{02}(s)z_2 + T_{03}(s)z_3 , \end{aligned} \tag{4.209}$$

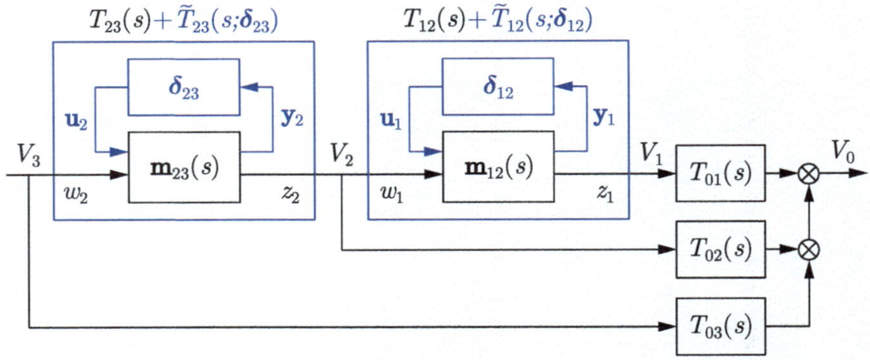

Fig. 4.38 Block diagram of the system in Fig. 4.11b under human uncertainties

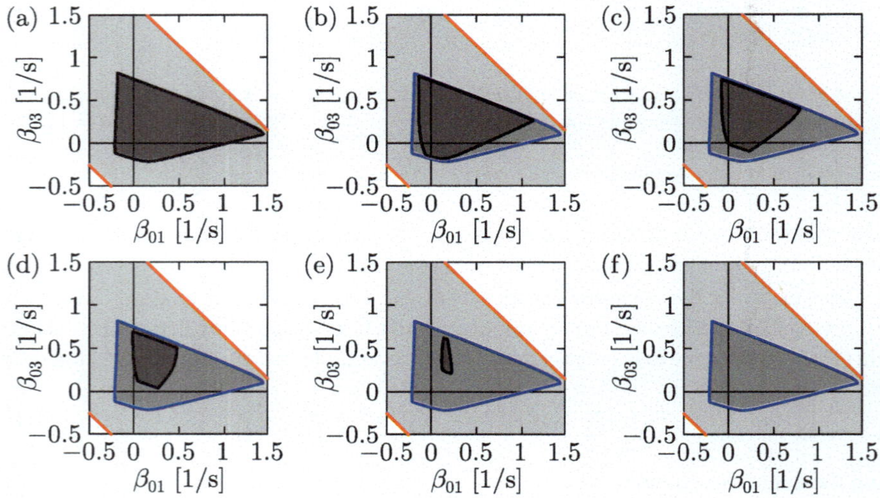

Fig. 4.39 Robust stability charts for the same setup as in Fig. 4.13c, i.e., $\beta_{02} = 0.5\,[1/\mathrm{s}]$. Both human drivers have identical nominal parameters and parameter uncertainty. The relative uncertainty of each human driver parameter is assumed to be the same, $\rho_\kappa/\kappa_\mathrm{h} = \rho_\alpha/\alpha_\mathrm{h} = \rho_\beta/\beta_\mathrm{h} = \rho_\tau/\tau =: p$, where **a** $p = 0\%$, **b** $p = 5\%$, **c** $p = 10\%$, **d** $p = 15\%$, **e** $p = 20\%$, and **f** $p = 25\%$

where $w_1 = z_2$, $w_2 = z_3 = w$, $m_{12}^{zw}(s) = T_{12}(s)$, and $m_{23}^{zw}(s) = T_{23}(s)$. This can be rearranged to (4.198) by eliminating z_1 and z_2, where the coefficient matrix $\mathbf{M}(s)$ becomes

$$\mathbf{M}(s) = \begin{bmatrix} \mathbf{m}_{12}^{yu}(s) & \mathbf{m}_{12}^{yw}(s)\mathbf{m}_{23}^{zu}(s) & \mathbf{m}_{12}^{yw}(s)T_{23}(s) \\ \mathbf{0} & \mathbf{m}_{23}^{yu}(s) & \mathbf{m}_{23}^{yw}(s) \\ \mathbf{m}_{12}^{zu}(s)T_{01}(s) & \mathbf{m}_{23}^{zu}(s)(T_{01}(s)T_{12}(s) + T_{02}(s)) & G_{03}(s) \end{bmatrix}. \quad (4.210)$$

Figure 4.38 depicts the block diagram corresponding to Fig. 4.11b and illustrates how the **m**-δ framework (4.200) is embedded in the **M**-$\mathbf{\Delta}$ framework (4.198); cf. Fig. 4.34b. Again, the structure of the nominal, black part corresponds to the network structure in Fig. 4.11b and the blue parts are added to represent the uncertainties of the two human drivers.

The robust stability charts are shown for different levels of uncertainty in Fig. 4.39. Note that in this case there are $2 \cdot 4 = 8$ uncertain human driver parameters, and consequently "brute force" computation of the robust charts is not feasible any more. Again, the black robustly head-to-tail string stable domain shrinks with increasing levels of uncertainty (even disappears for 25% uncertainty), and the most robust (β_{01}, β_{03}) parameter combination is at the top part of the nominal head-to-tail string stable domain.

Chapter 5
Experiments with Connected Automated Vehicles

In this chapter we present experimental results to demonstrate the applicability of the theoretical results presented in the previous chapters in real-world scenarios. Initial experiments were carried out with ground robots [175, 176], which acted as scaled vehicles, and these allowed us to test the controllers in laboratory environment. Here we present measurements that were taken using full-size connected automated vehicles (CAVs): a passenger vehicle and a class-8 truck. The parameters of these CAVs are shown in Table 3.1—with the exception that in case of the truck only the tractor was used, reducing the mass to 9000 kg. When performing experiments with these CAVs, we also utilize multiple connected human-driven vehicles (CHVs) as will be explained below. More details about the experiments with automobiles can be found in [7, 20, 61, 79, 153], while the ones involving the class-8 truck are detailed in [2, 3, 84]. Some experiments were performed with CHVs only [64, 134, 139].

Figure 5.1 depicts two CHVs and one CAV utilized in the experiments. All were equipped with the V2X kit shown below, which consists of a V2X on-board unit (OBU), an antenna used for GPS and V2X communication, and a laptop. The OBU is connected to the laptop via Ethernet, and it is powered by 12 V through the cigar lighter. This way vehicles were able to share motion information with each other via V2X communication, including GPS position, GPS-based speed, and heading angle 10 Hz update rate. In some of the experiments we utilized WiFi-based dedicated short range communication (DSRC), while in others we used LTE-based cellular V2X (CV2X). The tested control algorithms, however, are agnostic to the change in communication technology.

The CAV shown in the top right of Fig. 5.1 (KIA Soul) was a production vehicle which was modified such that we were able to control its longitudinal and lateral motions. In particular, an electronic control unit (ECU), which was connected to the V2X OBU via Ethernet, was used to provide the vehicle with pedal (gas and brake) commands and steering angle commands, which were converted to torque commands using lower-level controllers. In some of the experiments steering was done by a human driver and in some others the vehicle was driven by a human,

G. Orosz and T. G. Molnár, *Dynamics and Control of Connected Vehicles*, Surveys and Tutorials in the Applied Mathematical Sciences 15,
https://doi.org/10.1007/978-3-031-94555-7_5

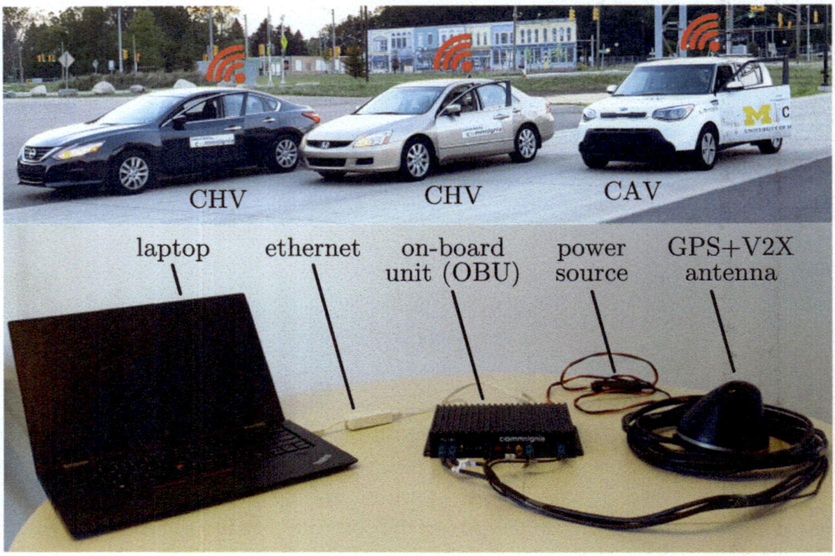

Fig. 5.1 Two connected human-driven vehicles (CHVs) and a connected automated vehicle (CAV) utilized in the experiments are shown in the top, and a V2X kit (used to make vehicles connected) is displayed at the bottom

essentially degrading to a CHV. Even in such human-driven cases, we collected motion data through the GPS as well as the vehicle's internal CAN network.

5.1 Response to a Single Vehicle

The first set of experiments involves two vehicles: a CAV and a CHV. We utilize the controller

$$u = \alpha\big(V(h) - v\big) + \beta\big(W(v_1) - v\big)\,, \tag{5.1}$$

cf. (3.30) and the first line of (4.41). Recall that V is the range policy (3.32), whose gradient is denoted by κ and W is the speed policy (3.33); see Fig. 3.7. We also test a variant with acceleration feedback, that is,

$$u = \alpha\big(V(h) - v\big) + \beta\big(W(v_1) - v\big) + \gamma\,\dot{v}_1\,, \tag{5.2}$$

cf. (4.6). We study the dynamical behavior of the CAV experimentally for different values of α, β, γ, and κ.

For simplicity, the resistance terms, like those in (3.10), are neglected when designing the controllers, but these are incorporated when evaluating the performance (e.g., energy consumption). In terms of implementation, we use V2X connectivity rather than range sensors; that is, the distances between vehicles are calculated using the Haversine formula (4.2), and we implement CCC rather than ACC even when the vehicle ahead is within the line of sight. Further experiments [2] (not detailed in this book) justified that indeed the V2X-based distance matches well with the radar-based distance.

Below we present the results of three sets of experiments of increasing complexity. We start with an experiment where a stationary vehicle is stopped on the road beyond the line of sight, and we compare the response of a human-driven vehicle with that of a connected automated vehicle. We continue with tuning CCC gains when responding to a moving vehicle ahead which follows a velocity profile consisting of acceleration, braking, and constant-speed cruising sections. We end with highlighting the effects of acceleration feedback in scenarios when a connected automated truck responds to a connected human-driven vehicle performing an emergency braking.

5.1.1 Response to a Stationary Vehicle

Figure 5.2 shows the map of the Mcity test track at the University of Michigan where many of our closed track experiments were carried out. A stationary vehicle is shown hidden behind a curve while another vehicle is traveling north along Wolverine Ave. Due to road elevation and vegetation (green patch) the stationary vehicle is beyond the line of sight of the moving vehicle until the latter one reaches the red segment of the trajectory.

Figure 5.3 shows the scenario when the moving vehicle is driven by a human driver. Apart from a bird's eye view picture highlighting the line of sight, we plot the headway, speed, and acceleration profiles of the moving vehicle. We remark that the acceleration is obtained from the velocity while applying a Savitzky–Golay filter with third-order polynomials and a frame length of 21 data points (corresponding to a window size of 2 [s]). Such filter was utilized for the rest of the experimental results presented in this chapter. The vehicle accelerates until it reaches a speed around 15 [m/s] which is close to the speed limit (35 [mph] $\approx$ 15.6 [m/s]) along this road section. Around $t \approx 13$ [s], that is, when $h \approx 25$ [m], the driver notices the stationary vehicle, applies the brakes, and brings the vehicle to a halt. To avoid collision in this safety-critical situation, the deceleration reaches -10 [m/s^2], which is often referred to as 1 g deceleration. In this case, the occupants of the vehicle experience forces from their safety belt equal to their weight.

The driver performing these experiments is a highly trained individual such that the overall reaction time (from the visual stimuli appearing until the brakes are clamped) was around $\tau \approx 0.8$ [s], leaving enough time for the vehicle to slow down. Such a maneuver was also made feasible by the adequate road friction on dry asphalt, and it may not be possible when having poor road surface conditions, e.g., if the road was covered with ice or snow. Moreover, if the approaching vehicle was an automated vehicle relying on range sensors only (i.e., without V2X connectivity), it would face

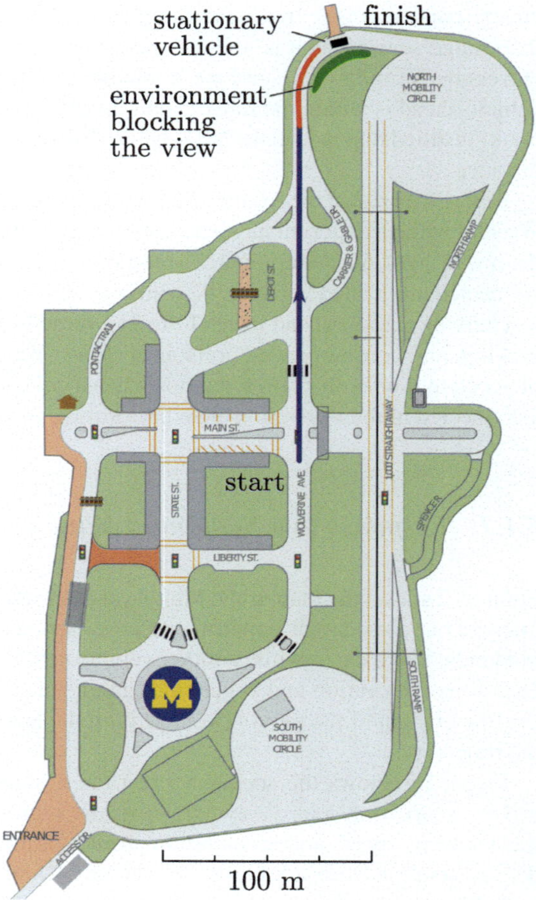

Fig. 5.2 Experimental setup where a vehicle is driving north along Wolverine Ave, as highlighted by the colored trajectory, while a stationary vehicle is hidden behind the curve by road elevation and vegetation (green patch). The stationary vehicle is within the line of sight of the moving vehicle only along the red section of the trajectory

the same issues. Due to the limited line of sight it would also run into a safety-critical situation, and whether it could resolve the problem and avoid an accident would depend on the available road friction.

Figure 5.4 shows the same experiment when the approaching vehicle is a CAV, which is made aware of the presence of the stationary CHV via V2X connectivity, thereby eliminating the limitations posed by the line of sight. Initially the CAV applies cruise control of the form (3.11) with $\alpha = 0.9\,[1/\mathrm{s}]$, $\eta = 0$ in order to reach the reference speed $v_\mathrm{r} = 15\,[\mathrm{m/s}]$. Since the proportional gain is chosen large enough, the steady-state error of the velocity can be reduced below $0.2\,[\mathrm{m/s}]$ even without using an integral gain. Once the headway drops below $70\,[\mathrm{m}]$, the CAV switches to connected cruise control using the parameters listed in the caption of Fig. 5.4. In this case, the braking starts around $t \approx 11\,[\mathrm{s}]$ and the deceleration only reaches $-2.5\,[\mathrm{m/s^2}]$, i.e., a quarter g, which is so mild that the occupants of the vehicle

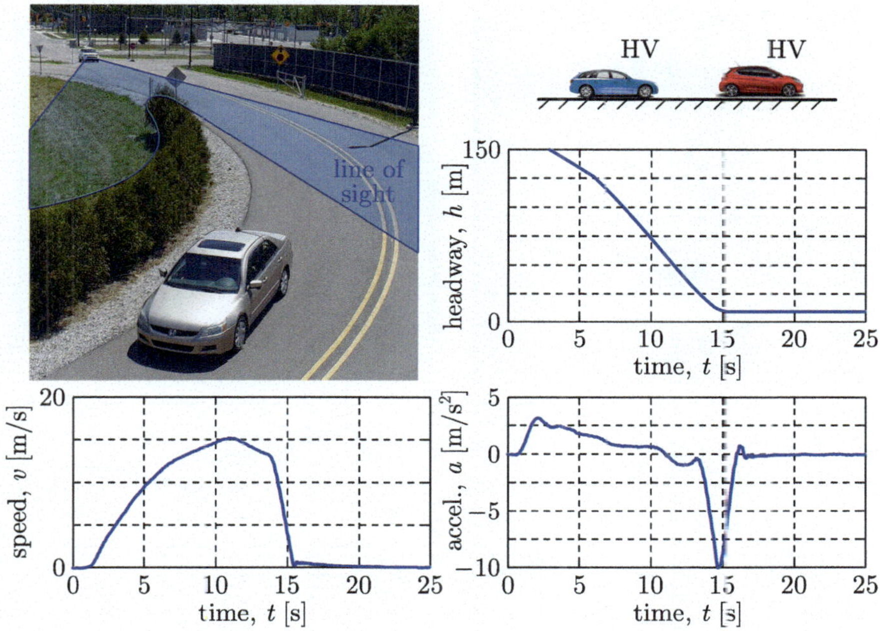

Fig. 5.3 Human-driven vehicle stopping behind a stationary vehicle that is beyond the line of sight until the headway reaches about 25 [m]. The human driver performs an emergency braking with the deceleration reaching 1 g

could hold their beverages without spilling them. These results demonstrate that V2X connectivity may benefit the safety of road participants in limited line-of-sight situations, which will be further reinforced by other experiments described below.

According to our measurements, the overall delay (from the time when GPS data are transmitted by the CHV until the time the CAV activates its brakes) is around 0.6–0.7 [s]. Most of this delay (about 0.5 [s]) is due to the actuation delay in the CAV's controller, and only about 0.1–0.2 [s] is from V2X communication because no significant amount of packet drops occurred during the experiments.

Note that the algorithms of the cruise controller and the connected cruise controller can in fact be integrated, which we also exploited in other experiments performed on public roads [2]. In the experiments of Figs. 5.3 and 5.4 the gains of the controller (5.1) were varied as function of the headway h. Namely, the constant gains α and β were changed to

$$\bar{\alpha}(h) = \begin{cases} \alpha & \text{if} \quad h \leq h_{\rm sw}\,, \\ \alpha_{\rm cc} & \text{if} \quad h > h_{\rm sw}\,, \end{cases} \tag{5.3}$$

and

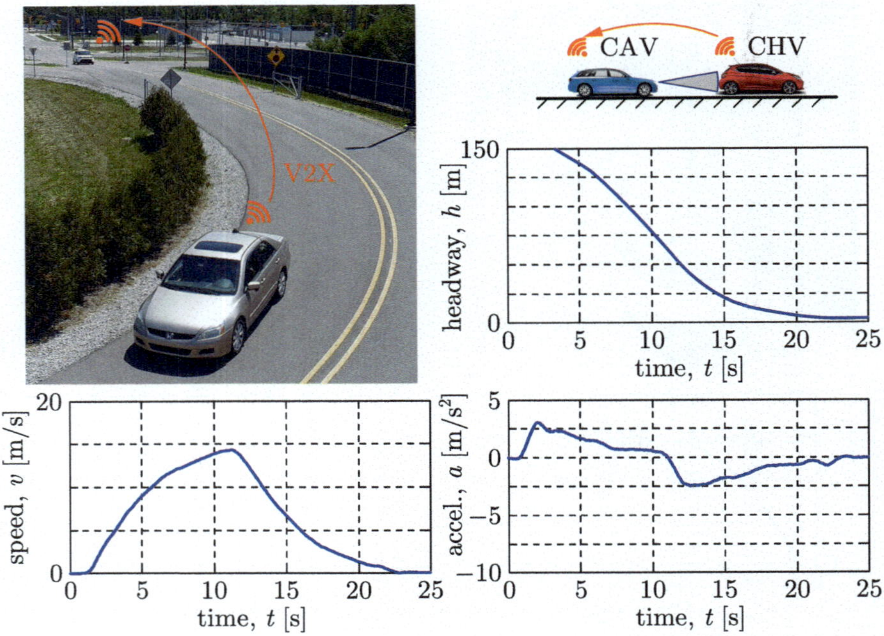

Fig. 5.4 CAV stopping behind a stationary CHV utilizing the CCC (5.1), (3.32), (3.33) with parameters $v_{\max} = 15\,[\mathrm{m/s}]$, $h_{\mathrm{st}} = 5\,[\mathrm{m}]$, $h_{\mathrm{go}} = 30\,[\mathrm{m}]$ yielding $\kappa = 0.6\,[1/\mathrm{s}]$. The gain parameters are set to $\alpha = 0.4\,[1/\mathrm{s}]$ and $\beta = 0.5\,[1/\mathrm{s}]$. This results in a mild braking where the deceleration only reaches a quarter g

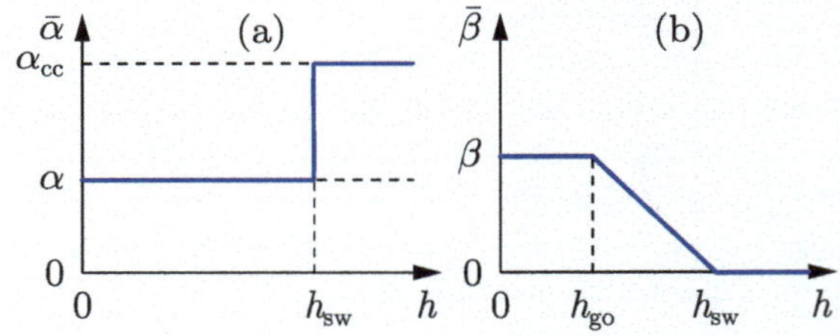

Fig. 5.5 Control gains implemented in the experiments of Sect. 5.1.1. Panel **a** shows the gain $\bar{\alpha}$ given by (5.3), whereas panel **b** plots the gain $\bar{\beta}$ defined by (5.4) as a function of the headway

$$\bar{\beta}(h) = \begin{cases} \beta & \text{if} \quad h \le h_{\mathrm{go}}\,, \\ \beta \dfrac{h_{\mathrm{sw}} - h}{h_{\mathrm{sw}} - h_{\mathrm{go}}} & \text{if} \quad h_{\mathrm{go}} < h \le h_{\mathrm{sw}}, \\ 0 & \text{if} \quad h > h_{\mathrm{sw}}, \end{cases} \tag{5.4}$$

which are depicted in Fig. 5.5a, b.

Combining these with the range policy (3.32) and speed policy (3.33) depicted in Fig. 3.7, one may observe that for $h > h_{\mathrm{sw}}$ (5.1) degrades to the cruise controller

$$u = \alpha_{cc}(v_{max} - v) , \tag{5.5}$$

cf. (3.11) with $\eta = 0$. In the experiments we used $\alpha_{cc} = 0.9\,[1/s]$. Between h_{go} and h_{sw} the α gain is set to its regular value while the β gain is decreased linearly to zero. For $h \leq h_{go}$ we recover the controller (5.1). In Figs. 5.3 and 5.4, we used $h_{sw} - h_{go} = 40\,[m]$, while in other experiments (like those in [2]) we used $h_{sw} - h_{go} = 20\,[m]$.

5.1.2 Response to a Moving Vehicle

When the CHV ahead is stationary, one may apply a large set of controllers to bring the CAV to a stop. However, once the CHV starts to move, it is crucial to tune the CCC algorithms so that certain performance requirements, e.g., plant stability and string stability, are satisfied. String stability can be evaluated by studying the response to harmonic excitations of different frequencies, but such experiments are only feasible when using uncrewed ground robots [175, 176]. When experimenting with full-sized vehicles that carry human occupants, a suitable trajectory should be chosen for the vehicle ahead, which should be rich enough to excite multiple frequencies. We utilize the trajectory shown in Fig. 5.6, referred to as "grand circle," which contains acceleration (green), braking (red), and constant-speed cruising (blue) sections. As the CHV ahead is human driven, each run is slightly different but they have rich enough frequency content to evaluate the performance of the controllers.

Figures 5.7 and 5.8 show two examples of the "grand circle" experiments. The velocity, acceleration, and headway profiles are shown in the top, middle, and bottom panels, respectively. Starting from standstill, the vehicles accelerate to 15 [mph] $\approx$ 6.7 [m/s] and maintain this speed for a while. Then they accelerate to 30 [mph] $\approx$ 13.4 [m/s] and cruise at that speed before reducing their speed to 10 [mph] $\approx$ 4.5 [m/s]. After driving at that speed for a while, they accelerate to 20 [mph] $\approx$ 8.9 [m/s] and cruise with that speed before coming to a stop at that end of the run. The recorded CVH profiles are shown as red while the CAV profiles are shown as blue. The dashed green trajectories are simulations of the model (3.8)–(3.10) with parameters in Table 3.1 such that the CAV controller (5.1), (3.32), (3.33) responds to the CHV's data. The differences between the blue and green trajectories show the difference between the real vehicle and the mathematical model. For example, during acceleration, the experimental headway is larger than the one obtained by simulation as the real powertrain may deliver less torque (depending on the state of the engine) compared to the model, making the vehicle to "lag behind." Still, there is good correspondence between the experiments and simulations despite the unmodeled powertrain dynamics.

When comparing the profiles in Figs. 5.7 and 5.8 one may notice the high-frequency oscillations in the latter case. In order to quantify this phenomenon, and to relate it to the string stability definitions discussed in the previous chapters, the fast Fourier transforms (FFT) of the velocity signals are computed. The results are fil-

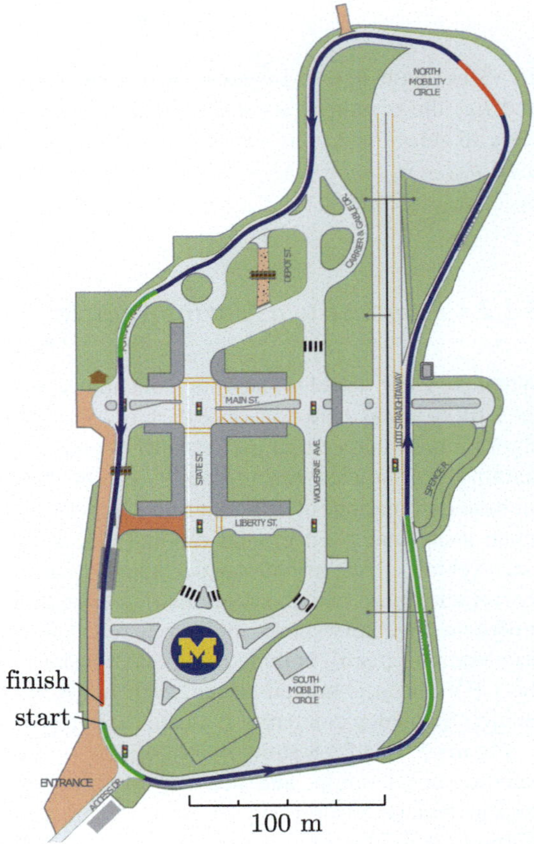

Fig. 5.6 Experimental setup referred to as the "grand circle" which is utilized tune the CCC algorithms when the CAV was following a CHV. The colored trajectory indicates the path of the vehicles such that green, red, and blue sections correspond to acceleration, braking, and constant-speed cruising, respectively

tered using the Savitzky–Golay filter with third-order polynomials and frame length of 31 data points, and these are shown in the left column of Fig. 5.9. Notice that the frequency f (measured in Hz) is plotted in the horizontal axis rather than the angular frequency ω (measured in 1/s) and that the vertical axis has logarithmic scale.

The right column in Fig. 5.9 shows the ratio $|V(f)|/|V_1(f)|$ of the FFT signals. This can serve as the experimental frequency response curve (cf. Fig. 2.20), and can be used to evaluate string stability experimentally. However, according to condition (2.70), both cases in Fig. 5.9 would be categorized as string unstable, due to the apparent noise in the experimental signals. In contrast, the FFT of the simulation signal would categorize case T (panels a, b) as string stable and would categorize case Z (panels c, d) as string unstable. In order to quantify the difference between the experimental frequency response curves, we introduce the string instability index

$$C_S = \frac{1}{f_{\text{end}}} \int_0^{f_{\text{end}}} g\left(\frac{|V(f)|}{|V_1(f)|} - 1\right) \mathrm{d}f \,, \tag{5.6}$$

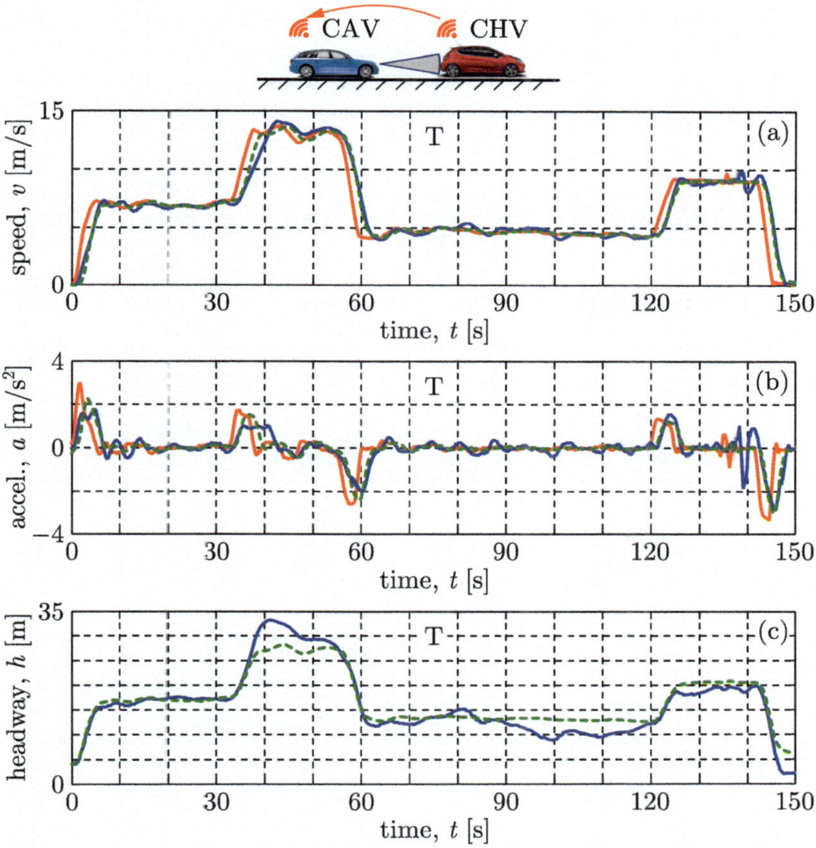

Fig. 5.7 CAV following a CHV along the "grand circle" while utilizing the CCC (5.1), (3.32), (3.33) with parameters $v_{\max} = 30\,[\mathrm{m/s}]$, $h_{\mathrm{st}} = 5\,[\mathrm{m}]$, $h_{\mathrm{st}} = 55\,[\mathrm{m}]$ yielding $\kappa = 0.6\,[1/\mathrm{s}]$. The gain parameters are set to $\alpha = 0.4\,[1/\mathrm{s}]$ and $\beta = 0.5\,[1/\mathrm{s}]$ which correspond to point T in Fig. 5.10. The red trajectories depict the speed and acceleration profiles of the CHV while the blue trajectories show the response of the CAV. The green dashed trajectories show the simulation results with the vehicle model (3.8)–(3.10) and the parameters in Table 3.1

where g denotes the ReLU function defined in (4.112) and depicted in Fig. 4.26. This index is proportional to the area under the experimental frequency response curve that is above 1, as illustrated by red shading in the right column of Fig. 5.9. We chose the frequency limit $f_{\mathrm{end}} = 1$ [Hz] corresponding to the range of characteristic frequencies of the observed velocity fluctuations.

Figure 5.10a visualizes the string instability index with red dots of different sizes for experiments with different gain combinations. The index is large when the gains are either small or when they are large, while smaller dots occur for mid-range gains. To unveil the reason behind this behavior, we overlay the theoretical string

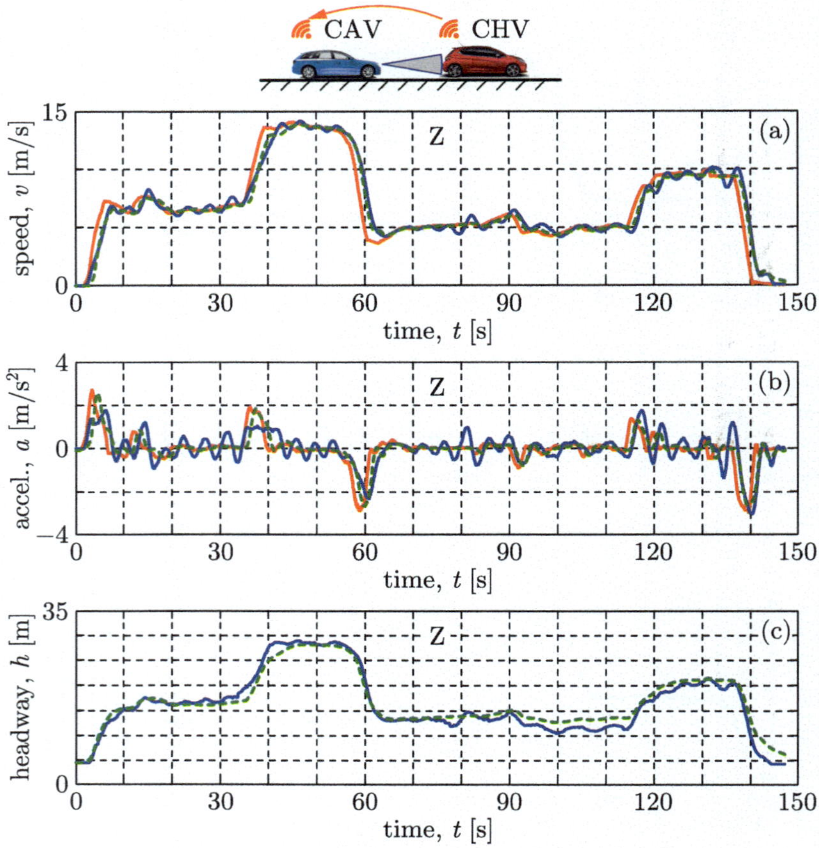

Fig. 5.8 CAV following a CHV along the "grand circle" while utilizing the CCC (5.1), (3.32), (3.33) with parameters $v_{\max} = 30\,[\mathrm{m/s}]$, $h_{\mathrm{st}} = 5\,[\mathrm{m}]$, $h_{\mathrm{st}} = 55\,[\mathrm{m}]$ yielding $\kappa = 0.6\,[1/\mathrm{s}]$. The gain parameters are set to $\alpha = 0.6\,[1/\mathrm{s}]$ and $\beta = 0.8\,[1/\mathrm{s}]$ which correspond to point Z in Fig. 5.10. The red trajectories depict the speed and acceleration profiles of the CHV while the blue trajectories show the response of the CAV. The green dashed trajectories show the simulation results with the vehicle model (3.8)–(3.10) and the parameters in Table 3.1

stability chart on the top of the figure; see (3.82), (3.83) and Fig. 3.11b. The string stable domain is depicted for three different delay values: $\sigma = 0.6\,[\mathrm{s}]$, $\sigma = 0.65\,[\mathrm{s}]$ and $\sigma = 0.7\,[\mathrm{s}]$. Choosing the gains corresponding to point T results in small string instability index and guarantees string stability theoretically for this range of delay values.

Apart from string stability, it is also important to consider safety when selecting the gains in the CCC algorithm. There exist multiple different metrics to evaluate safety [137], and one of the most commonly used one is the time to collision. This is typically calculated by assuming that the vehicles keep traveling with their current

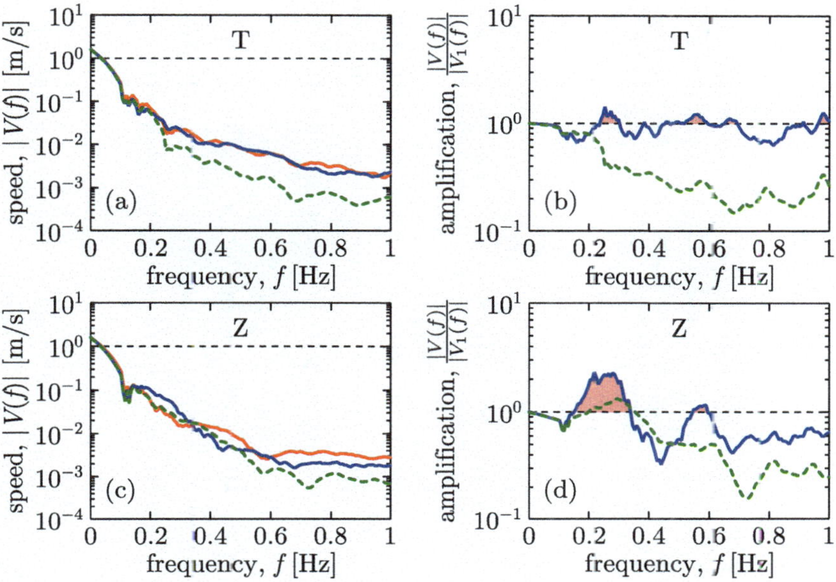

Fig. 5.9 Panels **a**, **c** on the left show the fast Fourier transforms of the velocity signals in Figs. 5.7a and 5.8a, respectively. Panels **b** and **d** on the right show the ratios of the FFT signals, and the red shaded area is utilized in the string instability index (5.6)

speed, that is, $h(t)/\big(v(t) - v_1(t)\big)$. Keeping this quantity above a given value $T_{\rm cr}$ is desired as it helps prevent front-end collisions. This inspires the definition of the time to collision index:

$$C_{\rm T} = \frac{1}{t_{\rm end}} \int_0^{t_{\rm end}} g\left(T_{\rm cr} - g\left(\frac{h(t)}{v(t) - v_1(t)}\right)\right) {\rm d}t \, , \tag{5.7}$$

where g still denotes the ReLU function (4.112) shown in Fig. 4.26. This is also applied twice to ensure that the integrand only counts for the cases when the time to collision is between zero and $T_{\rm cr}$. We chose the value $T_{\rm cr} = 2\,[\rm s]$ based on our driving experience.

Figure 5.10b visualizes the time to collision index with yellow dots of different sizes for the different gain combinations. Larger gains, in particular larger β values, result in smaller indices, indicating higher levels of safety. We would like to relate this behavior to plant stability, but the whole parameter domain shown in the figure is plant stable, cf. Fig. 3.11b. Thus, we consider a generalized notion of plant stability where we wish to keep the real part of the characteristic roots smaller than a given number $\delta \leq 0$ (rather than just keeping them in the left half of the complex plane).

Considering the characteristic equation (3.78), one may substitute $s = \delta$ and obtain

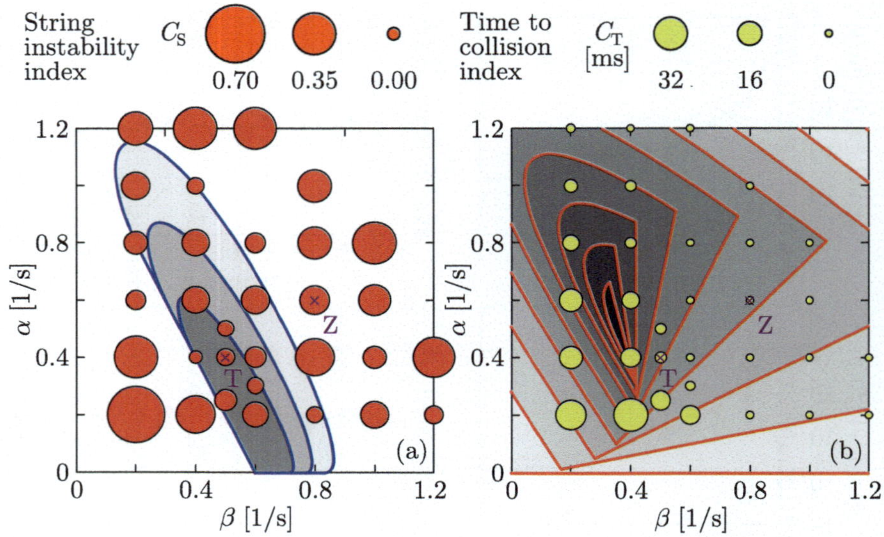

Fig. 5.10 Panel **a** shows the string instability index (5.6) for experiments with different gain combinations. The blue curves are the string stability boundaries (3.82) and (3.83) for $\kappa = 0.6\,[1/\text{s}]$ and delays $\sigma = 0.6\,[\text{s}]$, $\sigma = 0.65\,[\text{s}]$ and $\sigma = 0.7\,[\text{s}]$. Panel **b** shows the time to collision index (5.7) for different gain combinations. The red curves depict the generalized plant stability boundaries (5.8) and (5.9) for $\kappa = 0.6\,[1/\text{s}]$, $\sigma = 0.6\,[\text{s}]$ and different values of δ selected from $\delta \in [-0.8, 0]$

$$\alpha = -\frac{\delta}{\delta + \kappa}\left(\beta + \delta \mathrm{e}^{\delta\sigma}\right), \tag{5.8}$$

which gives a straight line in the (β, α) plane for a fixed value of δ. Similarly, substituting $s = \delta + \mathrm{j}\Omega$ and separating the real and imaginary parts result in

$$\begin{aligned} \alpha &= \frac{\delta^2 + \Omega^2}{\kappa\Omega}\mathrm{e}^{\delta\sigma}\left(\Omega\cos(\Omega\sigma) + \delta\sin(\Omega\sigma)\right), \\ \beta &= -\frac{1}{\Omega}\mathrm{e}^{\delta\sigma}\left(2\delta\Omega\cos(\Omega\sigma) + \left(\delta^2 - \Omega^2\right)\sin(\Omega\sigma)\right) - \alpha, \end{aligned} \tag{5.9}$$

which gives a curve in the (β, α) plane that is parametrized by $\Omega > 0$ (when δ is fixed). Indeed, (5.8) and (5.9) simplify to (3.80) and (3.81), respectively, for $\delta = 0$.

Figure 5.10b shows the generalized plant stability curves (5.8) and (5.9) for $\sigma = 0.6\,[\text{s}]$ and different values of δ. As δ is decreased the generalized plant stability region shrinks, and it becomes a point $(\beta_{\text{fast}}, \alpha_{\text{fast}})$ at a critical value δ_{fast}. This value can be obtained by noticing that at the critical case the gradient of (5.8), i.e., $-\frac{\delta}{\delta+\kappa}$, is equal to the gradient of (5.9) at its starting point. The latter can be obtained by calculating $\frac{\mathrm{d}\alpha}{\mathrm{d}\Omega} / \frac{\mathrm{d}\beta}{\mathrm{d}\Omega}$ and taking the limit $\Omega \to 0$. This yields the critical value

$$\delta_{\text{fast}} = \frac{\sqrt{2}-2}{\sigma} \,. \tag{5.10}$$

Substituting this and using $\Omega \to 0$ in (5.9) results in the corresponding gains

$$\begin{aligned} \alpha_{\text{fast}} &= \frac{10\sqrt{2}-14}{\kappa\sigma^2} \mathrm{e}^{\sqrt{2}-2} \approx \frac{0.0791}{\kappa\sigma^2} \,, \\ \beta_{\text{fast}} &= \frac{2\sqrt{2}-2}{\sigma} \mathrm{e}^{\sqrt{2}-2} - \alpha_{\text{fast}} \approx \frac{0.4612}{\sigma} - \alpha_{\text{fast}} \,, \end{aligned} \tag{5.11}$$

see [20, 99]. In particular, for parameters $\kappa = 0.6\,[1/\text{s}]$, $\sigma = 0.6\,[\text{s}]$, we have $\delta_{\text{fast}} = -0.9763\,[1/\text{s}]$, $\alpha_{\text{fast}} = 0.3663\,[1/\text{s}]$ and $\beta_{\text{fast}} = 0.4023\,[1/\text{s}]$ which is close to point T selected based on string stability. Thus, choosing the corresponding gains benefits both string stability and plant stability that is related to safety.

5.1.3 Utilizing Acceleration Feedback

One may improve the safety of CCC algorithms by incorporating acceleration feedback as in (5.2). This was demonstrated in Fig. 4.2 using numerical simulations. In Fig. 5.11 we show related experimental results where a connected automated truck (CAT) responds to a connected human-driven vehicle ahead while performing an emergency braking. When performing these experiments, we exploited a unique property of V2X connectivity. Namely, by recording the GPS data of the CHV, we could play it back and test different controllers with the same input signal (shown in red), thereby eliminating the variations in the lead vehicle's motion in different experiments. Thus, the difference between the responses is solely due to the differences in the applied controllers.

Figure 5.11 compares experiments without acceleration feedback ($\gamma = 0$, thin purple curves) and with acceleration feedback ($\gamma > 0$, thick purple curves). One may notice that in the latter case the truck brakes earlier and thus is able to stop farther away from the CHV. In this case we set $\gamma = 0.25$. Increasing the acceleration gain further may result in additional improvements in terms of stopping distance but may also deteriorate the regular car-following performance and lead to "jerky rides."

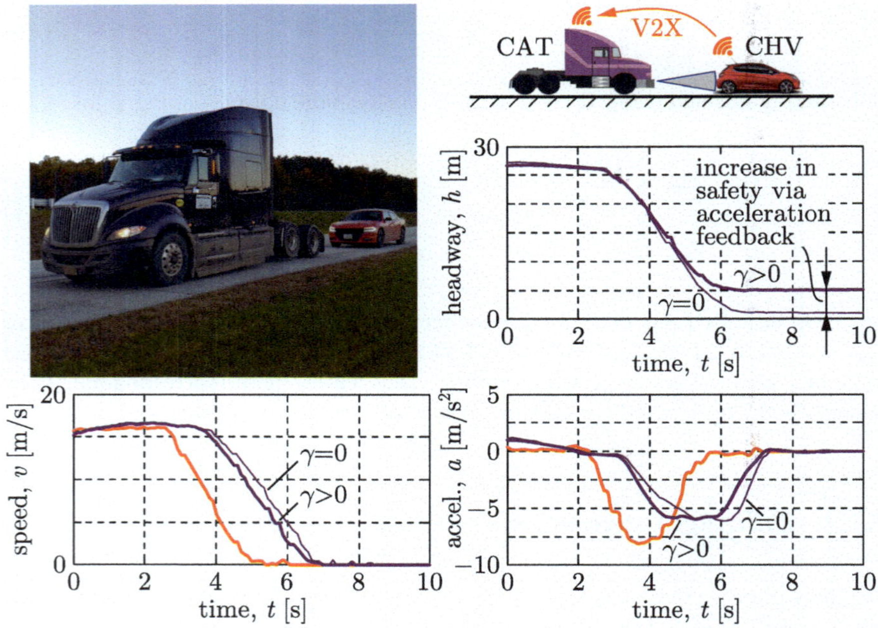

Fig. 5.11 CAT responding to a CHV during an emergency braking maneuver while utilizing the CCC (5.2), (3.32), (3.33) with parameters $v_{\max} = 20\,[\mathrm{m/s}]$, $h_{\mathrm{st}} = 5\,[\mathrm{m}]$, $h_{\mathrm{go}} = 30\,[\mathrm{m}]$ yielding $\kappa = 0.8\,[1/\mathrm{s}]$. Here we still used the gain parameters $\alpha = 0.4\,[1/\mathrm{s}]$ and $\beta = 0.5\,[1/\mathrm{s}]$ while the acceleration gain is either $\gamma = 0$ (thin) or $\gamma = 0.25$ (thick)

5.2 Response to Multiple Vehicles

The second set of experiments focuses on utilizing information from multiple human-driven vehicles ahead. In particular, we investigate the performance of the CCC algorithms

$$u = \alpha_{01}\big(V(h) - v\big) + \beta_{01}\big(W(v_1) - v\big) + \beta_{02}\big(W(v_2) - v_0\big)\,, \tag{5.12}$$

cf. (4.41) with $\alpha_{02} = 0$, and

$$u = \alpha_{01}\big(V(h) - v\big) + \beta_{01}\big(W(v_1) - v\big) + \beta_{02}\big(W(v_2) - v\big) + \beta_{03}\big(W(v_3) - v\big)\,, \tag{5.13}$$

using three vehicles and four vehicles, respectively. See also (5.1) used for two vehicles in the precious section. We still utilize the piecewise linear range policy V given in (3.32) with gradient κ and the speed policy W given in (3.33); see Fig. 3.7. We fix the gain parameter α_{01} and vary the gains β_{01}, β_{02} and β_{03}, as in Sect. 4.3.

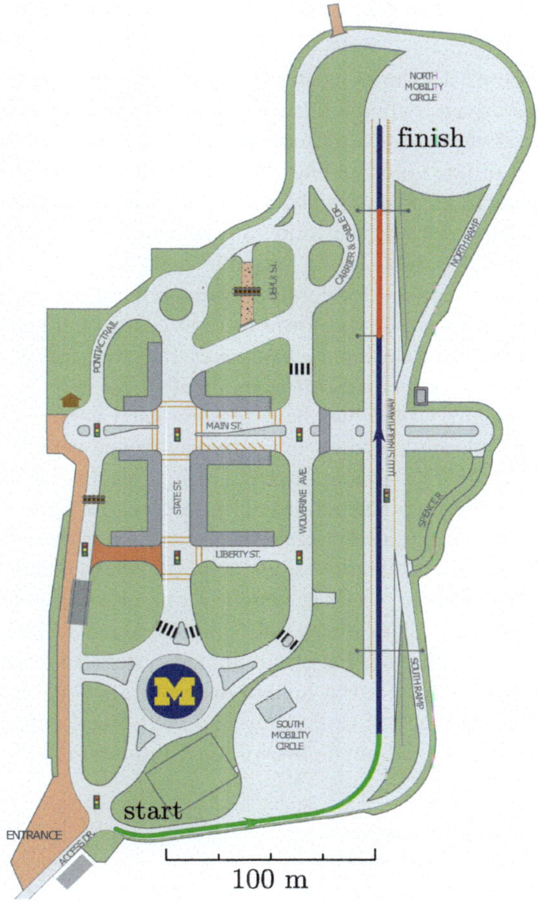

Fig. 5.12 Experimental setup when the CAV responds to multiple connected vehicles ahead using CCC algorithms. The colored trajectory indicates the path of the vehicles such that green, blue, and red sections correspond to acceleration, constant-speed cruising and braking, respectively

We study the dynamics when the CAV responds to a sequence of brakings of the vehicles ahead. The trajectories of the vehicles during the experiments are shown in Fig. 5.12, where, again, green, blue, and red colors correspond to acceleration, constant-speed cruising and braking, respectively. Here we only focus on the latter, which happens at the northern part of the highway. We remark that the primary goal of these experiments is to demonstrate the safety benefits of connectivity. A larger variety of real-world scenarios will be presented in the next section to demonstrate congestion mitigation and energy efficiency.

5.2.1 Experiments with Three Vehicles

Figure 5.13 shows car-following experiments with three vehicles where all of them are driven by human drivers. The time profiles of the velocities, accelerations, and headways are shown in the left, middle, and right panels at the bottom, respectively. We utilize color coding to differentiate the vehicles—a method that we will keep using in the rest of the book. The driver of the head vehicle (black) initiates a sudden braking, forcing the following vehicles (red and blue) to respond. These following vehicles exhibit string unstable behavior, i.e., they reduce their speed more and display harsher deceleration compared to their predecessors. As a consequence, the tail vehicle (blue) performs an emergency braking with 1 g deceleration and it comes to a full stop during the maneuver. In the meantime it also comes close to the vehicle ahead—its headway drops to about 6 [m].

Figure 5.14 shows how an automated vehicle (blue) handles the same situation when it only responds to the human-driven vehicle within its line of sight (red vehicle). The red vehicle displays string unstable behavior and amplifies the speed and acceleration perturbations of its predecessor significantly. Automation helps the blue vehicle to attenuate perturbations and reduce its velocity less and exhibit less harsh deceleration than the red vehicle. In fact, the deceleration of the automated vehicle reaches about a half g, which is similar to that of the head vehicle (black). At the same time, its headway only drops to about 13 [m]. Comparing Figs. 5.13 and 5.14, one may observe that an automated system relying on line-of-sight information may indeed outperform a human driver in a safety-critical situation, mainly due to its reduced reaction time delay and well-tuned controller. Yet, further improvements are possible when the line-of-sight limitation is lifted.

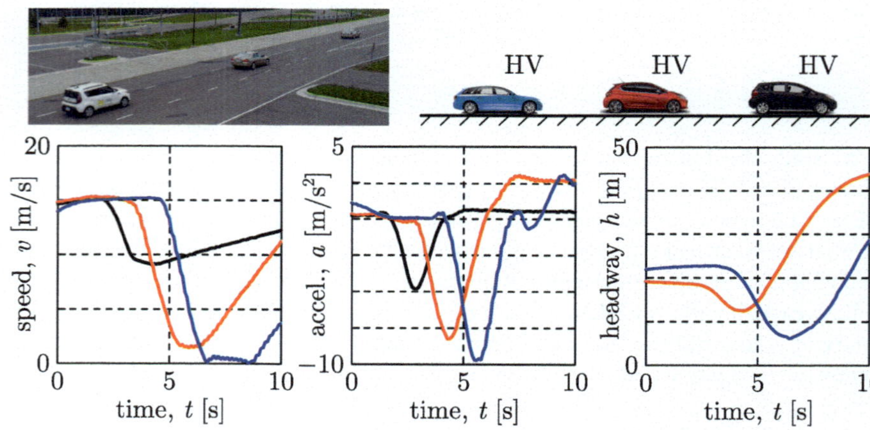

Fig. 5.13 A chain of human-driven vehicles responding to a sequence of brakings. During the maneuver the deceleration of the last vehicle reaches 1 g

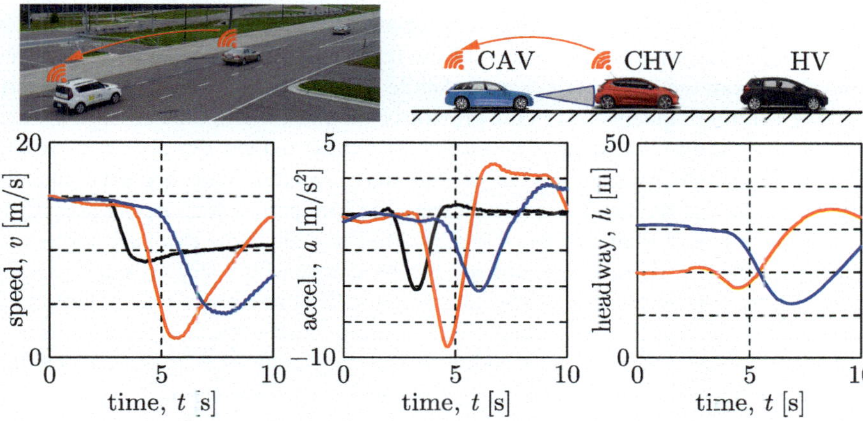

Fig. 5.14 An automated vehicle responding to a sequence of brakings performed by two human-driven vehicles ahead while utilizing the CCC (5.12), (3.32), (3.33) with parameters $v_{max} = 15$ [m/s], $h_{st} = 5$ [m], $h_{go} = 30$ [m] yielding $\kappa = 0.6$ [1/s]. The gain parameters are set to $\alpha_{01} = 0.4$ [1/s], $\beta_{01} = 0.5$ [1/s] and $\beta_{02} = 0$. During the maneuver the deceleration of the automated vehicle reaches a half g

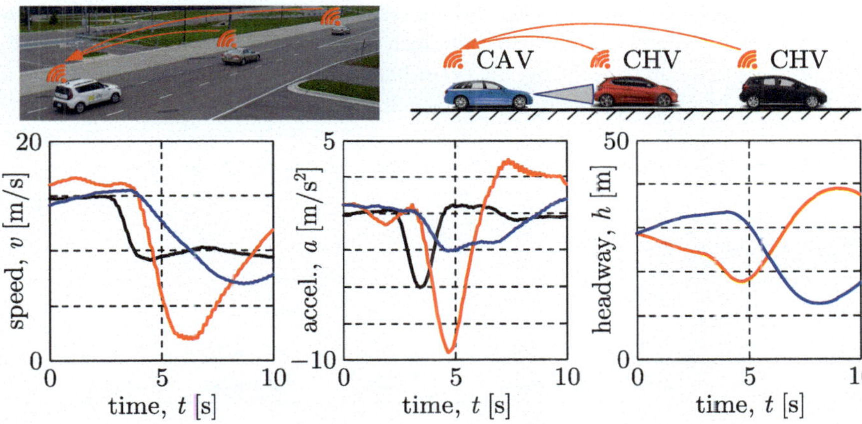

Fig. 5.15 A CAV responding to a sequence of brakings performed by two human-driven vehicles ahead while utilizing the CCC (5.12), (3.32), (3.33) with parameters $v_{max} = 15$ [m/s], $h_{st} = 5$ [m], $h_{go} = 30$ [m] yielding $\kappa = 0.6$ [1/s]. The gain parameters are set to $\alpha_{01} = 0.4$ [1/s], $\beta_{01} = 0.2$ [1/s] and $\beta_{02} = 0.6$ [1/s]. During the maneuver the deceleration of the CAV reaches a quarter g

Figure 5.15 highlights such benefits by showing the response of a CAV (blue) which utilizes information from beyond the line of sight, obtained via V2X communication from the head CHV (black). While the red human-driven vehicle ahead executes a similar maneuver as in the previous examples, V2X connectivity enables the blue vehicle to start to respond earlier, and thus, reduce its speed less while applying less harsh braking. Only about a quarter g deceleration is needed, which is only half of that of the black vehicle. Indeed, the gain parameters used in Fig. 5.15 are within the head-to-tail string stable region in Fig. 4.9b. Note that such improvement is achieved without compromising the headway, which still only reduces to about 13 [m]. This demonstrates that integrating V2X connectivity with automation can significantly improve safety and result in motions which can also be executed even under limited road friction. In addition, passenger comfort is improved, and the occupants can maintain their regular activities without interruption.

5.2.2 *Experiments with a Four-Vehicle Network*

Figures 5.16, 5.17, 5.18, 5.19 and 5.20 show a similar set of experiments for four vehicles as Figs. 5.13, 5.14 and 5.15 did for three vehicles. Comparing the human-driven chains in Figs. 5.13 and 5.16 shows that adding a vehicle can make the situation worse. When four vehicles are amplifying the perturbations, the last two vehicles come to a stop while the headway between them goes below 4 [m]. Adding automation, which relies on line-of-sight information, improves safety in both cases by decreasing the minimum deceleration and increasing the minimum headway of the automated vehi-

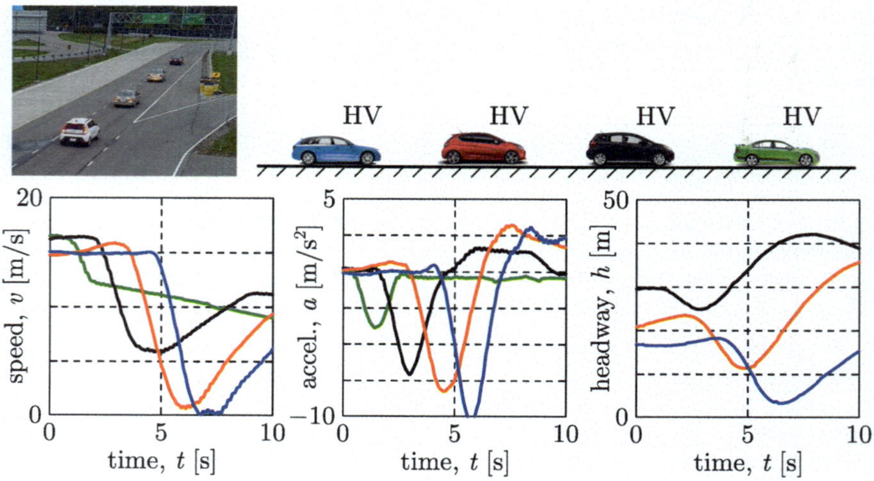

Fig. 5.16 A chain of human-driven vehicles responding to a sequence of brakings. The deceleration of the last vehicle reaches 1 g

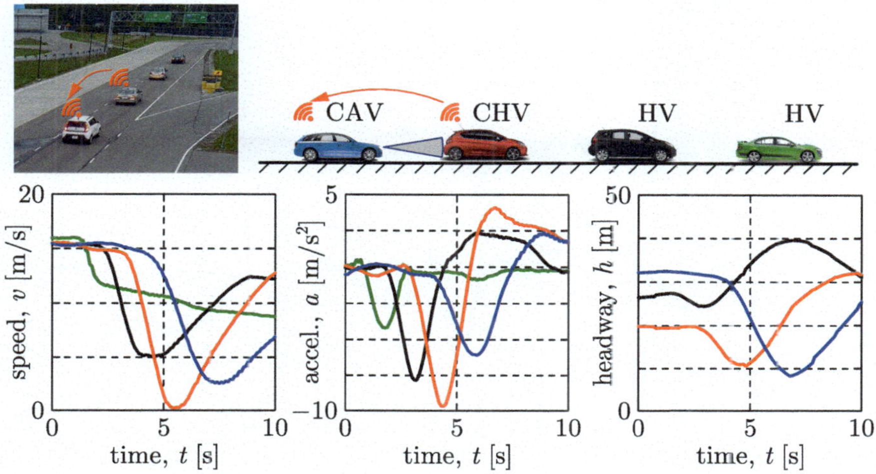

Fig. 5.17 An automated vehicle responding to a sequence of brakings performed by three human-driven vehicles ahead while utilizing the CCC (5.13), (3.32), (3.33) with parameters $v_{\max} = 15\,[\mathrm{m/s}]$, $h_{\mathrm{st}} = 5\,[\mathrm{m}]$, $h_{\mathrm{go}} = 30\,[\mathrm{m}]$ yielding $\kappa = 0.6\,[1/\mathrm{s}]$. The gain parameters are set to $\alpha_{01} = 0.4\,[1/\mathrm{s}]$, $\beta_{01} = 0.5\,[1/\mathrm{s}]$ and $\beta_{02} = \beta_{03} = 0$. During the maneuver the deceleration of the automated vehicle reaches 0.6 g

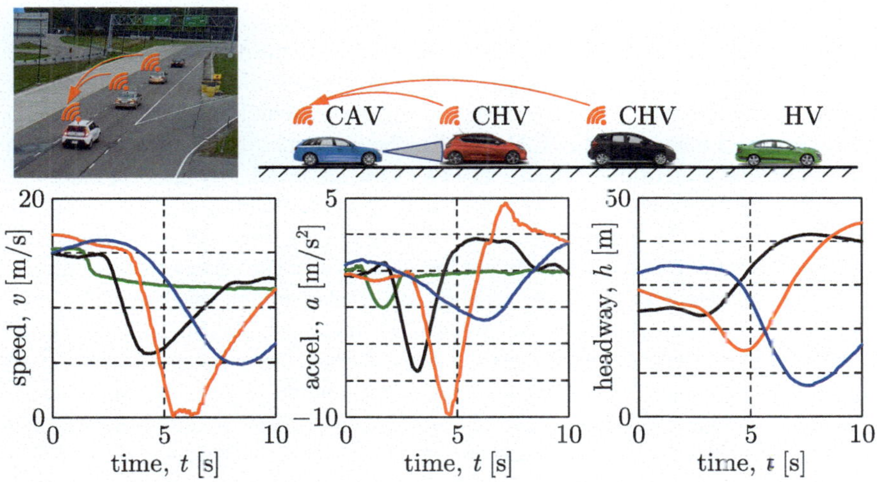

Fig. 5.18 A CAV responding to a sequence of brakings performed by the human-driven vehicles ahead while utilizing the CCC (5.13), (3.32), (3.33) with parameters $v_{\max} = 15\,[\mathrm{m/s}]$, $h_{\mathrm{st}} = 5\,[\mathrm{m}]$, $h_{\mathrm{go}} = 30\,[\mathrm{m}]$ yielding $\kappa = 0.6\,[1/\mathrm{s}]$. The gain parameters are set to $\alpha_{01} = 0.4\,[1/\mathrm{s}]$, $\beta_{01} = 0.2\,[1/\mathrm{s}]$, $\beta_{02} = 0.3\,[1/\mathrm{s}]$ and $\beta_{03} = 0$. During the maneuver the deceleration of the CAV reaches 0.3 g

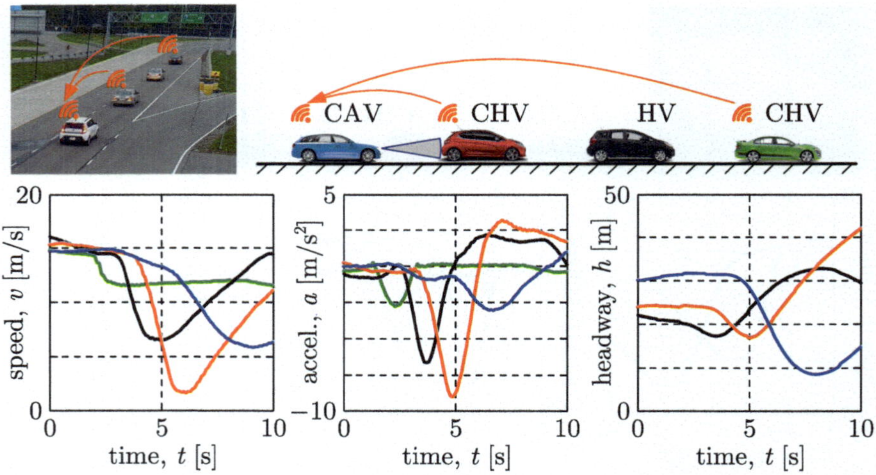

Fig. 5.19 A CAV responding to a sequence of brakings performed by the human-driven vehicles ahead while utilizing the CCC (5.13), (3.32), (3.33) with parameters $v_{\max} = 15\,[\mathrm{m/s}]$, $h_{\mathrm{st}} = 5\,[\mathrm{m}]$, $h_{\mathrm{go}} = 30\,[\mathrm{m}]$ yielding $\kappa = 0.6\,[1/\mathrm{s}]$. The gain parameters are set to $\alpha_{01} = 0.4\,[1/\mathrm{s}]$, $\beta_{01} = 0.2\,[1/\mathrm{s}]$, $\beta_{02} = 0$ and $\beta_{03} = 0.3\,[1/\mathrm{s}]$. During the maneuver the deceleration of the CAV reaches 0.3 g

cle (blue), as can be observed in Figs. 5.14 and 5.17. Still, the deceleration of the blue vehicle is larger in the four-vehicle case than in the three-vehicle case, as the vehicle shall compensate for the negative effects of one more vehicle. Similar phenomena can be observed when the blue CAV receives beyond-line-of-sight information via V2X communication from the black CHV two vehicles ahead, cf. Figs. 5.15 and 5.18. Utilizing V2X information improves the safety of the CAV, but this improvement is less pronounced in the four-vehicle case compared to the three-vehicle case.

However, in the four-vehicle network there exists a larger variety of connectivity topologies compared to the three-vehicle case. For example, the blue CAV may obtain motion information from the green CHV three vehicles ahead, as depicted in Fig. 5.19. As a matter of fact, the same long-distance gains are used in Fig. 5.18 ($\beta_{02} = 0.3\,[1/\mathrm{s}]$) and in Fig. 5.19 ($\beta_{03} = 0.3\,[1/\mathrm{s}]$) and these yield similar behavior in terms of speed, acceleration and headway of the CAV. This demonstrates that the CCC algorithm is (to some level) agnostic to which distant vehicle the velocity information comes from. On the other hand, when the CAV responds to all three vehicles ahead, as in Fig. 5.20, further improvements of safety can be observed: the CAV's deceleration only reaches a quarter g while its headway stays above 12 [m]. These experiments demonstrate that beyond-line-of-sight information can improve the response of the CAV in safety-critical scenarios and that string stability charts, like that in Fig. 4.14c, can be utilized to select how to weigh the V2X information obtained from distant vehicles.

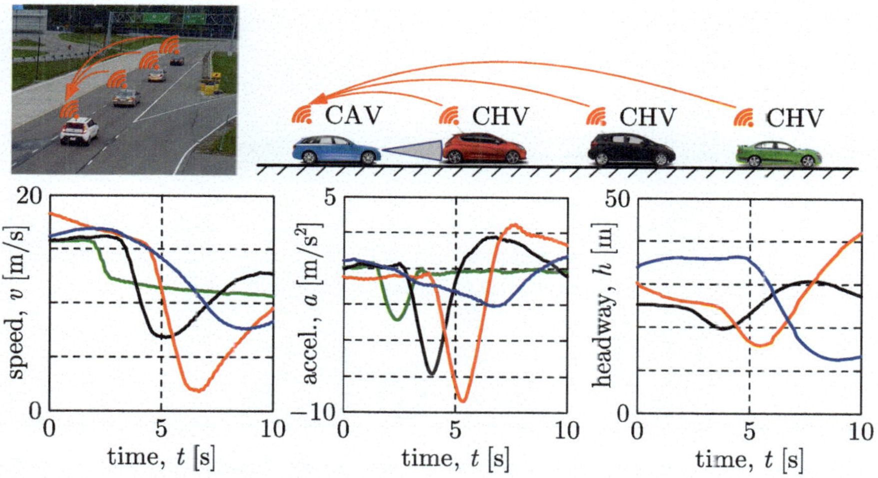

Fig. 5.20 A CAV responding to a sequence of brakings performed by the human-driven vehicles ahead while utilizing the CCC (5.13), (3.32), (3.33) with parameters $v_{max} = 15\,[\mathrm{m/s}]$, $h_{st} = 5\,[\mathrm{m}]$, $h_{go} = 30\,[\mathrm{m}]$ yielding $\kappa = 0.6\,[1/\mathrm{s}]$. The gain parameters are set to $\alpha_{01} = 0.4\,[1/\mathrm{s}]$, $\beta_{01} = 0.2\,[1/\mathrm{s}]$, $\beta_{02} = 0.3\,[1/\mathrm{s}]$ and $\beta_{03} = 0.3\,[1/\mathrm{s}]$. During the maneuver the deceleration of the CAV reaches a quarter g

This is exploited further in [7] where a weighted average of the velocity of connected vehicles ahead is utilized in the CCC algorithm. We also mention that using information from distant vehicles further downstream may require one to delay the signals, so that they are not only utilized with the right weight but also at the right time in the controller [191]. The ultimate goal of the CAV may be to "ride the traffic waves" (triggered by human-driven vehicles downstream) so that it improves the safety and comfort of its occupants. At the same time, the CAV may also mitigate these congestion waves and benefit vehicles traveling upstream as demonstrated in the next section.

5.3 Experiments on Public Roads

Figure 5.21 highlights the Howell/Mason Road near Ann Arbor, Michigan on the map. A 8.4 mile (13.5 km) long section of this road was selected for the next experiments, where driving is not interrupted by cross traffic. In these experiments, V2X-equipped personal vehicles are driven along the road such that a connected automated vehicle (CAV) responds to the motion of multiple connected human-driven vehicles (CHVs) using the connected cruise control (CCC) algorithms (5.12) and (5.13). The CAV utilizes the piecewise linear range policy (3.32) with gradient κ and the speed

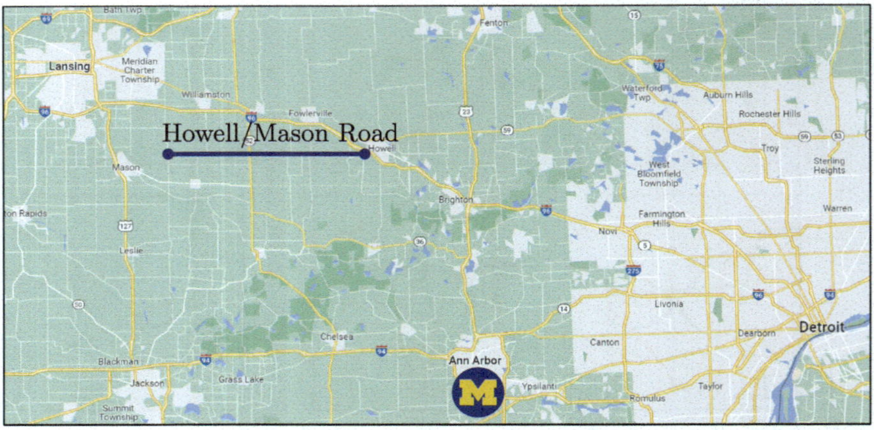

Fig. 5.21 Location of the Howell/Mason Road near Ann Arbor, Michigan. This road includes a 8.4 mile (13.5 km) long section of uninterrupted driving

policy (3.33), see Fig. 3.7. The parameters α_{01}, β_{0j}, and κ are chosen based on the on-track experiments presented in the previous section.

One set of experiments was carried out in the open chain configuration which allows us to study how the CAV can exploit connectivity to mitigate congestion waves triggered by a human driver downstream. The impacts on string stability and on energy efficiency are highlighted through these experiments. In another set of experiments we created a so-called virtual ring [133] to evaluate the impact of automation and connectivity at the nonlinear level and to investigate the bistability between equilibrium and traveling waves.

5.3.1 *Open Chain Experiments*

In the first set of experiments eight vehicles travel along the road while the head vehicle of the chain imposes velocity perturbations by reducing its speed and then accelerating back to its cruising speed repeatedly. Compared to the previous section these perturbations are less abrupt and each of these lasts around 30 s in order to mimic velocity fluctuations observed in highway traffic. The results are shown in Figs. 5.22, 5.23, 5.24, 5.25 and 5.26 where the same controllers are utilized as in Figs. 5.16, 5.17, 5.18, 5.19 and 5.20, respectively, except for the speed limit which is increased to highway level. Note that the range policy gradient at $\kappa = 0.6\,[1/\text{s}]$ is maintained by simultaneously increasing the speed limit to $v_{\max} = 30\,[\text{m/s}]$ and the free-flow headway to $h_{\text{go}} = 55\,[\text{m}]$.

Figure 5.22 shows the experimental results when all vehicles are driven by human drivers. The top panel depicts the speed of the head vehicle (brown) and that of the

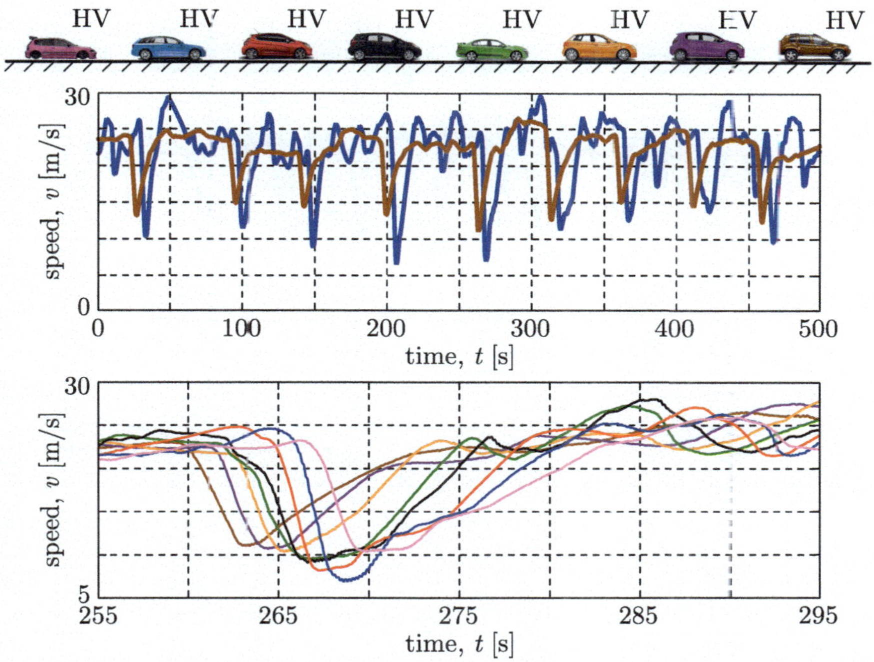

Fig. 5.22 Human-driven experiment with eight vehicles where the first vehicle imposes perturbations on the chain

blue vehicle as the latter will be automated in later experiments. The bottom panel zooms into one of the braking maneuvers and displays the velocity profiles of all vehicles. Due to the string instability of the human drivers the velocity perturbations are amplified as they propagate backward along the chain. This happens for each braking maneuver as seen when comparing the nearby minima of the brown and blue curves on the top panel, indicating a strong brown-to-blue string instability.

When the blue vehicle is automated, as in Fig. 5.23, it can ensure its string stability with respect to the red vehicle immediately in front and attenuate the perturbations. This is useful in mitigating congestion waves as indicated by the smaller differences between the nearby minima of the brown and blue curves. Still, all braking maneuvers are brown-to-blue string unstable. This demonstrates the benefit of inserting automated vehicles (with well-designed longitudinal controllers) into the traffic flow in an effort to mitigate congestion. The limitations of this strategy are also clear: when the penetration of automation is low, the ability of automated vehicles to eliminate congestion may be limited.

When the blue vehicle also utilizes V2X signals from the black vehicle (see Fig. 5.24) or from the green vehicle (see Fig. 5.25), it can improve its ability of mitigating congestion waves as indicated by the increasing minima of the blue curve (compared to the minima of the brown curve). This is improved even further when

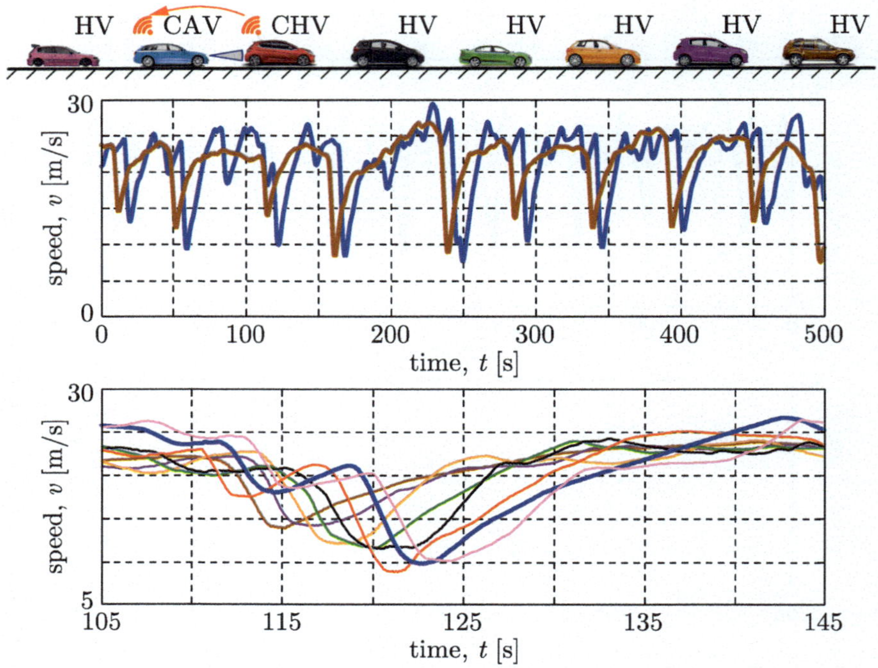

Fig. 5.23 An automated vehicle is embedded in the chain of seven human-driven vehicles using the same CCC algorithm as in Fig. 5.17 but with $v_{max} = 30\,[\text{m/s}]$ and $h_{go} = 55\,[\text{m}]$, which still yields $\kappa = 0.6\,[1/\text{s}]$. The gains are set to $\alpha_{01} = 0.4\,[1/\text{s}]$, $\beta_{01} = 0.5\,[1/\text{s}]$, $\beta_{02} = \beta_{03} = 0$

the blue CAV utilizes motion information from both the black CHV and the green CHV, as shown in Fig. 5.26. These experimental results demonstrate the benefits of V2X connectivity in congestion mitigation: having access to beyond-line-of-sight information enhances the ability of automated vehicles to mitigate congestion even for low penetration of automation.

We remark that while in Figs. 5.22 and 5.23 every braking maneuver exhibits brown-to-blue string instability, not all braking maneuvers show brown-to-blue string instability in Figs. 5.24, 5.25 and 5.26. This is in agreement with the stability chart in Fig. 4.14c used for the CCC design, even though that only guarantees green-to-blue string stability. Guaranteeing brown-to-blue string stability may in fact require larger penetration of connectivity and automation.

Apart from string stability we also evaluate the different CCC strategies based on the energy efficiency of the participating vehicles. Namely, we calculate the energy consumption per unit mass (4.111), that is,

$$w_i(t) = \int_0^t v_i(\xi) g\big(\dot{v}_i(\xi) + p(v_i(\xi))\big) \mathrm{d}\xi \ , \tag{5.14}$$

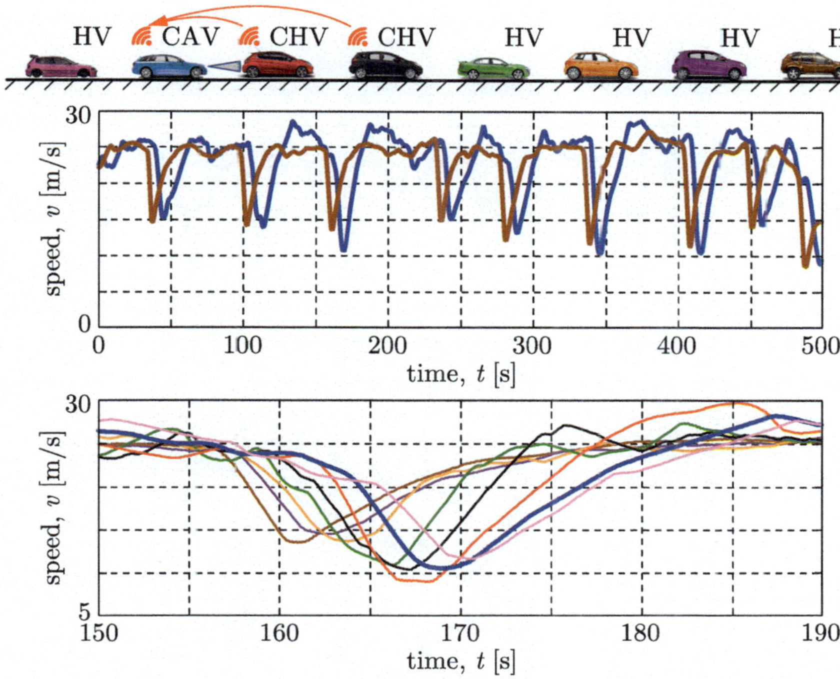

Fig. 5.24 A CAV is embedded in the chain of seven human-driven vehicles using the same CCC algorithm as in Fig. 5.18 but with $v_{max} = 30\,[\mathrm{m/s}]$ and $h_{go} = 55\,[\mathrm{m}]$, which still yields $\kappa = 0.6\,[1/\mathrm{s}]$. The gains are set to $\alpha_{01} = 0.4\,[1/\mathrm{s}]$, $\beta_{01} = 0.2\,[1/\mathrm{s}]$, $\beta_{02} = 0.3\,[1/\mathrm{s}]$ and $\beta_{03} = 0$

for each vehicle i. Here g still denotes the ReLU function (4.112) shown in Fig. 4.26. This enables us to compare the energy efficiency of different vehicles participating in an experiment, despite having different masses. The loss terms are calculated using the parameters in the first column of Table 3.1 for all vehicles.

The energy consumption results corresponding to Figs. 5.22, 5.23, 5.24, 5.25 and 5.26 are shown in Fig. 5.27a–e, respectively. Figure 5.27a shows the human-driven scenario, cf. Fig. 5.22. Moving upstream along the chain, vehicles consume more and more energy corresponding to the increasing velocity fluctuations. Notice that the consumption of the blue vehicle is very similar to that of the red vehicle immediately ahead. This is related to the fact that the blue vehicle was driven by a highly trained individual who decelerated and accelerated in an energy efficient a way when responding to the perturbations of the red vehicle. Moreover, the last vehicle (pink) in the chain has smaller energy consumption than the blue vehicle. This is because this vehicle had a very cautious driver who kept large distances and was able to respond mildly to the velocity perturbations of the blue vehicle.

In Fig. 5.27b, which corresponds to Fig. 5.23, the automated vehicle (blue) improves its energy consumption compared to the red vehicle ahead, demonstrat-

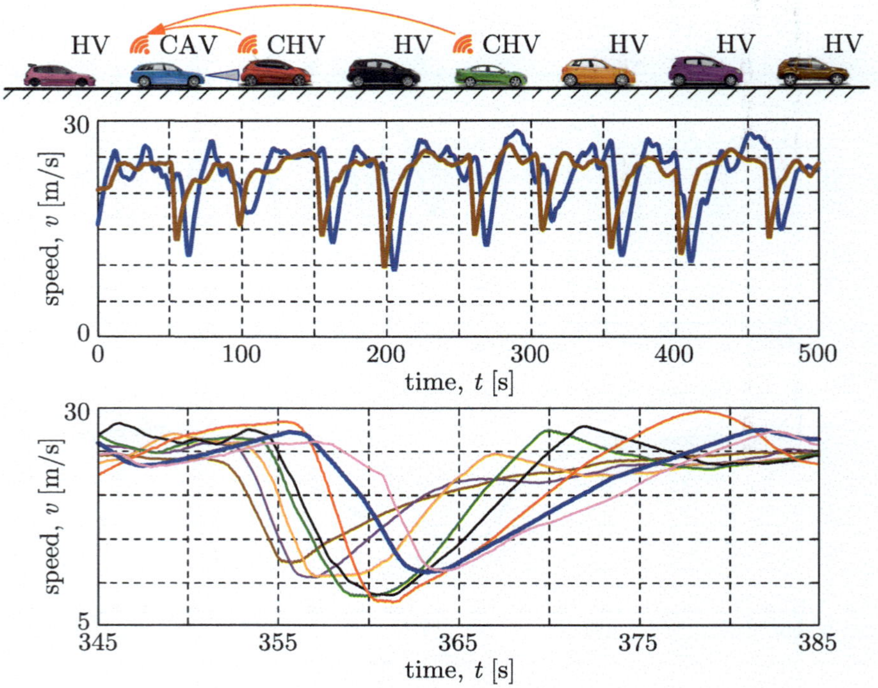

Fig. 5.25 A CAV is embedded in the chain of seven human-driven vehicles using the same CCC algorithm as in Fig. 5.19 but with $v_{\max} = 30\,[\mathrm{m/s}]$ and $h_{\mathrm{go}} = 55\,[\mathrm{m}]$, which still yields $\kappa = 0.6\,[1/\mathrm{s}]$. The gains are set to $\alpha_{01} = 0.4\,[1/\mathrm{s}]$, $\beta_{01} = 0.2\,[1/\mathrm{s}]$, $\beta_{02} = 0$ and $\beta_{03} = 0.3\,[1/\mathrm{s}]$

ing that automation can indeed improve energy efficiency. Figure 5.27c–e, which correspond to Figs. 5.24, 5.25 and 5.26, show that increasing the penetration of connectivity increases the energy savings of the CAV, as the blue curve moves away from the red curve. Indeed, the absolute energy consumption differs in each experiment as it depends on the perturbations imposed by the brown head vehicle.

To quantify the improvements of energy efficiency between different experiments, we scale the energy consumption with the consumption of the vehicle which consumes the most energy in the experiment (typically the red vehicle). That is, we define

$$\hat{w}_i = \frac{w_i}{\max_k\{w_k\}}\,, \tag{5.15}$$

where w_i denotes the energy consumption per unit mass of the i-th vehicle at the end of the experiment, i.e., at $t = 500\,[\mathrm{s}]$. The results are collected in Table 5.1 as percentages. For each experiment, i.e., for each column, the numbers increase when moving upstream along the chain from the brown vehicle to the red vehicle, corresponding to the string unstable behavior of human drivers. The blue vehicle, by using automation and connectivity, is able to reduce the energy consumption

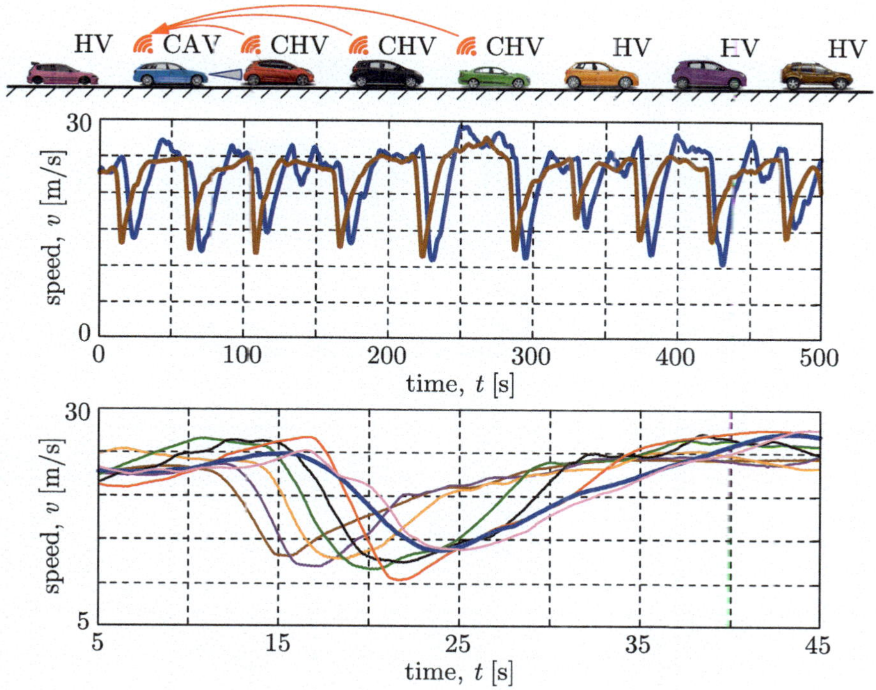

Fig. 5.26 A CAV is embedded in the chain of seven human-driven vehicles using the same CCC algorithm as in Fig. 5.20 but with $v_{max} = 30\,[\mathrm{m/s}]$ and $h_{go} = 55\,[\mathrm{m}]$, which still yields $\kappa = 0.6\,[1/\mathrm{s}]$. The gains are set to $\alpha_{01} = 0.4\,[1/\mathrm{s}]$, $\beta_{01} = 0.2\,[1/\mathrm{s}]$, $\beta_{02} = 0.3\,[1/\mathrm{s}]$ and $\beta_{03} = 0.3\,[1/\mathrm{s}]$

compared to the red vehicle: this improvement is 7% when it responds to the motion of the red vehicle only (second column), and it increases to 25% when it uses information from three vehicles ahead (last column). As a matter of fact, even the pink vehicle at the tail benefits from the smoother driving of the blue vehicle, as shown by the decreasing numbers of energy consumption in the last row.

As discussed in Sect. 4.6, one may optimize the gains in the CCC algorithm to benefit the energy efficiency of the CAV (and consequently the energy efficiency of the human-driven vehicles upstream). The data collected from the human-driven vehicles can be utilized in high-fidelity simulations to validate the effectiveness of this approach [85]. In addition, more sophisticated CCC algorithms can be designed, e.g., by utilizing model predictive control [190], to further reduce the energy consumption, and these can be tested using the collected data.

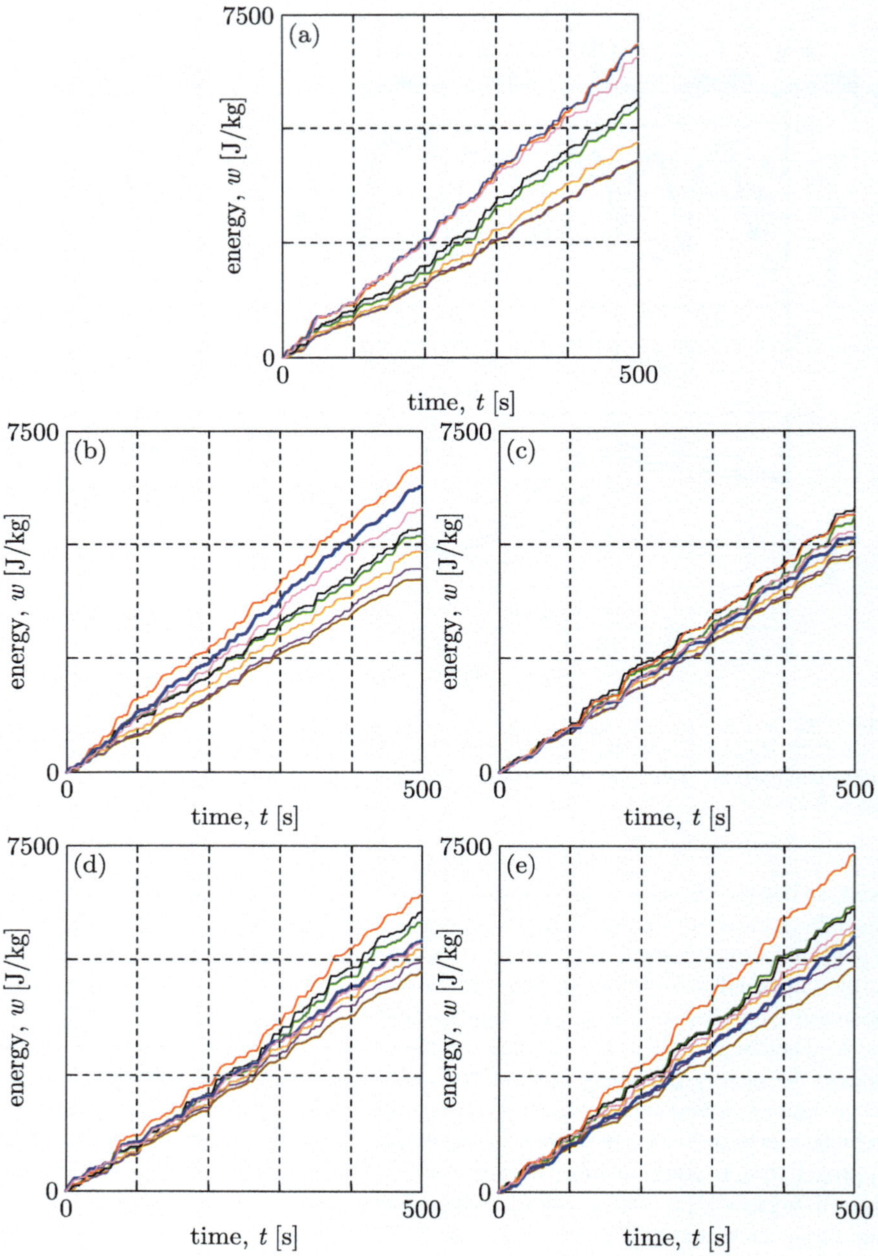

Fig. 5.27 The energy consumption per unit mass (5.14) shown for the open chain experiments. Namely, **a** Fig. 5.22, **b** Fig. 5.23, **c** Fig. 5.24, **d** Fig. 5.25 and **e** Fig. 5.26

Table 5.1 The total energy consumption per unit mass (5.15) of vehicles scaled by the consumption of the most consuming vehicle in each experiment. The numbers denote percentages

Car	Fig. 5.22	Fig. 5.23	Fig. 5.24	Fig. 5.25	Fig. 5.26
Brown	62	62	81	73	65
Purple	62	66	83	77	71
Orange	68	72	87	81	77
Green	79	77	96	91	85
Black	82	79	100	93	84
Red	100	100	97	100	100
Blue	100	93	88	84	75
Pink	96	86	90	83	80

5.3.2 *Virtual Ring Experiments*

While the experiments above highlight the rise (and possible suppression) of congestion waves, the chain is not long enough to study fully grown nonlinear oscillations, like those shown by simulations in Figs. 2.31 and 4.23 in Sects. 2.4 and 4.5. In order to study such nonlinear phenomena, we take advantage of the ring configuration. In this case, after congestion waves are triggered, they keep traveling along the ring, and the state of the system converges to limit cycle oscillations.

Realizing the ring configuration experimentally carries a significant challenge. Namely, the ring needs to be large enough so that the lateral dynamics can be neglected and drivers can focus on the car-following task. For example, in prior experiments [198, 199] rings with radii of about 40 meters were used with 22 vehicles. Still, the speed had to be kept low in order to limit the lateral acceleration—keeping the speed below 12 [m/s] limited the lateral acceleration around 3.5 [m/s^2], corresponding to the comfort level of human drivers. This speed range is, however, much lower than what human drivers experience in highway driving scenarios.

To overcome this challenge, we put forward the concept of virtual ring which is depicted in Fig. 5.28 for three vehicles. The blue CAV is leading a chain of vehicles while it pretends to follow the red CHV at the tail. Such configuration is enabled by V2X communication which allows the CAV to respond to multiple vehicles "ahead." In the example shown, the red vehicle is "immediately ahead" of the blue vehicle while the black vehicle is "two vehicles ahead" of the blue vehicle. This setup enables vehicles to drive at realistic speed and facilitates the study of their behavior at the nonlinear level.

The experimental realization of the virtual ring requires the definition of the headway of the CAV. Similarly to Sects. 2.2.2, 2.3.2, 2.4, 4.2.2, and 4.5, consider that N vehicles are placed on a (virtual) ring of length $L + Nl$ where L is the net ring length and l is the length of the vehicles (considered to be the same for all vehicles for simplicity). We number the vehicles from the tail CHV to the head CAV as $i = 1, \ldots, N$, see Fig. 5.28a for $N = 3$. According to (2.92), the sum of the

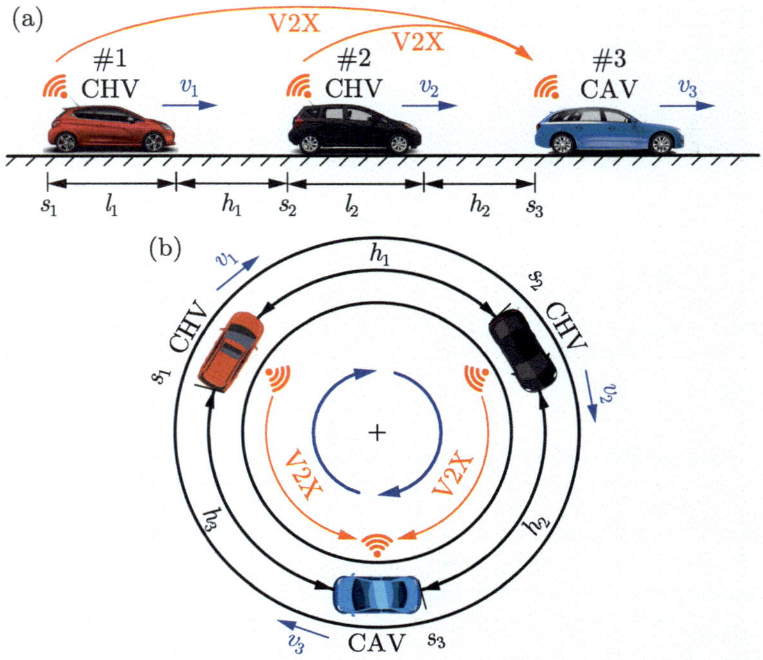

Fig. 5.28 The concept of virtual ring utilized in the experiments. The head CAV responds to the CHVs behind while it pretends to follow the tail CHV on a virtual ring

headways gives the net ring length which allows us to express the CAV's headway as

$$\sum_{i=1}^{N} h_i = L \quad \Rightarrow \quad h_N = L - \sum_{i=1}^{N-1} h_i \, . \tag{5.16}$$

As discussed in Sect. 4.1, V2X communication and the Haversine formula (4.2) can be used to obtain the distance between the rear bumpers of the head CAV and the tail CHV:

$$s_N - s_1 = \sum_{i=1}^{N-1} h_i + (N-1)l \, , \tag{5.17}$$

cf. Fig. 5.28b for $N = 3$. Substituting this into (5.16) results in

$$h_N = L + (N-1)l - (s_N - s_1) \, . \tag{5.18}$$

In accordance with the periodic boundary conditions (2.91), we can define the shifted position $s_{N+1} = s_1 + L + Nl$ and write

$$h_N = s_{N+1} - s_N - l \,, \tag{5.19}$$

cf. definition (4.1).

Note that the above arguments can easily be generalized for vehicles of different lengths which leads to the generalizations of (5.17)–(5.19) as

$$s_N - s_1 = \sum_{i=1}^{N-1} (h_i + l_i) \,, \tag{5.20}$$

$$h_N = L + \sum_{i=1}^{N-1} l_i - (s_N - s_1) \,, \tag{5.21}$$

$$h_N = s_{N+1} - s_N - l_N \,, \tag{5.22}$$

where we used the definition $s_{N+1} = s_1 + L + \sum_{i=1}^{N} l_i$.

For the experimental results presented below we utilize $N = 3$ vehicles of length $l = 5\,[\mathrm{m}]$ on a virtual ring of length $L + 3l = 90\,[\mathrm{m}]$. Figures 5.29, 5.30, 5.31, and 5.32 depict the measured speed, acceleration, and headway of all three vehicles in the top, middle, and bottom panels, respectively. Observe that the speed reaches $25\,[\mathrm{m/s}]$, which is a realistic speed for highway driving. If one wanted to carry out these experiments on a physical ring road, lateral accelerations would have reached $43.5\,[\mathrm{m/s^2}]$, that is, around 4.5 g. This is more than ten times larger than what is considered to be comfortable in normal driving, and only race car drivers can venture into this territory. This demonstrates the necessity of the virtual ring setup.

In Fig. 5.29 the CAV utilizes the same controller as in Fig. 5.14, that is, the blue CAV only responds to the red CHV "ahead." Again, we increase the speed limit to $v_{\max} = 30\,[\mathrm{m/s}]$ and the free-flow headway to $h_{\mathrm{go}} = 55\,[\mathrm{m}]$ simultaneously to maintain the range policy gradient $\kappa = 0.6\,[1/\mathrm{s}]$. After some initial transients the system approaches an equilibrium where the vehicles maintain the same constant speed and constant (but not uniform) headways. Around $t = 50\,[\mathrm{s}]$ the red CHV imposes a perturbation by tapping the brake. This perturbation propagates along the ring as the blue CAV and then the black HV respond to it, and thus, it reaches the red CHV again. The consecutive braking events become weaker and weaker, and after these transient oscillations decay, the system returns to the equilibrium. Around $t = 70\,[\mathrm{s}]$ the CHV applies another perturbation which also leads to transient oscillations. In this case it takes longer time for the transients to decay but eventually the system returns to the equilibrium. Based on these experiments, one may declare that the equilibrium is globally stable for the chosen connected cruise controller, which we also found to be the case for other values of the net ring length L.

Note that the equilibrium is nonuniform (i.e., the equilibrium headways differ from each other) since the vehicles are not identical. That is, instead of the uniform flow equilibrium (2.93), we have

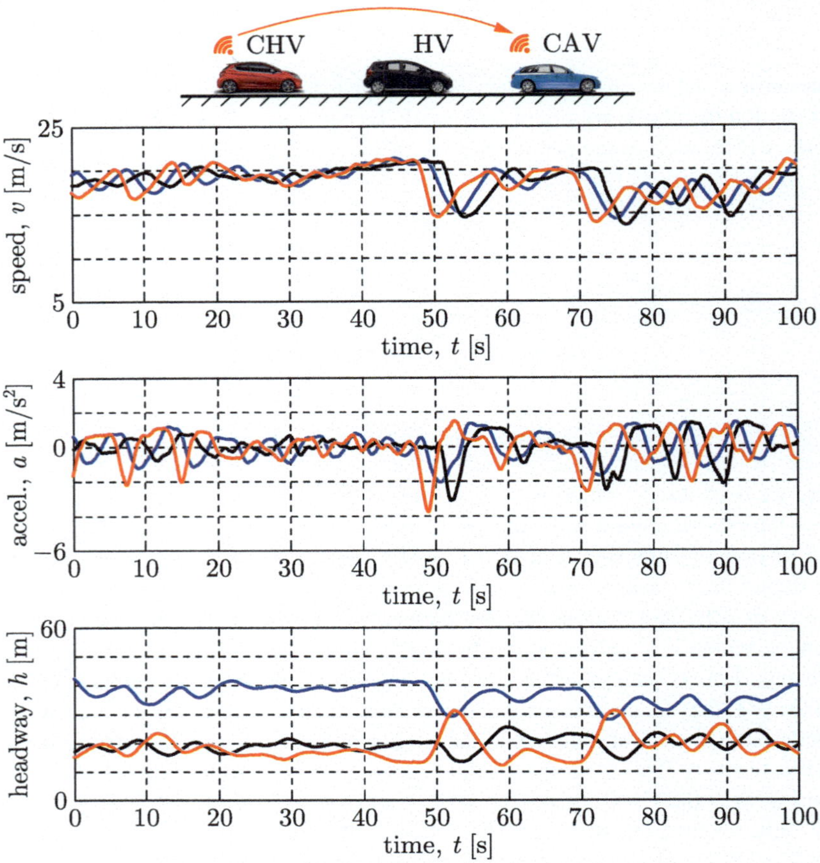

Fig. 5.29 Virtual ring experiment for virtual ring length $L + 3l = 90$ [m] where the CAV responds to one CHV "ahead" while utilizing the CCC (5.12), (3.32), (3.33) with parameters $v_{\max} = 30$ [m/s], $h_{\rm st} = 5$ [m], $h_{\rm go} = 55$ [m] yielding $\kappa = 0.6$ [1/s]. The gain parameters are set to $\alpha_{01} = 0.4$ [1/s], $\beta_{01} = 0.5$ [1/s], $\beta_{02} = 0$ as in Fig. 5.14. The red CHV imposes perturbations around $t = 50$ [s] and $t = 70$ [s] and in both cases the system returns to the equilibrium after some transient oscillations

$$\sum_{i=1}^{N} h_i^* = L\,, \quad v^* = V_1(h_1^*) = V_2(h_2^*) = \ldots = V_N(h_N^*)\,, \tag{5.23}$$

where we only designed V_N (cf. (3.32)) while V_i, $i = 1, \ldots, N-1$ were provided by the human drivers (cf. (2.5)). This results in the nonuniform equilibrium $h_1^* \neq h_2^* \neq \ldots \neq h_N^*$. In particular, in Fig. 5.29 we have $h_1^* \approx h_2^* < h_3^*$, that is, the CAV maintains a larger headway compared to those of the human drivers.

In Fig. 5.30 the same control strategy is used as in Fig. 5.29, i.e., the blue CAV only responds to the red CHV "ahead," but the gradient of the CAV's range policy is increased to $\kappa = 0.8$ [1/s] by changing the free-flow headway to $h_{\rm go} = 42.5$ [m]. This results in a smaller equilibrium headway for the CAV, and in turn, larger head-

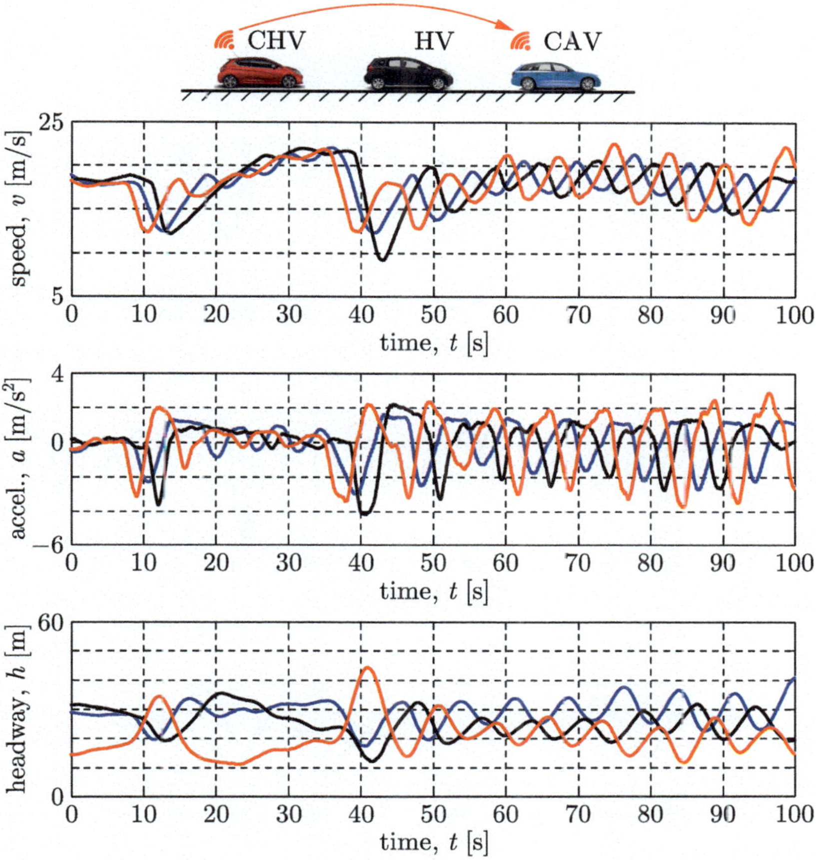

Fig. 5.30 Virtual ring experiment for virtual ring length $L + 3l = 90$ [m] where the CAV responds to one CHV "ahead" while utilizing the CCC (5.12), (3.32), (3.33) with parameters $v_{\max} = 30$ [m/s], $h_{st} = 5$ [m], $h_{go} = 42.5$ [m] yielding $\kappa = 0.8$ [1/s]. The gain parameters are set to $\alpha_{01} = 0.4$ [1/s], $\beta_{01} = 0.5$ [1/s], $\beta_{02} = 0$ as in Fig. 5.14. The red CHV imposes perturbations around $t = 10$ [s] and $t = 35$ [s]. In the former case the system returns to the equilibrium while in the latter case limit cycle oscillations develop

ways for the human-driven vehicles. Initially the system stays close to the equilibrium until around $t = 10$ [s] when the red CHV imposes a perturbation. The corresponding transients decay and the system returns to the equilibrium. This means that the equilibrium is locally stable. However, when around $t = 35$ [s] the red CHV imposes another, slightly larger, perturbation, the related transients grow into sustained oscillations. According to Sects. 2.4 and 4.5, the experiments reveal bistable behavior: the system either converges to the locally stable equilibrium or to the locally stable limit cycle depending on the applied perturbations.

In Fig. 5.31 the gradient of the range policy is increased further to $\kappa = 1$ [1/s], by decreasing the free-flow headway to $h_{go} = 35$ [m], while the CAV still utilizes the same control strategy as in Figs. 5.29 and 5.30. This further decreases the equilibrium

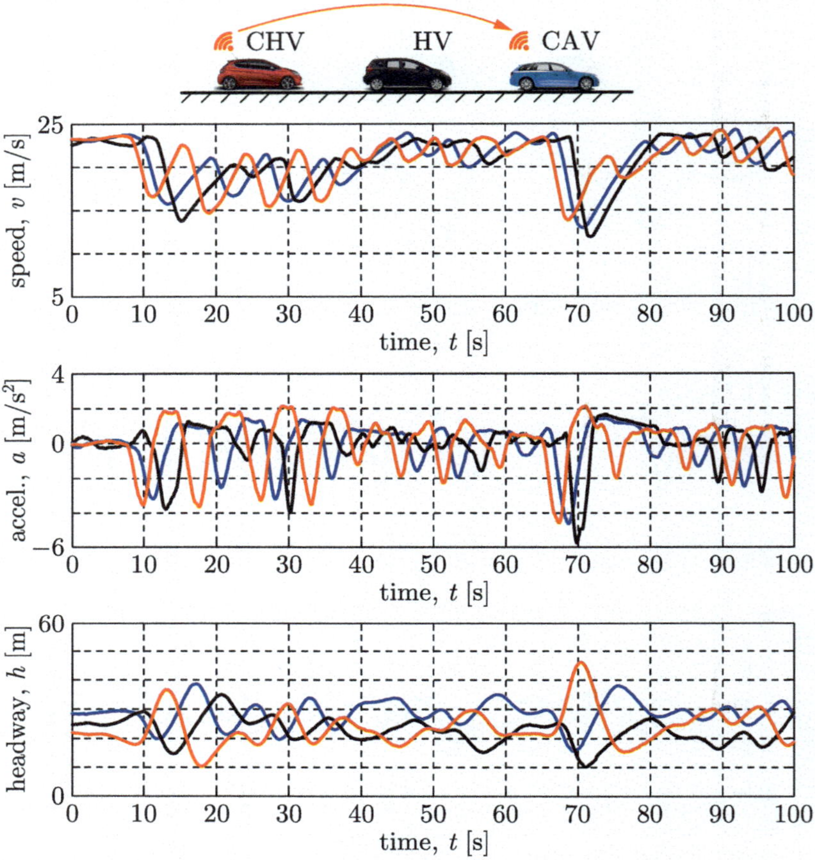

Fig. 5.31 Virtual ring experiment for virtual ring length $L + 3l = 90$ [m] where the CAV responds to one CHV "ahead" while utilizing the CCC (5.12), (3.32), (3.33) with parameters $v_{\max} = 30$ [m/s], $h_{\text{st}} = 5$ [m], $h_{\text{go}} = 35$ [m] yielding $\kappa = 1$ [1/s]. The gain parameters are set to $\alpha_{01} = 0.4$ [1/s], $\beta_{01} = 0.5$ [1/s], $\beta_{02} = 0$ as in Fig. 5.14. The red CHV imposes perturbations around $t = 10$ [s] and $t = 65$ [s] and in both cases limit cycle oscillations develop

headway of the CAV, and in turn, further increases the equilibrium headways of the human-driven vehicles. Again, perturbations, like those applied by the red CHV around $t = 10$ [s] and $t = 65$ [s], lead to sustained oscillations, while the system can still return to the equilibrium from time to time, indicating bistability. The fact that even smaller perturbations can trigger sustained oscillations suggests that the region of attraction of the equilibrium is smaller than it was in the $\kappa = 0.8$ [1/s] case.

Finally, to eliminate bistability, we change the control strategy as demonstrated in Fig. 5.32. Here the CAV responds to the red CHV "ahead" as well as to the black CHV "two vehicles ahead," using the control strategy as in Fig. 5.15, but with parameters $\kappa = 1$ [1/s], $\beta_{01} = 0.2$ [1/s] and $\beta_{02} = 0.3$ [1/s]. In this case the equilibrium

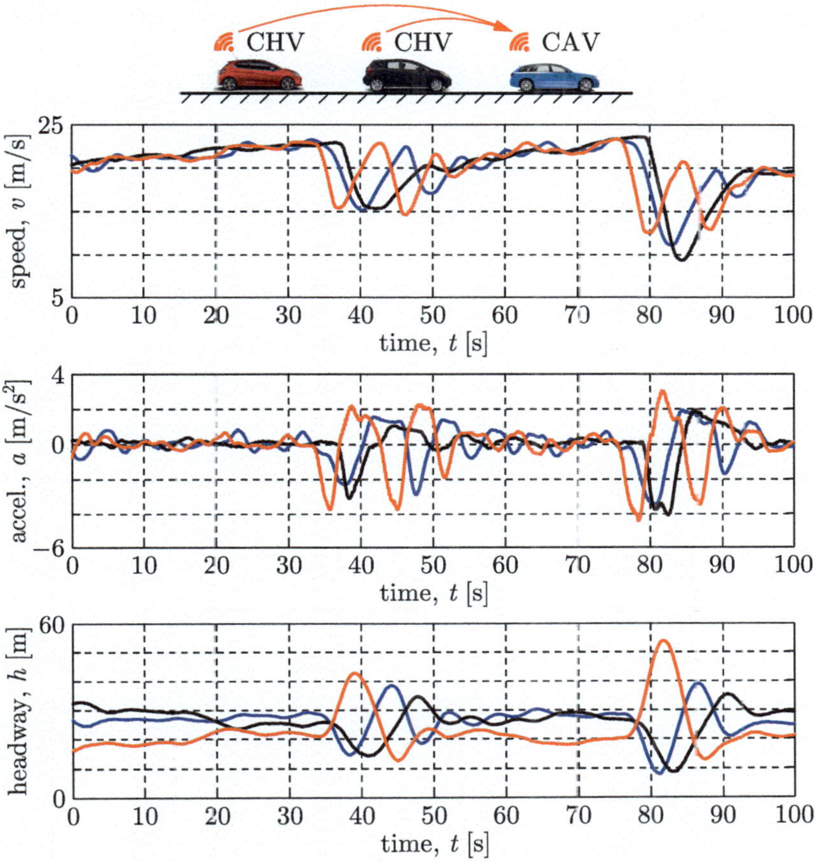

Fig. 5.32 Virtual ring experiment for virtual ring length $L + 3l = 90\,[\mathrm{m}]$ where the CAV responds to two CHVs "ahead" while utilizing the CCC (5.12), (3.32), (3.33) with parameters $v_{\max} = 30\,[\mathrm{m/s}]$, $h_{\mathrm{st}} = 5\,[\mathrm{m}]$, $h_{\mathrm{go}} = 35\,[\mathrm{m}]$ yielding $\kappa = 1\,[1/\mathrm{s}]$. The gain parameters are set to $\alpha_{01} = 0.4\,[1/\mathrm{s}]$, $\beta_{01} = 0.2\,[1/\mathrm{s}]$, $\beta_{02} = 0.3\,[1/\mathrm{s}]$ similar to Fig. 5.15. The red CHV imposes perturbations around $t = 35\,[\mathrm{s}]$ and $t = 75\,[\mathrm{s}]$, and in both cases the system returns to the equilibrium after some transient oscillations

becomes globally stable as shown by the decay of the oscillations triggered by the red CHV around $t = 35\,[\mathrm{s}]$ and $t = 75\,[\mathrm{s}]$. This demonstrates the benefits of V2X connectivity in stabilizing a system that is otherwise prone to limit cycle oscillations. Compared to the case in Fig. 5.29, where the equilibrium was also globally stable, here the CAV is able to maintain smaller headways which can lead to larger traffic flux.

Appendix A
Routh–Hurwitz Criteria

Consider an n-dimensional linear dynamical system given by the ordinary differential equation

$$\dot{\mathbf{x}} = \mathbf{A}\,\mathbf{x}\,, \tag{A.1}$$

with $\mathbf{x} \in \mathbb{R}^n$, $\mathbf{A} \in \mathbb{R}^{n\times n}$, and $n \geq 1$. The corresponding characteristic polynomial is given in the form

$$p(\lambda) = \det(\lambda\mathbf{I} - \mathbf{A}) = \sum_{k=0}^{n} a_k \lambda^k\,, \tag{A.2}$$

where $a_k \in \mathbb{R}$ for all $k \in \{0, \ldots, n\}$ with $a_0 > 0$. The linear system (A.1) is asymptotically stable if and only if $\mathrm{Re}(\lambda_k) < 0$ holds for the roots λ_k of the characteristic polynomial p in (A.2) for all $k \in \{0, \ldots, n\}$. Below we relate the stability of the linear system (A.1) to the coefficients of the characteristic polynomial (A.2) by revisiting the Routh–Hurwitz criteria given in, for example, [52, 96].

The following condition is necessary for the asymptotic stability of system (A.1).

Theorem A.1 *If system (A.1) is asymptotically stable (i.e., the roots of the characteristic polynomial (A.2) with $a_0 > 0$ have negative real parts) then $a_k > 0$ for all $k \in \{1, \ldots, n\}$. That is, the necessary condition for asymptotic stability is*

$$a_n, \ldots, a_2, a_1 > 0\,. \tag{A.3}$$

Let the Hurwitz matrix of (A.2) be defined as

$$\mathbf{H} = \begin{bmatrix} a_1 & a_0 & 0 & 0 & \ldots & 0 \\ a_3 & a_2 & a_1 & a_0 & \ldots & 0 \\ a_5 & a_4 & a_3 & a_2 & \ldots & 0 \\ \vdots & \vdots & \vdots & \vdots & \vdots & \vdots \\ a_{2n-1} & a_{2n-2} & a_{2n-3} & a_{2n-4} & \ldots & a_n \end{bmatrix}, \tag{A.4}$$

G. Orosz and T. G. Molnár, *Dynamics and Control of Connected Vehicles*, Surveys and Tutorials in the Applied Mathematical Sciences 15,
https://doi.org/10.1007/978-3-031-94555-7

where $a_k = 0$ if $k > n$. Then, the necessary and sufficient conditions for the asymptotic stability of system (A.1) are established by the Routh–Hurwitz criteria as follows.

Theorem A.2 (Routh–Hurwitz criteria) *System (A.1) is asymptotically stable (i.e., the roots of the characteristic polynomial (A.2) with $a_0 > 0$ have negative real parts) if and only if all leading principal minors of the Hurwitz matrix* **H** *are positive. That is, the condition for asymptotic stability is*

$$\begin{aligned} H_1 &= a_1 > 0\,, \\ H_2 &= \det\begin{bmatrix} a_1 & a_0 \\ a_3 & a_2 \end{bmatrix} > 0\,, \\ H_3 &= \det\begin{bmatrix} a_1 & a_0 & 0 \\ a_3 & a_2 & a_1 \\ a_5 & a_4 & a_3 \end{bmatrix} > 0\,, \\ &\vdots \\ H_n &= \det(\mathbf{H}) > 0\,. \end{aligned} \tag{A.5}$$

Notice that (A.5) with $a_0 > 0$ implies (A.3).

Furthermore, [52] highlighted that the Routh–Hurwitz stability conditions can be simplified as follows, in order to reduce the number of determinants to be calculated.

Theorem A.3 (Liénard–Chipart criteria) *System (A.1) is asymptotically stable (i.e., the roots of the characteristic polynomial (A.2) with $a_0 > 0$ have negative real parts) if and only if*

$$a_n, \ldots, a_2, a_1 > 0\,, \tag{A.6}$$

$$\begin{aligned} H_{n-1}, H_{n-3}, \ldots, H_3 > 0\,, &\quad \textit{for even } n\,, \\ H_{n-1}, H_{n-3}, \ldots, H_2 > 0\,, &\quad \textit{for odd } n\,, \end{aligned} \tag{A.7}$$

where (A.7) is dropped if $n \leq 2$.

These stability conditions become the following for low-dimensional systems. For $n = 1$ the characteristic polynomial, the Hurwitz matrix and the stability condition read

$$p(\lambda) = a_1\lambda + a_0\,, \quad \mathbf{H} = a_1\,, \quad a_1,\ a_0 > 0\,, \tag{A.8}$$

while for $n = 2$ these become

$$p(\lambda) = a_2\lambda^2 + a_1\lambda + a_0\,, \quad \mathbf{H} = \begin{bmatrix} a_1 & a_0 \\ 0 & a_2 \end{bmatrix}, \quad a_2,\ a_1,\ a_0 > 0\,. \tag{A.9}$$

For $n = 3$ we have

$$p(\lambda) = a_3\lambda^3 + a_2\lambda^2 + a_1\lambda + a_0 \,, \quad \mathbf{H} = \begin{bmatrix} a_1 & a_0 & 0 \\ a_3 & a_2 & a_1 \\ 0 & 0 & a_3 \end{bmatrix} , \tag{A.10}$$
$$a_3,\ a_2,\ a_1,\ a_0 > 0 \,, \quad H_2 = a_2a_1 - a_3a_0 > 0 \,,$$

whereas for $n = 4$ we get

$$p(\lambda) = a_4\lambda^4 + a_3\lambda^3 + a_2\lambda^2 + a_1\lambda + a_0 \,, \quad \mathbf{H} = \begin{bmatrix} a_1 & a_0 & 0 & 0 \\ a_3 & a_2 & a_1 & a_0 \\ 0 & a_4 & a_3 & a_2 \\ 0 & 0 & 0 & a_4 \end{bmatrix} , \tag{A.11}$$
$$a_4,\ a_3,\ a_2,\ a_1,\ a_0 > 0 \,, \quad H_3 = a_3(a_2a_1 - a_3a_0) - a_4a_1^2 > 0 \,.$$

Finally, we remark that some of the literature lists the Routh–Hurwitz criteria with the coefficients a_{n-k} instead of a_k, i.e., they consider the coefficients in the reverse order which corresponds to the polynomial $\tilde{p}(\lambda) = \sum_{k=0}^{n} a_{n-k}\lambda^k$ instead of p in (A.2). Both formulations are correct, since polynomial p is stable if and only if polynomial $\tilde{p}$ is stable. This is proven as follows. Notice that $p(\lambda) = \lambda^n \tilde{p}(1/\lambda)$ and $\tilde{p}(\lambda) = \lambda^n p(1/\lambda)$ hold, while $\mathrm{Re}(\lambda) < 0$ if and only if $\mathrm{Re}(1/\lambda) < 0$. If p is stable with roots λ_k, then we have $p(\lambda_k) = 0$ and $\mathrm{Re}(\lambda_k) < 0$. Hence $\tilde{p}(1/\lambda_k) = 0$ and $\mathrm{Re}(1/\lambda_k) < 0$, that is, $\tilde{p}$ is also stable with roots $\tilde{\lambda}_k = 1/\lambda_k$. Similarly, if $\tilde{p}$ is stable with roots $\tilde{\lambda}_k$, then p is also stable with roots $\lambda_k = 1/\tilde{\lambda}_k$. Therefore, a_k can be swapped with a_{n-k} in the Hurwitz matrix without affecting the stability conditions.

Appendix B
Introduction to Linear Time Delay Systems

In this appendix, we give a brief overview on the properties and behavior of linear dynamical systems with time delays. We restrict ourselves to those properties which are relevant to the analysis presented in this book. For a more comprehensive description of time delay systems, please refer to the books [24, 43, 56, 73, 81, 96, 113, 130, 197].

B.1 Autonomous Systems with Point Delay

First, we discuss linear autonomous time delay systems, that is, linear delay differential equations (DDEs). Later, we address the effect of time-varying inputs (such as external excitation or control input) on the dynamics.

B.1.1 Scalar Example

First, consider one of the simplest time delay systems: the Hayes equation [83]. This equation is of the form

$$\dot{x}(t) = ax(t) + bx(t-\tau) \,, \tag{B.1}$$

where $x(t) \in \mathbb{R}$ is a scalar variable, $a, b \in \mathbb{R}$ are scalar parameters, and $\tau \geq 0$ is the delay. Below, we discuss the solution and stability properties of this equation with delay $\tau > 0$, and compare them to the delay-free case with $\tau = 0$. More details about this equation can be found, for example, in [96].

The initial conditions for the Hayes equation (B.1) are given by

$$x(t) = \varphi(t) \,, \quad t \in [-\tau, 0] \,, \tag{B.2}$$

G. Orosz and T. G. Molnár, *Dynamics and Control of Connected Vehicles*, Surveys and Tutorials in the Applied Mathematical Sciences 15,
https://doi.org/10.1007/978-3-031-94555-7

where $\varphi : [-\tau, 0] \to \mathbb{R}$ is a continuous function. Notice that, as opposed to the delay-free case ($\tau = 0$) where the initial conditions $x(0) = x_0$ are given by a single value $x_0 \in \mathbb{R}$ at $t = 0$, the initial conditions of DDEs ($\tau > 0$) are defined by the history of states over the interval $t \in [-\tau, 0]$.

Similar to delay-free systems, the solution of the initial value problem (B.1), (B.2) can be sought for in the exponential form

$$x(t) = c\mathrm{e}^{\lambda t}\ , \quad c, \lambda \in \mathbb{C}\ , \tag{B.3}$$

which allows us to analyze the stability of the system. Substituting the trial solution (B.3) into the Hayes equation (B.1) and considering $c \neq 0$ leads to the *characteristic equation*

$$D(\lambda) = \lambda - a - b\mathrm{e}^{-\lambda\tau} = 0\ , \tag{B.4}$$

where $D : \mathbb{C} \to \mathbb{C}$ is called the *characteristic function*. Notice that, due to the exponential term $\mathrm{e}^{-\lambda\tau}$, D is a transcendental function with infinitely many complex roots if $\tau > 0$. As opposed, the characteristic function of the delay-free case ($\tau = 0$) is a polynomial with finitely many roots (i.e., a single root for the scalar example (B.1)).

The solutions $\lambda_k \in \mathbb{C}$, $k \in \mathbb{N}$ of the characteristic equation (B.4) (i.e., the roots of the characteristic function D) are called *characteristic exponents* (or characteristic roots). The Hayes equation (B.1) is asymptotically stable if and only if $\mathrm{Re}(\lambda_k) < 0$ holds for all the characteristic exponents (for all $k \in \mathbb{N}$). Accordingly, we call a characteristic exponent as stable if $\mathrm{Re}(\lambda_k) < 0$ and unstable if $\mathrm{Re}(\lambda_k) > 0$. Note that, due to having infinitely many characteristic exponents, stability is more difficult to analyze than for delay-free systems, and stability conditions like the Routh–Hurwitz criteria in Appendix A are no longer applicable.

Yet, the *stability boundary*—where $\mathrm{Re}(\lambda_k) = 0$ holds for some $k \in \mathbb{N}$—can be identified algorithmically using the so-called *D-subdivision* method [96]. Specifically, $\mathrm{Re}(\lambda_k) = 0$ may occur if there is a real root at zero, $\lambda_k = 0$, or if there is a complex conjugate pair of roots on the imaginary axis, $\lambda_k = \pm\mathrm{j}\omega$ with $\mathrm{j}^2 = -1$, $\omega > 0$. Substituting $\lambda = 0$ into the characteristic equation (B.4) leads to

$$b = -a\ , \tag{B.5}$$

which is called a *D-curve* in the space (a, b) of the parameters. Substituting $\lambda = \mathrm{j}\omega$ into the characteristic equation (B.4) and separation into real and imaginary parts yields

$$\begin{aligned} -a - b\cos(\omega\tau) &= 0\ , \\ \omega + b\sin(\omega\tau) &= 0\ , \end{aligned} \tag{B.6}$$

which can be rearranged to

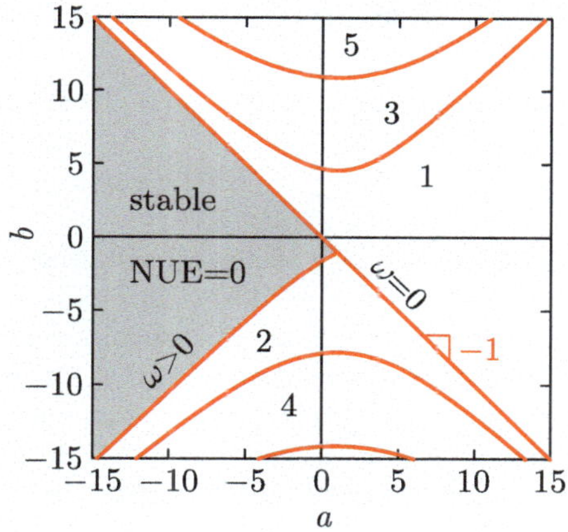

Fig. B.1 Stability chart of the Hayes equation (B.1) for $\tau = 1$ [96]. The number of unstable characteristic exponents (NUE) are shown in each region and the stable region is shaded

$$
\begin{aligned}
a &= \frac{\omega}{\tan(\omega\tau)}\,, \\
b &= -\frac{\omega}{\sin(\omega\tau)}\,.
\end{aligned}
\tag{B.7}
$$

This represents another D-curve in the space (a, b) that is parameterized by $\omega > 0$.

The D-curves are plotted in the space (a, b) in Fig. B.1 using red color. The D-curves separate those regions of the parameter space (a, b) that are associated with the same number of unstable characteristic exponents (NUE), and hence provide the same qualitative stability properties for the Hayes equation (B.1). When the parameters (a, b) are changed such that the D-curve (B.5) associated with $\lambda = 0$ is crossed, the NUE increases or decreases by one (i.e., a single real root crosses the imaginary axis at zero). When the parameters are changed across the D-curve (B.7) corresponding to $\lambda = j\omega$, the NUE changes by two (i.e., a complex conjugate pair of roots crosses the imaginary axis).

The plot of the D-curves in the parameter space (a, b) is called a *stability chart*. In this chart, the parameter region associated with zero unstable characteristic exponents is called the *stable region*; see the gray shaded domain in Fig. B.1. The segments of the D-curves that bound the stable region are referred to as *stability boundaries*. As such, stability charts help graphically identify parameters that yield stable (or unstable) dynamics, and hence these charts can be used as design tool to choose or tune parameters. This is leveraged throughout the book to characterize the stability of driving behavior and design controllers that drive vehicles in a stable manner.

B.1.2 Retarded Systems with Point Delay

Now we describe a more general class of DDEs, given by

$$\dot{\mathbf{x}}(t) = \mathbf{a}_0\,\mathbf{x}(t) + \mathbf{a}_\tau\,\mathbf{x}(t-\tau)\ , \tag{B.8}$$

where $\mathbf{x}(t) \in \mathbb{R}^n$ is a vector-valued variable, $\mathbf{a}_0, \mathbf{a}_\tau \in \mathbb{R}^{n\times n}$ are matrix-valued parameters, and $\tau \geq 0$ is the delay. This class of time delay systems is called as *retarded system*, because the rate of change of state depends on past states. We refer to τ as *point delay* because a single point $t-\tau$ in the history of the system determines the dynamics. Note that retarded systems with so-called distributed delays also exist, where time intervals distributed over the history of the system affect the dynamics.

The initial conditions corresponding to (B.8) are given by a continuous function $\boldsymbol{\varphi} : [-\tau, 0] \to \mathbb{R}^n$, in the form

$$\mathbf{x}(t) = \boldsymbol{\varphi}(t)\ , \quad t \in [-\tau, 0]\ . \tag{B.9}$$

The solution of the initial value problem (B.8), (B.9) can be found, for example, by the *method of steps* [96]. This method establishes the solution recursively over the intervals $t \in [0, \tau]$, $t \in [\tau, 2\tau]$, $t \in [2\tau, 3\tau]$, etc., in the form

$$\mathbf{x}(t) = \mathbf{x}_k(t)\ , \quad t \in [(k-1)\tau, k\tau]\ , \tag{B.10}$$

where $k \in \mathbb{N}$. Starting with $\mathbf{x}_0(t) = \boldsymbol{\varphi}(t)$, the solution $\mathbf{x}_{k+1}(t)$ over $t \in [k\tau, (k+1)\tau]$ is calculated from $\mathbf{x}_k(t)$ for each $k \in \mathbb{N}$ by the convolution integral

$$\mathbf{x}_{k+1}(t) = \mathrm{e}^{\mathbf{a}_0(t-k\tau)}\mathbf{x}_k(k\tau) + \int_{k\tau}^{t} \mathrm{e}^{\mathbf{a}_0(t-\theta)}\mathbf{a}_\tau \mathbf{x}_k(\theta-\tau)\mathrm{d}\theta\ . \tag{B.11}$$

It was shown in [130] that the solution gets smoother over time: $\mathbf{x}_k$ is k-times continuously differentiable over $t \in ((k-1)\tau, k\tau]$.

The convolution integral highlights that the dynamics of retarded systems are affected by their history over the delay interval. For this reason, the state of the time delay system (B.8) is considered to be the history of $\mathbf{x}$ over $[t-\tau, t]$ instead of the vector $\mathbf{x}(t)$ (that is the state of the corresponding delay-free system with $\tau = 0$). In particular, the state of the retarded system (B.8) is defined as the continuous function $\mathbf{x}_t : [-\tau, 0] \to \mathbb{R}^n$ given by

$$\mathbf{x}_t(\theta) = \mathbf{x}(t+\theta)\ , \quad \theta \in [-\tau, 0]\ , \tag{B.12}$$

see [43, 81, 197]. Since the state is not the vector $\mathbf{x}(t)$ in $\mathbb{R}^n$ but the continuous vector-valued function $\mathbf{x}_t$ mapping from $[-\tau, 0]$ to $\mathbb{R}^n$ (that includes the evaluation of infinitely many vector values), we say that DDEs and their state space are infinite dimensional.

The retarded system (B.8) admits the trial solution

$$\mathbf{x}(t) = \mathbf{c}\mathrm{e}^{\lambda t}\,, \quad \mathbf{c} \in \mathbb{C}^n\,, \quad \lambda \in \mathbb{C}\,. \tag{B.13}$$

Substitution into (B.8) and considering $\mathbf{c} \neq \mathbf{0}$ leads to the characteristic equation

$$D(\lambda) = \det\left(\lambda \mathbf{I} - \mathbf{a}_0 - \mathbf{a}_\tau \mathrm{e}^{-\lambda\tau}\right) = 0\,, \tag{B.14}$$

which has infinitely many roots $\lambda_k \in \mathbb{C}$, $k \in \mathbb{N}$. The infinitely many characteristic exponents also highlight the infinite-dimensional nature of time delay systems (whereas in the delay-free case with $\tau = 0$ we only have n roots for the underlying n-dimensional system).

It was shown in [130] for systems of retarded type that the characteristic exponents λ_k continuously depend on the parameters τ, $\mathbf{a}_0$, and $\mathbf{a}_\tau$. Furthermore, for retarded systems there are finitely many characteristic exponents with real parts larger than any constant $\delta \in \mathbb{R}$ (i.e., $\mathrm{Re}(\lambda_k) > \delta$). This implies that even though the number of characteristic exponents is infinite, the number of *unstable* characteristic exponents (satisfying $\mathrm{Re}(\lambda_k) > 0$) is finite.

The retarded system (B.8) is asymptotically stable if and only if all the infinitely many characteristic exponents have negative real parts, i.e., $\mathrm{Re}(\lambda_k) < 0$ holds for all $k \in \mathbb{N}$. The stability boundaries can be found by evaluating

$$D(0) = 0\,, \tag{B.15}$$

that corresponds to $\lambda = 0$, and

$$\begin{aligned} \mathrm{Re}\big(D(\mathrm{j}\omega)\big) &= 0\,, \\ \mathrm{Im}\big(D(\mathrm{j}\omega)\big) &= 0\,, \end{aligned} \tag{B.16}$$

associated with $\lambda = \mathrm{j}\omega$, $\omega > 0$. These can be plotted as D-curves in the space of the parameters (i.e., τ or the elements of $\mathbf{a}_0$ or $\mathbf{a}_\tau$), ultimately allowing the construction of stability charts.

B.1.3 Neutral Systems with Point Delay

Retarded systems can be further extended to a wider class of DDEs, called *neutral systems*. In neutral systems, the rate of change of state not only depends on past states but also on past state derivatives. For example, linear neutral systems with a single point delay are of the form

$$\dot{\mathbf{x}}(t) = \mathbf{a}_0\,\mathbf{x}(t) + \mathbf{a}_\tau\,\mathbf{x}(t-\tau) + \hat{\mathbf{a}}_\tau\,\dot{\mathbf{x}}(t-\tau)\,, \tag{B.17}$$

cf. (B.8), where an additional delayed state derivative is included with coefficient $\hat{\mathbf{a}}_\tau \in \mathbb{R}^{n\times n}$. This additional term leads to the following changes in properties compared to retarded systems.

The initial condition (B.9) applies where the initial state $\boldsymbol{\varphi} : [-\tau, 0] \to \mathbb{R}^n$ is continuously differentiable almost everywhere on $[-\tau, 0]$. The corresponding solution can be obtained by the method of steps in (B.10) with the underlying convolution integral being

$$\mathbf{x}_{k+1}(t) = \mathrm{e}^{\mathbf{a}_0(t-k\tau)}\mathbf{x}_k(k\tau) + \int_{k\tau}^{t} \mathrm{e}^{\mathbf{a}_0(t-\theta)} \left(\mathbf{a}_\tau \mathbf{x}_k(\theta-\tau) + \hat{\mathbf{a}}_\tau \dot{\mathbf{x}}_k(\theta-\tau)\right) \mathrm{d}\theta \,, \tag{B.18}$$

cf. (B.11). This solution does not necessarily get smoother over time.

The state space of the neutral system (B.17) is infinite dimensional, and its state $\mathbf{x}_t$ is defined by (B.12). Using the trial solution (B.13), the underlying characteristic equation becomes

$$D(\lambda) = \det\left(\lambda\mathbf{I} - \mathbf{a}_0 - \mathbf{a}_\tau \mathrm{e}^{-\lambda\tau} - \hat{\mathbf{a}}_\tau \lambda \mathrm{e}^{-\lambda\tau}\right) = 0 \,, \tag{B.19}$$

which has infinitely many roots $\lambda_k \in \mathbb{C}$, $k \in \mathbb{N}$. As opposed to retarded systems, the characteristic exponents of neutral systems may not depend continuously on the delay τ, and there can be infinitely many characteristic exponents with real parts larger than a constant $\delta \in \mathbb{R}$ (i.e., $\mathrm{Re}(\lambda_k) > \delta$). Accordingly, the number of unstable characteristic exponents ($\mathrm{Re}(\lambda_k) > 0$) can potentially be infinite.

Furthermore, [130] also highlighted that if the neutral system (B.17) is asymptotically stable, then the associated delay difference equation

$$\mathbf{x}(t) = \hat{\mathbf{a}}_\tau \mathbf{x}(t-\tau) \,, \tag{B.20}$$

is also asymptotically stable. In other words, the stability of (B.20) is a necessary condition for the stability of (B.17). The delay difference equation (B.20) is associated with the characteristic equation

$$\det\left(\mathbf{I} - \hat{\mathbf{a}}_\tau \mathrm{e}^{-\lambda\tau}\right) = 0 \,, \tag{B.21}$$

which, by introducing $\mu = \mathrm{e}^{\lambda\tau}$, can also be written as

$$\mathcal{D}(\mu) = \det\left(\mu\mathbf{I} - \hat{\mathbf{a}}_\tau\right) = 0 \,. \tag{B.22}$$

The solutions $\mu_\ell \in \mathbb{C}$, $\ell \in \{1, 2, \ldots, n\}$ of this equation are called *characteristic multipliers*.

Note that each characteristic multiplier μ_ℓ corresponds to a sequence of infinitely many characteristic exponents with the same real parts:

$$\lambda_{\ell,m} = \frac{\ln|\mu_\ell|}{\tau} + \mathrm{j}\frac{\angle\mu_\ell + m2\pi}{\tau} \,, \quad m \in \mathbb{Z} \,. \tag{B.23}$$

The delay difference equation (B.20) is asymptotically stable if and only if all characteristic multipliers are located within the unit circle, i.e., $|\mu_\ell| < 1$ holds for all $\ell \in \{1, 2, \ldots, n\}$ (which implies $\mathrm{Re}(\lambda_{\ell,m}) < 0$ for all $\ell \in \{1, 2, \ldots, n\}$ and $m \in \mathbb{Z}$). If a multiplier μ_ℓ is outside the unit circle (i.e., $|\mu_\ell| > 1$ for some $\ell \in \{1, 2, \ldots, n\}$), then it gives rise to an infinite sequence of unstable characteristic exponents $\lambda_{\ell,m}$. In such a case, the characteristic equation (B.19) of the neutral system (B.17) also features infinitely many unstable characteristic exponents λ_k.

Similar to the retarded system (B.8), we can identify the stability boundaries of the delay difference equation (B.20). Specifically, $|\mu_k| = 1$ may occur if there is a real multiplier of unit magnitude, $\mu_\ell = 1$ or $\mu_\ell = -1$, or if there is a complex conjugate pair of multipliers on the unit circle, $\mu_\ell = \mathrm{e}^{\pm\mathrm{j}\omega\tau}$ with $\omega \in (0, \pi/\tau)$. Substituting $\mu = 1$ into the characteristic equation (B.22) leads to the D-curve

$$\mathcal{D}(1) = 0\,, \tag{B.24}$$

while $\mu = -1$ yields

$$\mathcal{D}(-1) = 0\,, \tag{B.25}$$

and $\mu = \mathrm{e}^{\mathrm{j}\omega\tau}$ gives

$$\begin{aligned} \mathrm{Re}\big(\mathcal{D}(\mathrm{e}^{\mathrm{j}\omega\tau})\big) &= 0\,, \\ \mathrm{Im}\big(\mathcal{D}(\mathrm{e}^{\mathrm{j}\omega\tau})\big) &= 0\,. \end{aligned} \tag{B.26}$$

Along these three D-curves the number of unstable multipliers changes by one, one and two, respectively.

Ultimately, the stability charts of the neutral system (B.17) can be constructed from the D-curves given by (B.15), (B.16) and (B.19) as well as from the D-curves of the associated delay difference equation (B.20) given by (B.22), (B.24)–(B.26).

B.2 Input-Output Systems with Point Delay

Next, we discuss the behavior of input-output time delay systems of retarded-type:

$$\begin{aligned} \dot{\mathbf{x}}(t) &= \mathbf{a}_0\,\mathbf{x}(t) + \mathbf{a}_\tau\,\mathbf{x}(t-\tau) + \mathbf{b}_0\,\mathbf{u}(t) + \mathbf{b}_\tau\,\mathbf{u}(t-\tau) + \hat{\mathbf{b}}_\tau\,\dot{\mathbf{u}}(t-\tau)\,, \\ \mathbf{y}(t) &= \mathbf{c}\,\mathbf{x}(t)\,, \end{aligned} \tag{B.27}$$

cf. (B.8), with the input $\mathbf{u}(t) \in \mathbb{R}^{n_u}$, the output $\mathbf{y}(t) \in \mathbb{R}^{n_y}$, and the coefficients $\mathbf{b}_0, \mathbf{b}_\tau, \hat{\mathbf{b}}_\tau \in \mathbb{R}^{n\times n_u}$ and $\mathbf{c} \in \mathbb{R}^{n_y\times n}$.

We study the response of this system to the input in Laplace domain, considering zero initial conditions to eliminate their effect on the output. Then, the Laplace transform of (B.27) leads to

$$\begin{aligned} s\mathbf{X}(s) &= \left(\mathbf{a}_0 + \mathbf{a}_\tau \mathrm{e}^{-s\tau}\right)\mathbf{X}(s) + \left(\mathbf{b}_0 + \mathbf{b}_\tau \mathrm{e}^{-s\tau} + \hat{\mathbf{b}}_\tau s \mathrm{e}^{-s\tau}\right)\mathbf{U}(s) \,, \\ \mathbf{Y}(s) &= \mathbf{c}\mathbf{X}(s) \,, \end{aligned} \tag{B.28}$$

where $\mathbf{X}(s) \in \mathbb{C}^n$, $\mathbf{U}(s) \in \mathbb{C}^{n_u}$ and $\mathbf{Y}(s) \in \mathbb{C}^{n_y}$ are the Laplace transforms of $\mathbf{x}(t)$, $\mathbf{u}(t)$ and $\mathbf{y}(t)$, respectively, with $s \in \mathbb{C}$. This can be rearranged to

$$\mathbf{Y}(s) = \mathbf{T}(s)\mathbf{U}(s) \,, \tag{B.29}$$

where $\mathbf{T}(s) \in \mathbb{C}^{n_y \times n_u}$ is the *transfer function* given by

$$\mathbf{T}(s) = \mathbf{c}\left(s\mathbf{I} - \mathbf{a}_0 - \mathbf{a}_\tau \mathrm{e}^{-s\tau}\right)^{-1}\left(\mathbf{b}_0 + \mathbf{b}_\tau \mathrm{e}^{-s\tau} + \hat{\mathbf{b}}_\tau s \mathrm{e}^{-s\tau}\right) . \tag{B.30}$$

Throughout the book, transfer functions are constructed for the purpose of analyzing stability and frequency response.

The stability of the system (B.27) is determined by the denominator of the transfer function. Notice that the denominator

$$D(s) = \det\left(s\mathbf{I} - \mathbf{a}_0 - \mathbf{a}_\tau \mathrm{e}^{-s\tau}\right) , \tag{B.31}$$

of the transfer function (B.30) matches the characteristic function given in (B.14), hence stability analysis can be done by following the procedure discussed for the retarded system (B.8).

The frequency response of the system (B.27) can be characterized using the transfer function and considering $s = \mathrm{j}\omega$ with some $\omega > 0$. In particular, if the components $u_j(t)$ ($j \in \{1, \ldots, n_u\}$) of the input $\mathbf{u}(t)$ are harmonic:

$$u_j(t) = u_{j,\mathrm{amp}} \cos(\omega_j t + \varphi_j) \,, \tag{B.32}$$

with amplitude $u_{j,\mathrm{amp}}$, angular frequency $\omega_j > 0$ and phase φ_j, the transfer function establishes the steady-state response

$$y_i^{(\mathrm{s})}(t) = \sum_{j=1}^{n_u} M_{i,j} u_{j,\mathrm{amp}} \cos\left(\omega_j t + \varphi_j + \theta_{i,j}\right) \,, \tag{B.33}$$

where $y_i(t)$ ($i \in \{1, \ldots, n_y\}$) is the i-th component of the output $\mathbf{y}(t)$ and superscript (s) refers to steady state. The amplification $M_{i,j}$ and the phase shift $\theta_{i,j}$ are given by the elements $T_{i,j}$ of the transfer function as

$$M_{i,j} = \left|T_{i,j}(\mathrm{j}\omega_j)\right|, \quad \theta_{i,j} = \angle T_{i,j}(\mathrm{j}\omega_j) \,. \tag{B.34}$$

For single input-single output systems, we have $n_u = n_y = 1$, $u(t) \in \mathbb{R}$, $y(t) \in \mathbb{R}$, and $T(s) \in \mathbb{C}$. Hence the harmonic input

$$u(t) = u_{\mathrm{amp}} \cos(\omega t + \varphi) \,, \tag{B.35}$$

induces the steady-state response

$$y^{(s)}(t) = M u_{\text{amp}} \cos(\omega t + \varphi + \theta) , \tag{B.36}$$

with amplification M and phase θ given by

$$M = |T(\mathrm{j}\omega)| , \quad \theta = \angle T(\mathrm{j}\omega) . \tag{B.37}$$

These relations are used throughout the book to establish the response of vehicles (considered as output) to the motion of their preceding vehicle (considered as input).

B.3 Semi-discretization

Finally, we outline a numerical method called *semi-discretization* to analyze the stability of linear DDEs. This method was introduced in [95] and described in detail in [96]. We discuss the simplest instantiation of this method for retarded systems of the form (B.8) with $\tau > 0$. We remark, however, that semi-discretization is applicable to a much wider class of time delay systems, including linear time-periodic systems with multiple delays or distributed delays. The simplicity and the wide applicability of semi-discretization makes it an extremely powerful method for stability analysis.

Semi-discretization is based on the approximation of the DDE (B.8) by a sequence of ordinary differential equations (ODEs) defined over the time intervals $t \in [t_k, t_{k+1})$, where $t_k = k\Delta t$ are discrete time moments, $k \in \mathbb{N}$, and $\Delta t > 0$ is a time step. The choice of this time step affects the accuracy of the stability results obtained from this method. For simplicity of exposition, we assume that the time step is chosen such that the delay is an integer multiple of the time step, i.e., $\tau = r\Delta t$ with $r \in \mathbb{N}$, $r \geq 1$.

The heart of semi-discretization is the approximation of the delayed term $\mathbf{x}(t - \tau)$ without modifying the delay-free terms (hence the name "semi"). The simplest instantiation of semi-discretization approximates $\mathbf{x}(t - \tau)$ by the constant $\mathbf{x}(t_{k-r})$ as

$$\mathbf{x}(t - \tau) \approx \mathbf{x}(t_{k-r}) , \quad t \in [t_k, t_{k-1}) . \tag{B.38}$$

This yields the following sequence of ODEs approximating the retarded system (B.8)

$$\dot{\mathbf{x}}(t) = \mathbf{a}_0\, \mathbf{x}(t) + \mathbf{a}_\tau\, \mathbf{x}(t_{k-r}) , \quad t \in [t_k, t_{k+1}) . \tag{B.39}$$

By considering $\mathbf{x}(t_k)$ as initial condition, each ODE can be solved in closed form, and as a result $\mathbf{x}(t_{k+1})$ can be expressed:

$$\mathbf{x}(t_{k+1}) = \mathbf{P}\, \mathbf{x}(t_k) + \mathbf{R}\, \mathbf{x}(t_{k-r}) . \tag{B.40}$$

The coefficients $\mathbf{P}$ and $\mathbf{R}$ read

$$\mathbf{P} = \mathrm{e}^{\mathbf{a}_0 \Delta t}\,, \quad \mathbf{R} = \int_0^{\Delta t} \mathrm{e}^{\mathbf{a}_0(\Delta t - s)} \mathrm{d}s\, \mathbf{a}_\tau = \mathbf{a}_0^{-1} (\mathbf{P} - \mathbf{I})\, \mathbf{a}_\tau\,, \tag{B.41}$$

where the last equality holds only if $\mathbf{a}_0$ is invertible.

According to [96], the solution of the discrete-time map (B.40) approximates that of the DDE (B.8) at the discrete time moments t_k, while the stability properties (i.e., stability charts) of (B.40) also approximate those of (B.8). Notice that by augmenting the state to

$$\mathbf{X}(t_k) = \begin{bmatrix} \mathbf{x}(t_k) \\ \mathbf{x}(t_{k-1}) \\ \vdots \\ \mathbf{x}(t_{k-r}) \end{bmatrix}, \tag{B.42}$$

one may represent an approximation of the state $\mathbf{x}_t$ of the DDE as given in (B.12). With the augmented state $\mathbf{X}$, the discrete-time dynamics (B.40) can be written equivalently as

$$\begin{bmatrix} \mathbf{x}(t_{k+1}) \\ \mathbf{x}(t_k) \\ \vdots \\ \mathbf{x}(t_{k-r+1}) \end{bmatrix} = \begin{bmatrix} \mathbf{P} & \mathbf{0} & \cdots & \mathbf{0} & \mathbf{R} \\ \mathbf{I} & \mathbf{0} & \ldots & \mathbf{0} & \mathbf{0} \\ \vdots & & \ddots & & \vdots \\ \mathbf{0} & \mathbf{0} & \ldots & \mathbf{I} & \mathbf{0} \end{bmatrix} \begin{bmatrix} \mathbf{x}(t_k) \\ \mathbf{x}(t_{k-1}) \\ \vdots \\ \mathbf{x}(t_{k-r}) \end{bmatrix}, \tag{B.43}$$

or more compactly, as

$$\mathbf{X}(t_{k+1}) = \mathbf{A}\mathbf{X}(t_k)\,, \tag{B.44}$$

with

$$\mathbf{A} = \begin{bmatrix} \mathbf{P} & \mathbf{0} & \cdots & \mathbf{0} & \mathbf{R} \\ \mathbf{I} & \mathbf{0} & \ldots & \mathbf{0} & \mathbf{0} \\ \vdots & & \ddots & & \vdots \\ \mathbf{0} & \mathbf{0} & \ldots & \mathbf{I} & \mathbf{0} \end{bmatrix}. \tag{B.45}$$

Note that as the discretization is refined, i.e., $\Delta t \to 0$ and $r \to \infty$, the dimension $(r+1)n$ of the state $\mathbf{X}$ and matrix $\mathbf{A}$ increases to infinity. This captures the infinite-dimensional nature of DDEs.

The linear discrete-time map (B.44) is asymptotically stable if and only if the eigenvalues of matrix $\mathbf{A}$ are of magnitude less than one. The eigenvalues are also called as *characteristic multipliers*, and they are found as the solutions of the characteristic equation

$$\det(\mu\mathbf{I} - \mathbf{A}) = 0\,. \tag{B.46}$$

In other words, the condition for asymptotic stability is that all characteristic multipliers $\mu_\ell \in \mathbb{C}$ are located within the unit circle of the complex plane, i.e., $|\mu_\ell| < 1$ for all $\ell \in \{1, \ldots, (r+1)n\}$.

Ultimately, the stability analysis of the DDE (B.8) can be approximated using numerical eigenvalue computation for the large but finite dimensional matrix $\mathbf{A}$ constructed in (B.45). With this approach, stability charts can be constructed by executing the eigenvalue analysis for various combinations of parameters, e.g., by considering a grid in the parameter space. Stability boundaries can be found by searching for the parameters where an eigenvalue is of unit magnitude. This can be done efficiently, for example, by bisection methods such as the one established in [11].

Finally, we remark that the accuracy of semi-discretization can be improved by higher-order approximations. Equation (B.38) represents a *zeroth-order* approximation where the delayed term $\mathbf{x}(t-\tau)$ is approximated by a constant (i.e., zeroth-order polynomial) over $t \in [t_k, t_{k+1})$. One could also approximate the delayed term, for example, by a linear function of time as

$$\mathbf{x}(t-\tau) \approx \frac{t-t_k}{\Delta t}\mathbf{x}(t_{k-r+1}) + \left(1 - \frac{t-t_k}{\Delta t}\right)\mathbf{x}(t_{k-r})\,, \quad t \in [t_k, t_{k+1})\,, \tag{B.47}$$

leading to *first-order* semi-discretization with higher accuracy. The corresponding ODE sequence is

$$\dot{\mathbf{x}}(t) = \mathbf{a}_0\,\mathbf{x}(t) + \frac{t-t_k}{\Delta t}\mathbf{a}_\tau\,\mathbf{x}(t_{k-r+1}) + \left(1 - \frac{t-t_k}{\Delta t}\right)\mathbf{a}_\tau\,\mathbf{x}(t_{k-r})\,, \quad t \in [t_k, t_{k+1})\,, \tag{B.48}$$

whose solution becomes

$$\mathbf{x}(t_{k+1}) = \mathbf{P}\,\mathbf{x}(t_k) + \mathbf{R}_1\,\mathbf{x}(t_{k-r+1}) + \mathbf{R}_0\,\mathbf{x}(t_{k-r})\,, \tag{B.49}$$

with $\mathbf{P}$ and $\mathbf{R}$ defined in (B.41) and

$$\begin{aligned} \mathbf{R}_1 &= \int_0^{\Delta t} \mathrm{e}^{\mathbf{a}_0(\Delta t - s)}\frac{s}{\Delta t}\mathrm{d}s\,\mathbf{a}_\tau = \mathbf{a}_0^{-1}\left(\frac{1}{\Delta t}\mathbf{R} - \mathbf{a}_\tau\right), \\ \mathbf{R}_0 &= \int_0^{\Delta t} \mathrm{e}^{\mathbf{a}_0(\Delta t - s)}\left(1 - \frac{s}{\Delta t}\right)\mathrm{d}s\,\mathbf{a}_\tau = \mathbf{R} - \mathbf{R}_1\,, \end{aligned} \tag{B.50}$$

where the integrals are evaluated explicitly only for the case where $\mathbf{a}_0$ is invertible.

System (B.49) can be written compactly using the augmented state $\mathbf{X}$ as (B.44) with

$$\mathbf{A} = \begin{bmatrix} \mathbf{P} & \mathbf{0} & \cdots & \mathbf{R}_1 & \mathbf{R}_0 \\ \mathbf{I} & \mathbf{0} & \ldots & \mathbf{0} & \mathbf{0} \\ \vdots & & \ddots & & \vdots \\ \mathbf{0} & \mathbf{0} & \ldots & \mathbf{I} & \mathbf{0} \end{bmatrix}, \tag{B.51}$$

cf. (B.45). That is, only the top right corner of matrix $\mathbf{A}$ is modified compared to the zeroth-order method, and the stability analysis is conducted the same way through computing the magnitudes of eigenvalues. For the problems presented in this book, first-order semi-discretization with a small time step Δt is capable of accurately capturing the stability properties within reasonable computation time.

References

1. Alam, A., Besselink, B., Turri, V., Mårtensson, J., Johansson, K.H.: Heavy-duty vehicle platooning towards sustainable freight transportation: A cooperative method to enhance safety and efficiency. IEEE Control Systems Magazine **36**(6), 34–56 (2015)
2. Alan, A., He, C.R., Molnar, T.G., Mathew, J.C., Bell, A.H., Orosz, G.: Integrating safety with performance in connected automated truck control: Experimental validation. IEEE Transactions on Intelligent Vehicles **9**(1), 3075–3088 (2024)
3. Alan, A., Taylor, A.J., He, C.R., Ames, A.D., Orosz, G.: Control barrier functions and input-to-state safety with application to automated vehicles. IEEE Transactions on Control Systems Technology **31**(6), 2744–2759 (2023)
4. Ames, A.D., Grizzle, J.W., Tabuada, P.: Control barrier function based quadratic programs with application to adaptive cruise control. In: 53rd IEEE Conference on Decision and Control, pp. 6271–6278 (2014)
5. Ard, T., Dollar, R.A., Vahidi, A., Zhang, Y., Karbowski, D.: Microsimulation of energy and flow effects from optimal automated driving in mixed traffic. Transportation Research Part C **120**, 102806 (2020)
6. Asadi, B., Vahidi, A.: Predictive cruise control: Utilizing upcoming traffic signal information for improving fuel economy and reducing trip time. IEEE Transactions on Control Systems Technology **19**(3), 707–714 (2011)
7. Avedisov, S.S., Bansal, G., Orosz, G.: Impacts of connected automated vehicles on freeway traffic patterns at different penetration levels. IEEE Transactions on Intelligent Transportation Systems **23**(5), 4305–4318 (2022)
8. Avedisov, S.S., Orosz, G.: Nonlinear network modes in cyclic systems with applications to connected vehicles. Journal of Nonlinear Science **25**(4), 1015–1049 (2015)
9. Avedisov, S.S., Orosz, G.: Analysis of connected vehicle networks using network-based perturbation techniques. Nonlinear Dynamics **89**, 1651–1672 (2017)
10. Aw, A., Rascle, M.: Resurrection of "second order" models of traffic flow. SIAM Journal on Applied Mathematics **60**(3), 916–938 (2000)
11. Bachrathy, D., Stépán, G.: Bisection method in higher dimensions and the efficiency number. Periodica Polytechnica - Mechanical Engineering **56**(2), 81–86 (2012)
12. Bae, S., Kim, Y., Choi, Y., Guanetti, J., Gill, P., Borrelli, F., Moura, S.J.: Ecological adaptive cruise control of plug-in hybrid electric vehicle with connected infrastructure and on-road experiments. Journal of Dynamic Systems, Measurement, and Control **144**(1), 011109 (2022)

G. Orosz and T. G. Molnár, *Dynamics and Control of Connected Vehicles*, Surveys and Tutorials in the Applied Mathematical Sciences 15, https://doi.org/10.1007/978-3-031-94555-7

13. Balas, G.J., Doyle, J.C., Glover, K., Packard, A., Smith, R.: μ-Analysis and Synthesis Toolbox for Use with MATLAB: User's Guide. The MathWorks (1993)
14. Bando, M., Hasebe, K., Nakanishi, K., Nakayama, A.: Analysis of optimal velocity model with explicit delay. Physical Review E **58**(5), 5429–5435 (1998)
15. Bando, M., Hasebe, K., Nakayama, A., Shibata, A., Sugiyama, Y.: Dynamical model of traffic congestion and numerical simulation. Physical Review E **51**(2), 1035–1042 (1995)
16. Bekiaris-Liberis, N.: Robust string stability and safety of CTH predictor-feedback CACC. IEEE Transactions on Intelligent Transportation Systems **24**(8), 8209–8221 (2023)
17. Bekiaris-Liberis, N.: Nonlinear predictor-feedback cooperative adaptive cruise control of vehicles with nonlinear dynamics and input delay. International Journal of Robust and Nonlinear Control **34**(10), 6683–6698 (2024)
18. Bekiaris-Liberis, N., Roncoli, C., Papageorgiou, M.: Predictor-based adaptive cruise control design. IEEE Transactions on Intelligent Transportation Systems **19**(10), 3181–3195 (2017)
19. Bellman, R., Cooke, K.L.: Differential-Difference Equations. Academic Press (1963)
20. Beregi, S., Avedisov, S.S., He, C.R., Takács, D., Orosz, G.: Connectivity-based delay-tolerant control of automated vehicles: Theory and experiments. IEEE Transactions on Intelligent Vehicles **8**(1), 275–289 (2023)
21. Berg, P., Mason, A.D., Woods, A.W.: Continuum approach to car-following models. Physical Review E **61**(2), 1056–1066 (2000)
22. Besselink, B., Johansson, K.H.: String stability and a delay-based spacing policy for vehicle platoons subject to disturbances. IEEE Transactions on Automatic Control **62**(9), 4376–4391 (2017)
23. Biron, Z.A., Dey, S., Pisu, P.: Real-time detection and estimation of denial of service attack in connected vehicle systems. IEEE Transactions on Intelligent Transportation Systems **19**(12), 3893–3902 (2018)
24. Breda, D., Maset, S., Vermiglio, R.: Stability of Linear Delay Differential Equations. Springer (2015)
25. Brown, R.C., Fink, A.M., Hinton, D.B.: Some Opial, Lyapunov, and De la Valée Poussin inequalities with nonhomogeneous boundary conditions. Journal of Inequalities and Applications **5**, 11–37 (2000)
26. Cai, K.X., Phan-Minh, T., Chung, S.J., Murray, R.M.: Rules of the road: formal guarantees for autonomous vehicles with behavioral contract design. IEEE Transactions on Robotics **39**(3), 1853–1872 (2023)
27. Campbell, M., Egerstedt, M., How, J.P., Murray, R.M.: Autonomous driving in urban environments: approaches, lessons and challenges. Philosophical Transactions of the Royal Society A **368**(1928), 4649–4672 (2010)
28. Caveney, D.: Cooperative vehicular safety applications. IEEE Control Systems Magazine **30**(4), 38–53 (2010)
29. Chandler, R.E., Herman, R., Montroll, E.W.: Traffic dynamics: studies in car following. Operations Research **6**(2), 165–184 (1958)
30. Chen, N., Wang, M., Alkim, T., Van Arem, B.: A robust longitudinal control strategy of platoons under model uncertainties and time delays. Journal of Advanced Transportation **2018**(1), 9852721 (2018)
31. Chen, Y., Orosz, G., Molnar, T.G.: Safety-critical connected cruise control: leveraging connectivity for safe and efficient longitudinal control of automated vehicles. In: 27th IEEE International Conference on Intelligent Transportation Systems, pp. 110–115 (2024)
32. Chen, Z., Park, B.B., Hu, J.: Design and evaluation of a human-in-the-loop connected cruise control. IEEE Transactions on Vehicular Technology **71**(8), 8104–8115 (2022)
33. Čičić, M., Johansson, K.H.: Traffic regulation via individually controlled automated vehicles: a cell transmission model approach. In: 21st IEEE International Conference on Intelligent Transportation Systems, pp. 766–771 (2018)
34. Ciuffo, B., Mattas, K., Makridis, M., Albano, G., Anesiadou, A., He, Y., Jósvai, S., Komnos, D., Pataki, M., Vass, S., Szalay, Z.: Requiem on the positive effects of commercial adaptive cruise control on motorway traffic and recommendations for future automated driving systems. Transportation Research Part C **130**, 103305 (2021)

35. Darbha, S., Konduri, S., Pagilla, P.R.: Benefits of V2V communication for autonomous and connected vehicles. IEEE Transactions on Intelligent Transportation Systems **20**(5), 1954–1963 (2019)
36. Davis, L.C.: Comment on "Analysis of optimal velocity model with explicit delay". Physical Review E **66**(3), 038101 (2002)
37. Davis, L.C.: Modification of the optimal velocity traffic model to include delay due to driver reaction time. Physica A **319**, 557–567 (2003)
38. Davis, L.C.: Effect of adaptive cruise control systems on traffic flow. Physical Review E **69**(6), 066110 (2004)
39. De Haan, R., van der Sande, T., Lefeber, E., Besselink, I.: Cooperative adaptive cruise control for heterogeneous platoons with delays: Controller design and experiments. IEEE Transactions on Control Systems Technology (2024)
40. Delle Monache, M.L., Sprinkle, J., Vasudevan, R., Work, D.: Autonomous vehicles: From vehicular control to traffic control. In: 58th IEEE Conference on Decision and Control, pp. 4680–4696 (2019)
41. Dey, K.C., Yan, L., Wang, X., Wang, Y., Shen, H., Chowdhury, M., Yu, L., Qiu, C., Soundararaj, V.: A review of communication, driver characteristics, and controls aspects of cooperative adaptive cruise control (CACC). IEEE Transactions on Intelligent Transportation Systems **17**(2), 491–509 (2015)
42. di Bernardo, M., Salvi, A., Santini, S.: Distributed consensus strategy for platooning of vehicles in the presence of time varying heterogeneous communication delays. IEEE Transaction on Intelligent Transportation Systems **16**(1), 102–112 (2015)
43. Diekmann, O., van Gils, S.A., Verduyn Lunel, S.M., Walther, H.O.: Delay Equations: Functional-, Complex-, and Nonlinear Analysis. Springer (1995)
44. Donà, R., Mattas, K., Albano, G., Vass, S., Ciuffo, B.: Stability issues in adaptive cruise control systems and traffic implication. In: Advanced Vehicle Control Symposium, pp. 961–967. Springer (2024)
45. Donà, R., Mattas, K., Albano, G., Vass, S., Ciuffo, B.: Towards automated driving: findings and comparison with ADAS. In: Advanced Vehicle Control Symposium, pp. 954–960. Springer (2024)
46. Doyle, J.: Analysis of feedback systems with structured uncertainties. IEE Proceedings D: Control Theory and Applications **129**(6), 242–250 (1982)
47. Dunbar, W.B., Caveney, D.S.: Distributed receding horizon control of vehicle platoons: Stability and string stability. IEEE Transactions on Automatic Control **57**(3), 620–633 (2011)
48. Edie, L.C.: Car-following and steady-state theory for noncongested traffic. Operations Research **9**(1), 66–76 (1961)
49. Engelborghs, K., Luzyanina, T., Roose, D.: Numerical bifurcation analysis of delay differential equations using DDE-BIFTOOL. ACM Transactions on Mathematical Software **28**(1), 1–21 (2002)
50. Ersal, T., Kolmanovsky, I., Masoud, N., Ozay, N., Scruggs, J., Vasudevan, R., Orosz, G.: Connected and automated road vehicles: state of the art and future challenges. Vehicle System Dynamics **58**(5), 672–704 (2020)
51. Fan, M.K.H., Tits, A.L., Doyle, J.C.: Robustness in the presence of mixed parametric uncertainty and unmodeled dynamics. IEEE Transactions on Automatic Control **36**(1), 25–38 (1991)
52. Farkas, H., Simon, P.L.: Use of the parametric representation method in revealing the root structure and Hopf bifurcation. Journal of Mathematical Chemistry **9**, 323–339 (1992)
53. Fayazi, S.A., Vahidi, A.: Mixed-integer linear programming for optimal scheduling of autonomous vehicle intersection crossing. IEEE Transactions on Intelligent Vehicles **3**(3), 287–299 (2018)
54. Feng, S., Sun, H., Yan, X., Zhu, H., Zou, Z., Shen, S., Liu, H.X.: Dense reinforcement learning for safety validation of autonomous vehicles. Nature **615**(7953), 620–627 (2023)
55. Feng, S., Zhang, Y., Li, S.E., Cao, Z., Liu, H.X., Li, L.: String stability for vehicular platoon control: Definitions and analysis methods. Annual Reviews in Control **47**, 81–97 (2019)

56. Fridman, E.: Introduction to Time-Delay Systems. Birkhäuser (2014)
57. Gao, F., Li, S.E., Zheng, Y., Kum, D.: Robust control of heterogeneous vehicular platoon with uncertain dynamics and communication delay. IET Intelligent Transport Systems **10**(7), 503–513 (2016)
58. Gasser, I., Sirito, G., Werner, B.: Bifurcation analysis of a class of 'car-following' traffic models. Physica D **197**(3-4), 222–241 (2004)
59. Gazis, D.C., Herman, R., Potts, R.B.: Car-following theory of steady-state traffic flow. Operations Research **7**(4), 499–505 (1959)
60. Gazis, D.C., Herman, R., Rothery, R.W.: Nonlinear follow-the-leader models of traffic flow. Operations Research **9**(4), 545–567 (1961)
61. Ge, J.I., Avedisov, S.S., He, C.R., Qin, W.B., Sadeghpour, M., Orosz, G.: Experimental validation of connected automated vehicle design among human-driven vehicles. Transportation Research Part C **91**, 335–352 (2018)
62. Ge, J.I., Orosz, G.: Dynamics of connected vehicle systems with delayed acceleration feedback. Transportation Research Part C **46**, 46–64 (2014)
63. Ge, J.I., Orosz, G.: Optimal control of connected vehicle systems with communication delay and driver reaction time. IEEE Transactions on Intelligent Transportation Systems **18**(8), 2056–2070 (2017)
64. Ge, J.I., Orosz, G.: Connected cruise control among human-driven vehicles: Experiment-based parameter estimation and optimal control design. Transportation Research Part C **95**, 445–459 (2018)
65. Ge, J.I., Orosz, G., Hajdu, D., Insperger, T., Moehlis, J.: To delay or not to delay—stability of connected cruise control. In: T. Insperger, T. Ersal, G. Orosz (eds.) Time Delay Systems: Theory, Numerics, Applications, and Experiments, pp. 263–282. Springer (2017)
66. Geddes, N.B.: Magic Motorways. Read Books Ltd (1940)
67. Gillespie, T.D.: Fundamentals of Vehicle Dynamics. SAE International (1992)
68. Gloudemans, D., Wang, Y., Ji, J., Zachar, G., Barbour, W., Hall, E., Cebelak, M., Smith, L., Work, D.B.: I-24 MOTION: An instrument for freeway traffic science. Transportation Research Part C **155**, 104311 (2023)
69. Gong, S., Zhou, A., Peeta, S.: Cooperative adaptive cruise control for a platoon of connected and autonomous vehicles considering dynamic information flow topology. Transportation Research Record **2673**(10), 185–198 (2019)
70. Gong, Y., Zhu, W.X.: Modeling and robust H_∞ control synthesis of the CAV-HDV heterogeneous traffic system with different car-following modes. IEEE Transactions on Intelligent Transportation Systems **25**(10), 12980–12998 (2024)
71. Greenshields, B.D.: A study of traffic capacity. In: 14th Annual Meeting of the Highway Research Board, pp. 448–477 (1935)
72. Greenshields, B.D.: Reaction time in automobile driving. Journal of Applied Psychology **20**(3), 353–358 (1936)
73. Gu, K., Kharitonov, V.L., Chen, J.: Stability of Time-Delay System. Springer (2003)
74. Guanetti, J., Kim, Y., Borrelli, F.: Control of connected and automated vehicles: State of the art and future challenges. Annual Reviews in Control **45**, 18–40 (2018)
75. Guckenheimer, J., Holmes, P.: Nonlinear Oscillations, Dynamical Systems, and Bifurcations of Vector Fields. Springer (1983)
76. Gunter, G., Gloudemans, D., Stern, R.E., McQuade, S., Bhadani, R., Bunting, M., Delle Monache, M.L., Lysecky, R., Seibold, B., Sprinkle, J., Piccoli, B., Work, D.B.: Are commercially implemented adaptive cruise control systems string stable? IEEE Transactions on Intelligent Transportation Systems **22**(11), 6992–7003 (2020)
77. Gunter, G., Janssen, C., Barbour, W., Stern, R.E., Work, D.B.: Model-based string stability of adaptive cruise control systems using field data. IEEE Transactions on Intelligent Vehicles **5**(1), 90–99 (2020)
78. Guo, S., Orosz, G., Molnar, T.G.: Connected cruise and traffic control for pairs of connected automated vehicles. IEEE Transactions on Intelligent Transportation Systems **24**(11), 12648–12658 (2023)

79. Hajdu, D., Ge, J.I., Insperger, T., Orosz, G.: Robust design of connected cruise control among human-driven vehicles. IEEE Transactions on Intelligent Transportation Systems **21**(2), 749–761 (2020)
80. Hale, J.: Functional Differential Equations. Springer (1971)
81. Hale, J.K., Verduyn Lunel, S.M.: Introduction to Functional Differential Equations. Springer (1993)
82. Hayat, A., Alanqary, A., Bhadani, R., Denaro, C., Weightman, R.J., Xiang, S., Lee, J.W., Bunting, M., Gollakota, A., Nice, M.W., Gloudemans, D., Zachár, G., Davis, J.F., Delle Monache, M.L., Seibold, B., Bayen, A.M., Sprinkle, J., Work, D.B., Piccoli, B.: Traffic smoothing using explicit local controllers: Experimental evidence for dissipating stop-and-go waves with a single automated vehicle in dense traffic. IEEE Control Systems **45**(1), 95–110 (2025)
83. Hayes, N.D.: Roots of the transcendental equation associated with a certain difference-differential equation. Journal of the London Mathematical Society **1**(3), 226–232 (1950)
84. He, C.R., Alan, A., Molnár, T.G., Avedisov, S.S., Bell, A.H., Zukouski, R., Hunkler, M., Yan, J., Orosz, G.: Improving fuel economy of heavy-duty vehicles in daily driving. In: American Control Conference, pp. 2306–2311 (2020)
85. He, C.R., Ge, J.I., Orosz, G.: Fuel efficient connected cruise control for heavy-duty trucks in real traffic. IEEE Transactions on Control Systems Technology **28**(6), 2474–2481 (2020)
86. He, C.R., Maurer, H., Orosz, G.: Fuel consumption optimization of heavy-duty vehicles with grade, wind, and traffic information. Journal of Computational and Nonlinear Dynamics **11**(6), 061011 (2016)
87. He, C.R., Orosz, G.: Safety guaranteed connected cruise control. In: 21st IEEE International Conference on Intelligent Transportation Systems, pp. 549–554 (2018)
88. Helbing, D.: Traffic and related self-driven many-particle systems. Reviews of Modern Physics **73**(4), 1067–1141 (2001)
89. Herman, R., Montroll, E.W., Potts, R.B., Rothery, R.W.: Traffic dynamics: analysis of stability in car following. Operations Research **7**(1), 86–106 (1959)
90. Hinrichsen, D., Pritchard, A.J.: Mathematical Systems Theory I: Modelling, State Space Analysis, Stability and Robustness. Springer (2006)
91. Huang, M., Gao, W., Jiang, Z.P.: Connected cruise control with delayed feedback and disturbance: An adaptive dynamic programming approach. International Journal of Adaptive Control and Signal Processing **33**(2), 356–370 (2019)
92. Inman, J.: Navigation and Nautical Astronomy: For the Use of British Seamen. C. W. Woodward and J. Rivington (1835)
93. Insperger, T.: On the approximation of delayed systems by Taylor series expansion. Journal of Computational and Nonlinear Dynamics **10**(2), 024503 (2015)
94. Insperger, T., Milton, J., Stépán, G.: Acceleration feedback improves balancing against reflex delay. Journal of The Royal Society Interface **10**(79), 20120763 (2013)
95. Insperger, T., Stépán, G.: Semi-discretization method for delayed systems. International Journal for Numerical Methods in Engineering **55**(5), 503–518 (2002)
96. Insperger, T., Stépán, G.: Semi-discretization for Time-delay Systems: Stability and Engineering Applications. Springer (2011)
97. Ioannou, P.A., Chien, C.C.: Autonomous intelligent cruise control. IEEE Transactions on Vehicular Technology **42**(4), 657–672 (1993)
98. Isermann, R.: Automotive Control. Springer (2022)
99. Ji, X.A., Avedisov, S.S., Khan, M.I., Lucas-Estan, M.C., Coll-Perales, B., Vörös, I., Altintas, O., Orosz, G.: Fundamental rules of teleoperated driving with network latency on curvy roads. In: IEEE Intelligent Vehicles Symposium, pp. 1841–1846 (2024)
100. Ji, X.A., Molnár, T.G., Gorodetsky, A.A., Orosz, G.: Bayesian inference for time delay systems with application to connected automated vehicles. In: 24th IEEE International Conference on Intelligent Transportation Systems, pp. 3259–3264 (2021)
101. Ji, X.A., Orosz, G.: Trainable delays in time delay neural networks for learning delayed dynamics. IEEE Transactions on Neural Networks and Learning Systems **36**(3), 5219–5229 (2025)

102. Jiang, L., Molnár, T.G., Orosz, G.: On the deployment of V2X roadside units for traffic prediction. Transportation Research Part C **129**, 103238 (2021)
103. Jin, L., Čičić, M., Amin, S., Johansson, K.H.: Modeling the impact of vehicle platooning on highway congestion: A fluid queuing approach. In: 21st International Conference on Hybrid Systems: Computation and Control, pp. 237–246 (2018)
104. Jin, W.L.: On the equivalence between continuum and car-following models of traffic flow. Transportation Research Part B **93**(A), 543–559 (2016)
105. Kerner, B.S.: The Physics of Traffic: Empirical Freeway Pattern Features, Engineering Applications, and Theory. Springer (2004)
106. Kerner, B.S.: Introduction to Modern Traffic Flow Theory and Control: The Long Road to Three-Phase Traffic Theory. Springer (2009)
107. Kesting, A., Treiber, M.: How reaction time, update time, and adaptation time influence the stability of traffic flow. Computer-Aided Civil and Infrastructure Engineering **23**(2), 125–137 (2008)
108. Kesting, A., Treiber, M., Helbing, D.: Enhanced intelligent driver model to access the impact of driving strategies on traffic capacity. Philosophical Transactions of the Royal Society A **368**(1928), 4607–4626 (2010)
109. Kesting, A., Treiber, M., Schönhof, M., Helbing, D.: Adaptive cruise control design for active congestion avoidance. Transportation Research Part C **16**(6), 668–683 (2008)
110. Kiss, A.K., Avedisov, S.S., Bachrathy, D., Orosz, G.: On the global dynamics of connected vehicle systems. Nonlinear Dynamics **96**(3), 1865–1877 (2019)
111. Kolmanovskii, V., Myshkis, A.: Applied Theory of Functional Differential Equations. Kluwer Academic Publishers (1992)
112. Kolmanovskii, V.B., Nosov, V.R.: Stability of Functional Differential Equations. Academic Press (1986)
113. Krstic, M.: Delay Compensation for Nonlinear, Adaptive, and PDE Systems. Birkhäuser (2003)
114. Laval, J.A., Leclercq, L.: A mechanism to describe the formation and propagation of stop-and-go waves in congested freeway traffic. Philosophical Transactions of the Royal Society A **368**(1928), 4519–4541 (2010)
115. Laval, J.A., Leclercq, L.: The Hamilton-Jacobi partial differential equation and the three representations of traffic flow. Transportation Research Part B **52**, 17–30 (2013)
116. Li, S., Li, K., Rajamani, R., Wang, J.: Model predictive multi-objective vehicular adaptive cruise control. IEEE Transactions on Control Systems Technology **19**(3), 556–566 (2010)
117. Li, S., Zhou, B., Xu, M.: Unraveling the platoon dynamics of connected automated vehicles with information errors. IEEE Transactions on Intelligent Transportation Systems **25**(11), 17260–17278 (2024)
118. Li, S.E., Gao, F., Li, K., Wang, L.Y., You, K., Cao, D.: Robust longitudinal control of multi-vehicle systems – A distributed H-infinity method. IEEE Transactions on Intelligent Transportation Systems **19**(9), 2779–2788 (2018)
119. Li, S.E., Zheng, Y., Li, K., Wu, Y., Hedrick, J.K., Gao, F., Zhang, H.: Dynamical modeling and distributed control of connected and automated vehicles: Challenges and opportunities. IEEE Intelligent Transportation Systems Magazine **9**(3), 46–58 (2017)
120. Liang, K.Y., Mårtensson, J., Johansson, K.H.: Heavy-duty vehicle platoon formation for fuel efficiency. IEEE Transactions on Intelligent Transportation Systems **17**(4), 1051–1061 (2015)
121. Lighthill, M.J., Whitham, G.B.: On kinematic waves II: A theory of traffic flow on long crowded roads. Proceedings of the Royal Society of London A **229**(1178), 317–345 (1955)
122. Limebeer, D.J.N., Massaro, M.: Dynamics and Optimal Control of Road Vehicles. Oxford University Press (2018)
123. Liu, C., Zheng, F., Li, R., Liu, X.: An anti-disturbance adaptive control approach for automated vehicles in mixed connected traffic environment. IEEE Transactions on Intelligent Transportation Systems **24**(12), 15274–15287 (2023)
124. Liu, D., Besselink, B., Baldi, S., Yu, W., Trentelman, H.L.: On structural and safety properties of head-to-tail string stability in mixed platoons. IEEE Transactions on Intelligent Transportation Systems **24**(6), 6614–6626 (2023)

125. Liu, D., Besselink, B., Baldi, S., Yu, W., Trentelman, H.L.: A reachability approach to disturbance and safety propagation in mixed platoons. IEEE Transactions on Automatic Control **69**(5), 3206–3213 (2023)
126. Luo, L., Liu, Y., Feng, Y., Liu, H.X., Ge, Y.E.: Stabilizing traffic flow by autonomous vehicles: Stability analysis and implementation considerations. Transportation Research Part C **158**, 104449 (2024)
127. Malikopoulos, A.A., Beaver, L., Chremos, I.V.: Optimal time trajectory and coordination for connected and automated vehicles. Automatica **125**, 109469 (2021)
128. Malikopoulos, A.A., Cassandras, C.G., Zhang, Y.J.: A decentralized energy-optimal control framework for connected automated vehicles at signal-free intersections. Automatica **93**, 244–256 (2018)
129. Mi, T., Takács, D., Liu, H., Orosz, G.: Capturing the true bounding boxes: vehicle kinematic data extraction using unmanned aerial vehicles. Journal of Intelligent Transportation Systems (2024)
130. Michiels, W., Niculescu, S.I.: Stability and stabilization of time delay systems - An eigenvalue based approach. SIAM (2007)
131. Milanes, V., Shladover, S.E., Spring, J., Nowakowski, C., Kawazoe, H., Nakamura, M.: Cooperative adaptive cruise control in real traffic situations. IEEE Transactions on Intelligent Transportation Systems **15**(1), 296–305 (2014)
132. Mirković, T.Z., Tričković, S.B., Stanković, M.S.: Opial inequality in q-calculus. Journal of Inequalities and Applications **2018**(1), 347 (2018)
133. Molnár, T.G., Hopka, M., Upadhyay, D., Van Nieuwstadt, M., Orosz, G.: Virtual rings on highways: Traffic control by connected automated vehicles. In: AI-enabled Technologies for Autonomous and Connected Vehicles, pp. 441–479. Springer (2023)
134. Molnár, T.G., Ji, X.A., Oh, S., Takács, D., Hopka, M., Upadhyay, D., Nieuwstadt, M.V., Orosz, G.: On-board traffic prediction for connected vehicles: Implementation and experiments on highways. In: American Control Conference, pp. 1036–1041 (2022)
135. Molnar, T.G., Kiss, A.K., Ames, A.D., Orosz, G.: Safety-critical control with input delay in dynamic environment. IEEE Transactions on Control Systems Technology **31**(4), 1507–1520 (2023)
136. Molnar, T.G., Orosz, G.: Destroying phantom jams with connectivity and automation: Nonlinear dynamics and control of mixed traffic. Transportation Science **58**(6), 1319–1334 (2024)
137. Molnar, T.G., Orosz, G., Ames, A.D.: On the safety of connected cruise control: Analysis and synthesis with control barrier functions. In: 62nd IEEE Conference on Decision and Control, pp. 1106–1111 (2023)
138. Molnár, T.G., Qin, W.B., Insperger, T., Orosz, G.: Application of predictor feedback to compensate time delays in connected cruise control. IEEE Transactions on Intelligent Transportation Systems **19**(2), 545–559 (2018)
139. Molnár, T.G., Upadhyay, D., Hopka, M., Van Nieuwstadt, M., Orosz, G.: Delayed Lagrangian continuum models for on-board traffic prediction. Transportation Research Part C **123**, 102991 (2021)
140. Montanino, M., Monteil, J., Punzo, V.: From homogeneous to heterogeneous traffic flows: L_p string stability under uncertain model parameters. Transportation Research Part B **146**, 136–154 (2021)
141. Montanino, M., Punzo, V.: On string stability of a mixed and heterogeneous traffic flow: A unifying modelling framework. Transportation Research Part B **144**, 133–154 (2021)
142. Monteil, J., Russo, G., Shorten, R.: On L_∞ string stability of nonlinear bidirectional asymmetric heterogeneous platoon systems. Automatica **105**, 198–205 (2019)
143. Nagel, K., Schreckenberg, M.: A cellular automaton model for freeway traffic. Journal de Physique I **12**(2), 2221–2229 (1992)
144. Nagel, K., Wagner, P., Woesler, R.: Still flowing: Approaches to traffic flow and traffic jam modeling. Operations Research **51**(5), 681–710 (2003)
145. Naus, G.J.L., Vugts, R.P.A., Ploeg, J., van de Molengraft, M.J.G., Steinbuch, M.: String-stable CACC design and experimental validation: A frequency-domain approach. IEEE Transactions on Vehicular Technology **59**(9), 4268–4279 (2010)

146. Newell, G.F.: Nonlinear effects in the dynamics of car following. Operations Research **9**(2), 209–229 (1961)
147. Newell, G.F.: A simplified car-following theory: a lower order model. Transportation Research Part B **36**(3), 195–205 (2002)
148. Nilsson, P., Hussien, O., Balkan, A., Chen, Y., Ames, A.D., Grizzle, J.W., Ozay, N., Peng, H., Tabuada, P.: Correct-by-construction adaptive cruise control: Two approaches. IEEE Transactions on Control Systems Technology **24**(4), 1294–1307 (2015)
149. Olgac, N., Sipahi, R.: An exact method for the stability analysis of time-delayed linear time-invariant (LTI) systems. IEEE Transactions on Automatic Control **47**(5), 793–797 (2002)
150. Olson, B.J., Shaw, S.W., Shi, C., Pierre, C., Parker, R.G.: Circulant matrices and their application to vibration analysis. Applied Mechanics Review **66**(4), 040803 (2014)
151. Öncü, S., Ploeg, J., van de Wouw, N., Nijmeijer, H.: Cooperative adaptive cruise control: network-aware analysis of string stability. IEEE Transaction on Intelligent Transportation Systems **15**(4), 1527–1537 (2014)
152. Orosz, G.: Connected cruise control: modelling, delay effects, and nonlinear behaviour. Vehicle System Dynamics **54**(8), 1147–1176 (2016)
153. Orosz, G., Ge, J.I., He, C.R., Avedisov, S.S., Qin, W.B., Zhang, L.: Seeing beyond the line of sight – controlling connected automated vehicles. ASME Mechanical Engineering Magazine **139**(12), S8–S12 (2017)
154. Orosz, G., Krauskopf, B., Wilson, R.E.: Bifurcations and multiple traffic jams in a car-following model with reaction-time delay. Physica D **211**(3–4), 277–293 (2005)
155. Orosz, G., Shah, S.P.: A nonlinear modeling framework for autonomous cruise control. In: ASME Dynamical Systems and Control Conference, pp. 467–471 (2012)
156. Orosz, G., Stépán, G.: Hopf bifurcation calculations in delayed systems with translational symmetry. Journal of Nonlinear Science **14**(6), 505–528 (2004)
157. Orosz, G., Stépán, G.: Subcritical Hopf bifurcations in a car-following model with reaction-time delay. Proceedings of the Royal Society A **462**(2073), 2643–2670 (2006)
158. Orosz, G., Wilson, R.E., Krauskopf, B.: Global bifurcation investigation of an optimal velocity traffic model with driver reaction time. Physical Review E **70**(2), 026207 (2004)
159. Orosz, G., Wilson, R.E., Stépán, G.: Traffic jams: dynamics and control. Philosophical Transactions of the Royal Society A **368**(1928), 4455–4479 (2010)
160. Orosz, G., Wilson, R.E., Szalai, R., Stépán, G.: Exciting traffic jams: Nonlinear phenomena behind traffic jam formation on highways. Physical Review E **80**(4), 046205 (2009)
161. Packard, A., Doyle, J.: The complex structured singular value. Automatica **29**(1), 71–109 (1993)
162. Payne, H.J.: Model of freeway traffic and control. Mathematical Model of Public System pp. 51–61 (1971)
163. Petrillo, A., Pescape, A., Santini, S.: A secure adaptive control for cooperative driving of autonomous connected vehicles in the presence of heterogeneous communication delays and cyberattacks. IEEE Transactions on Cybernetics **51**(3), 1134–1149 (2020)
164. Piacentini, G., Goatin, P., Ferrara, A.: Traffic control via moving bottleneck of coordinated vehicles. IFAC-PapersOnLine **51**(9), 13–18 (2018)
165. Pipes, L.A.: An operational analysis of traffic dynamics. Journal of Applied Physics **24**(3), 274–281 (1953)
166. Ploeg, J., Englund, C., Nijmeijer, H., Semsar-Kazerooni, E., Shladover, S.E., Voronov, A., van de Wouw, N.: Introduction to the special issue on the 2016 grand cooperative driving challenge. IEEE Transactions on Intelligent Transportation Systems **19**(4), 1208–1212 (2018)
167. Ploeg, J., Scheepers, B.T., van Nunen, E., van de Wouw, N., Nijmeijer, H.: Design and experimental evaluation of cooperative adaptive cruise control. In: 14th IEEE International Conference on Intelligent Transportation Systems, pp. 260–265 (2011)
168. Ploeg, J., Shukla, D.P., van de Wouw, N., Nijmeijer, H.: Controller synthesis for string stability of vehicle platoons. IEEE Transaction on Intelligent Transportation Systems **15**(2), 854–865 (2014)

169. Ploeg, J., van de Wouw, N., Nijmeijer, H.: L_p string stability of cascaded systems: application to vehicle platooning. IEEE Transactions on Control Systems Technology **22**(2), 786–793 (2014)
170. Popp, K., Schiehlen, W.: Ground Vehicle Dynamics. Springer (2010)
171. Powell, E.A., Poehler, E.: Rush hour in Pompeii: An archaeologist navigates first-century gridlock. Archaeology **61**(6), 18 (2008)
172. Qin, W.B., Gomez, M.M., Orosz, G.: Stability and frequency response under stochastic communication delays with applications to connected cruise control design. IEEE Transactions of Intelligent Transportation Systems **18**(2), 388–403 (2017)
173. Qin, W.B., Orosz, G.: Digital effects and delays in connected vehicles: linear stability and simulations. In: ASME Dynamical Systems and Control Conference, p. V003T30A001 (2013)
174. Qin, W.B., Orosz, G.: Scalable stability analysis on large connected vehicle systems subject to stochastic communication delays. Transportation Research Part C **83**, 39–60 (2017)
175. Qin, W.B., Orosz, G.: Experimental validation on connected cruise control with flexible connectivity topologies. IEEE/ASME Transactions on Mechatronics **24**(6), 2791–2802 (2019)
176. Qin, W.B., Orosz, G.: Experimental validation of string stability for connected vehicles subject to information delay. IEEE Transactions on Control Systems Technology **28**(4), 1203–1217 (2020)
177. Qin, Y., Wang, H.: Analytical framework of string stability of connected and autonomous platoons with electronic throttle angle feedback. Transportmetrica A **17**(1), 59–80 (2021)
178. Qin, Y., Wang, H., Ran, B.: Control design for stable connected cruise control systems to enhance safety and traffic efficiency. IET Intelligent Transport Systems **12**(8), 921–930 (2018)
179. Rajamani, R.: Vehicle Dynamics and Control. Springer (2012)
180. Rajamani, R., Tan, H.S., Law, B.K., Zhang, W.B.: Demonstration of integrated longitudinal and lateral control for the operation of automated vehicles in platoons. IEEE Transactions on Control Systems Technology **8**(4), 695–708 (2000)
181. Rakha, H., Crowther, B.: Comparison of Greenshields, Pipes, and Van Aerde car-following and traffic stream models. Transportation Research Record **1802**(1), 248–262 (2002)
182. Richards, P.I.: Shockwaves on the highway. Operations Research **4**(1), 42–51 (1956)
183. Rios-Torres, J., Malikopoulos, A.A.: Impact of partial penetrations of connected and automated vehicles on fuel consumption and traffic flow. IEEE Transactions on Intelligent Vehicles **3**(4), 453–462 (2018)
184. Rosen, K.: Discrete Mathematics and Its Applications. McGraw Hill (2019)
185. Samii, A., Bekiaris-Liberis, N.: Simultaneous compensation of actuation and communication delays for heterogeneous platoons via predictor-feedback CACC with integral action. IEEE Transactions on Intelligent Vehicles (2024)
186. Schramm, D., Hiller, M., Bardini, R.: Ground Vehicle Dynamics. Springer (2014)
187. Sciarretta, A., Vahidi, A.: Energy-Efficient Driving of Road Vehicles: Toward Cooperative, Connected, and Automated Mobility. Springer (2019)
188. Seiler, P., Pant, A., Hedrick, K.: Disturbance propagation in vehicle strings. IEEE Transactions on Automatic Control **49**(10), 1835–1842 (2004)
189. Shen, D., Karbowski, D., Rousseau, A.: A minimum principle-based algorithm for energy-efficient eco-driving of electric vehicles in various traffic and road conditions. IEEE Transactions on Intelligent Vehicles **5**(4), 725–737 (2020)
190. Shen, M., Dollar, R.A., Molnar, T.G., He, C.R., Vahidi, A., Orosz, G.: Energy-efficient reactive and predictive connected cruise control. IEEE Transactions on Intelligent Vehicles **9**(1), 944–957 (2024)
191. Shen, M., He, C.R., Molnar, T.G., Bell, A.H., Orosz, G.: Energy-efficient connected cruise control with lean penetration of connected vehicles. IEEE Transactions on Intelligent Transportation Systems **24**(4), 4320–4332 (2023)
192. Shladover, S.E., Nowakowski, C., Lu, X.Y., Ferlis, R.: Cooperative adaptive cruise control: Definitions and operating concepts. Transportation Research Record **2489**(1), 145–152 (2015)
193. Shladover, S.E., Su, D., Lu, X.Y.: Impacts of cooperative adaptive cruise control on freeway traffic flow. Transportation Research Record **2324**(1), 63–70 (2012)

194. Sieber, J., Engelborghs, K., Luzyanina, T., Samaey, G., Roose, D.: DDE-Biftool Manual - Bifurcation analysis of delay differential equations. arXiv preprint, arXiv:1406.7144 (2014)
195. Society of Automotive Engineers: Taxonomy and definitions for terms related to driving automation systems for on-road motor vehicles. SAE Standards (2021)
196. Society of Automotive Engineers: Application protocol and requirements for maneuver sharing and coordinating service. SAE Standards (2023)
197. Stépán, G.: Retarded Dynamical Systems: Stability and Characteristic Functions. Longman (1989)
198. Stern, R.E., Cui, S., Delle Monache, M.L., Bhadani, R., Bunting, M., Churchill, M., Hamilton, N., Haulcy, R., Pohlmann, H., Wu, F., Piccoli, B., Seibold, B., Sprinkle, J., Work, D.B.: Dissipation of stop-and-go waves via control of autonomous vehicles: Field experiments. Transportation Research Part C **89**, 205–221 (2018)
199. Sugiyama, Y., Fukui, M., Kikuchi, M., Hasebe, K., Nakayama, A., Nishinari, K., Tadaki, S.i., Yukawa, S.: Traffic jams without bottlenecks – experimental evidence for the physical mechanism of the formation of a jam. New Journal of Physics **10**(3), 033001 (2008)
200. Sun, C., Guanetti, J., Borrelli, F., Moura, S.J.: Optimal eco-driving control of connected and autonomous vehicles through signalized intersections. IEEE Internet of Things Journal **7**(5), 3759–3773 (2020)
201. Sun, J., Zheng, Z., Sharma, A., Sun, J.: Stability and extension of a car-following model for human-driven connected vehicles. Transportation Research Part C **155**, 104317 (2023)
202. Swaroop, D., Hedrick, J.K.: String stability of interconnected systems. IEEE Transactions on Automatic Control **41**(3), 349–357 (1996)
203. Swaroop, D., Hedrick, J.K., Chien, C.C., Ioannou, P.A.: A comparision of spacing and headway control laws for automatically controlled vehicles. Vehicle System Dynamics **23**(8), 597–625 (1994)
204. Sykora, H.T., Sadeghpour, M., Ge, J.I., Bachrathy, D., Orosz, G.: On the moment dynamics of stochastically delayed linear control systems. International Journal of Robust and Nonlinear Control **30**(18), 8074–8097 (2020)
205. Treiber, M., Hennecke, A., Helbing, D.: Congested traffic states in empirical observations and microscopic simulations. Physical Review E **62**(2), 1805–1824 (2000)
206. Treiber, M., Kesting, A.: Traffic Flow Dynamics: Data, Models and Simulation. Springer (2013)
207. Treiber, M., Kesting, A., Helbing, D.: Delays, inaccuracies and anticipation in microscopic traffic models. Physica A **360**(1), 71–80 (2006)
208. Turri, V., Besselink, B., Johansson, K.H.: Cooperative look-ahead control for fuel-efficient and safe heavy-duty vehicle platooning. IEEE Transactions on Control Systems Technology **25**(1), 12–28 (2016)
209. Ulsoy, A.G., Peng, H., Cakmakci, M.: Automotive Control Systems. Cambridge University Press (2012)
210. Vahidi, A., Eskandarian, A.: Research advances in intelligent collision avoidance and adaptive cruise control. IEEE Transactions on Intelligent Transportation Systems **4**(3), 143–153 (2003)
211. Vahidi, A., Sciarretta, A.: Energy saving potentials of connected and automated vehicles. Transportation Research Part C **95**, 822–843 (2018)
212. van Arem, B., van Driel, C.J.G., Visser, R.: The impact of cooperative adaptive cruise control on traffic-flow characteristics. IEEE Transactions on Intelligent Transportation Systems **7**(4), 429–436 (2006)
213. Vegamoor, V., Rathinam, S., Darbha, S.: String stability of connected vehicle platoons under lossy V2V communication. IEEE Transactions on Intelligent Transportation Systems **23**(7), 8834–8845 (2022)
214. Wang, C., Gong, S., Zhou, A., Li, T., Peeta, S.: Cooperative adaptive cruise control for connected autonomous vehicles by factoring communication-related constraints. Transportation Research Part C **113**, 124–145 (2020)
215. Wang, H.M., Avedisov, S.S., Altintas, O., Orosz, G.: Multi-vehicle conflict management with status and intent sharing under time delays. IEEE Transactions on Intelligent Vehicles **8**(2), 1624–1637 (2023)

216. Wang, H.M., Avedisov, S.S., Altintas, O., Orosz, G.: Intent sharing in cooperative maneuvering: Theory and experimental evaluation. IEEE Transactions on Intelligent Transportation Systems **25**(9), 12450–12463 (2024)
217. Wang, H.M., Avedisov, S.S., Altintas, O., Orosz, G.: Negotiation in cooperative maneuvering using conflict analysis: Theory and experimental evaluation. In: IEEE Intelligent Vehicles Symposium, pp. 2297–2302 (2024)
218. Wang, H.M., Avedisov, S.S., Molnár, T.G., Sakr, A.H., Altintas, O., Orosz, G.: Conflict analysis for cooperative maneuvering with status and intent sharing via V2X communication. IEEE Transactions on Intelligent Vehicles **8**(2), 1105–1118 (2023)
219. Wang, H.M., Orosz, G.: Improving the Efficiency of Trucks via CV2X Connectivity on Highways. Tech. rep., University of Michigan, Ann Arbor (2024). URL https://dx.doi.org/10.7302/21946
220. Wang, J., Zheng, Y., Chen, C., Xu, Q., Li, K.: Leading cruise control in mixed traffic flow: System modeling, controllability, and string stability. IEEE Transactions on Intelligent Transportation Systems **23**(8), 12861–12876 (2021)
221. Wang, J., Zheng, Y., Li, K., Xu, Q.: DeeP-LCC: Data-enabled predictive leading cruise control in mixed traffic flow. IEEE Transactions on Control Systems Technology **31**(6), 2760–2776 (2023)
222. Wang, M., Daamen, W., Hoogendoorn, S.P., van Arem, B.: Rolling horizon control framework for driver assistance systems. Part I: Mathematical formulation and non-cooperative systems. Transportation Research Part C **40**, 271–289 (2014)
223. Wang, M., Daamen, W., Hoogendoorn, S.P., van Arem, B.: Rolling horizon control framework for driver assistance systems. Part II: Cooperative sensing and cooperative control. Transportation Research Part C **40**, 290–311 (2014)
224. Wang, M., Hoogendoorn, S.P., Daamen, W., van Arem, B., Shyrokau, B., Happee, R.: Delay-compensating strategy to enhance string stability of adaptive cruise controlled vehicles. Transportmetrica B **6**(3), 211–229 (2018)
225. Wang, S., Levin, M.W., Stern, R.: Optimal feedback control law for automated vehicles in the presence of cyberattacks: A min-max approach. Transportation research part C **153**, 104204 (2023)
226. Wang, S., Shang, M., Levin, M.W., Stern, R.: A general approach to smoothing nonlinear mixed traffic via control of autonomous vehicles. Transportation Research Part C **146**, 103967 (2023)
227. Wang, X., Jiang, R., Li, L., Lin, Y., Zheng, X., Wang, F.Y.: Capturing car-following behaviors by deep learning. IEEE Transactions on Intelligent Transportation Systems **19**(3), 910–920 (2019)
228. Wang, Y., Delle Monache, M.L., Work, D.B.: Identifiability of car-following dynamics. Physica D **430**, 133090 (2022)
229. Wang, Y., Gunter, G., Nice, M., Delle Monache, M.L., Work, D.: Online parameter estimation methods for adaptive cruise control systems. IEEE Transactions on Intelligent Vehicles **6**(2), 288–298 (2021)
230. Wang, Y., Gunter, G., Work, D.B.: Online parameter estimation of adaptive cruise control models with delays and lags. In: 23rd IEEE International Conference on Intelligent Transportation Systems, pp. 1–6 (2020)
231. Wang, Y., Zhang, R., Masoud, N., Liu, H.X.: Anomaly detection and string stability analysis in connected automated vehicular platoons. Transportation Research Part C **151**, 104114 (2023)
232. Wang, Z., Bian, Y., Shladover, S.E., Wu, G., Li, S.E., Barth, M.J.: A survey on cooperative longitudinal motion control of multiple connected and automated vehicles. IEEE Intelligent Transportation Systems Magazine **12**(1), 4–24 (2019)
233. Wang, Z., Gao, Y., Fang, C., Liu, L., Guo, S., Li, P.: Optimal connected cruise control with arbitrary communication delays. IEEE Systems Journal **14**(2), 2913–2924 (2019)
234. Wang, Z., Gao, Y., Fang, C., Liu, L., Zhou, H., Zhang, H.: Optimal control design for connected cruise control with stochastic communication delays. IEEE Transactions on Vehicular Technology **69**(12), 15357–15369 (2020)

235. Wang, Z., Jin, S., Liu, L., Fang, C., Li, M., Guo, S.: Design of intelligent connected cruise control with vehicle-to-vehicle communication delays. IEEE Transactions on Vehicular Technology **71**(8), 9011–9025 (2022)
236. Wang, Z., Wu, G., Barth, M.J.: A review on cooperative adaptive cruise control (CACC) systems: Architectures, controls, and applications. In: 21st IEEE International Conference on Intelligent Transportation Systems, pp. 2884–2891 (2018)
237. Wang, Z., Wu, G., Barth, M.J.: Cooperative eco-driving at signalized intersections in a partially connected and automated vehicle environment. IEEE Transactions on Intelligent Transportation Systems **21**(5), 2029–2038 (2019)
238. Waqas, M., Murtaza, M.A., Nuzzo, P., Ioannou, P.: Correct-by-construction design of adaptive cruise control with control barrier functions under safety and regulatory constraints. In: American Control Conference, pp. 5140–5146 (2022)
239. Whitham, G.B.: Linear and Nonlinear Waves. Wiley-Interscience (1999)
240. Wilson, R.E.: Mechanisms for spatio-temporal pattern formation in highway traffic models. Philosophical Transactions of the Royal Society A **366**(1872), 2017–2032 (2008)
241. Wilson, R.E., Ward, J.A.: Car-following models: fifty years of linear stability analysis – a mathematical perspective. Transportation Planning and Technology **34**(1), 3–18 (2011)
242. Wong, S., Jiang, L., Walters, R., Molnár, T.G., Orosz, G., Yu, R.: Traffic forecasting using vehicle-to-vehicle communication. In: Learning for Dynamics and Control, pp. 917–929. PMLR (2021)
243. Wu, C., Kreidieh, A.R., Parvate, K., Vinitsky, E., Bayen, A.M.: Flow: A modular learning framework for mixed autonomy traffic. IEEE Transactions on Robotics **38**(2), 1270–1286 (2022)
244. Xing, H., Ploeg, J., Nijmeijer, H.: Smith predictor compensating for vehicle actuator delays in cooperative ACC systems. IEEE Transactions on Vehicular Technology **68**(2), 1106–1115 (2018)
245. Yanakiev, D., Kanellakopoulos, I.: Longitudinal control of automated CHVs with significant actuator delays. In: 36th IEEE Conference on Decision and Control, vol. 5, pp. 4756–4763 (1997)
246. Yanakiev, D., Kanellakopoulos, I.: Nonlinear spacing policies for automated heavy-duty vehicles. IEEE Transactions on Vehicular Technology **47**(4), 1365–1377 (1998)
247. Young, P., Newlin, M., Doyle, J.: μ analysis with real parametric uncertainty. In: 30th IEEE Conference on Decision and Control, vol. 2, pp. 1251–1256 (1991)
248. Yu, H., Koga, S., Krstic, M.: Stabilization of traffic flow with a leading autonomous vehicle. In: ASME Dynamic Systems and Control Conference, p. V002T22A006 (2018)
249. Yu, W., Ngoduy, D., Hua, X., Wang, W.: On the stability of a heterogeneous platoon-based traffic system with multiple anticipations in the presence of connected and automated vehicles. Transportation Research Part C **157**, 104389 (2023)
250. Zhang, H.M.: A non-equilibrium traffic model devoid of gas-like behavior. Transportation Research Part B **36**(3), 275–290 (2002)
251. Zhang, L., Orosz, G.: Motif-based design for connected vehicle systems in presence of heterogeneous connectivity structures and time delays. IEEE Transactions of Intelligent Transportation Systems **17**(6), 1638–1651 (2016)
252. Zhang, L., Orosz, G.: Beyond-line-of-sight identification by using vehicle-to-vehicle communication. IEEE Transactions on Intelligent Transportation Systems **19**(6), 1962–1972 (2017)
253. Zhang, L., Orosz, G.: Consensus and disturbance attenuation in multi-agent chains with nonlinear control and time delays. International Journal of Robust and Nonlinear Control **27**(5), 781–803 (2017)
254. Zhang, L., Stépán, G., Insperger, T.: Saturation limits the contribution of acceleration feedback to balancing against reaction delay. Journal of The Royal Society Interface **15**(138), 20170771 (2018)
255. Zhang, L., Sun, J., Orosz, G.: Hierarchical design of connected cruise control in the presence of information delays and uncertain vehicle dynamics. IEEE Transactions on Control Systems Technology **26**(1), 139–150 (2017)

256. Zhang, Y., Bai, Y., Hu, J., Cao, D., Wang, M.: Memory-anticipation strategy to compensate for communication and actuation delays for strings-stable platooning. IEEE Transactions on Intelligent Vehicles **8**(2), 1145–1155 (2022)
257. Zhang, Y., Bai, Y., Wang, M., Hu, J.: Cooperative adaptive cruise control with robustness against communication delay: An approach in the space domain. IEEE Transactions on Intelligent Transportation Systems **22**(9), 5496–5507 (2020)
258. Zhao, C., Yu, H.: Robust safety for mixed-autonomy traffic with delays and disturbances. IEEE Transactions on Intelligent Transportation Systems **25**(11), 16522–16535 (2024)
259. Zhao, C., Yu, H., Molnar, T.G.: Safety-critical traffic control by connected automated vehicles. Transportation Research Part C **154**, 104230 (2023)
260. Zheng, Y., Li, S.E., Li, K., Borrelli, F., Hedrick, J.K.: Distributed model predictive control for heterogeneous vehicle platoons under unidirectional topologies. IEEE Transactions on Control Systems Technology **25**(3), 899–910 (2017)
261. Zheng, Y., Li, S.E., Li, K., Ren, W.: Platooning of connected vehicles with undirected topologies: Robustness analysis and distributed H-infinity controller synthesis. IEEE Transactions on Intelligent Transportation Systems **19**(5), 1353–1364 (2018)
262. Zheng, Y., Li, S.E., Wang, J., Cao, D., Li, K.: Stability and scalability of homogeneous vehicular platoon: Study on the influence of information flow topologies. IEEE Transactions on Intelligent Transportation Systems **17**(1), 14–26 (2016)
263. Zhou, H., Zhou, A., Li, T., Chen, D., Peeta, S., Laval, J.: Congestion-mitigating MPC design for adaptive cruise control based on Newell's car following model: History outperforms prediction. Transportation Research Part C **142**, 103801 (2022)
264. Zhou, K., Doyle, J.C.: Essentials of Robust Control. Prentice Hall (1998)
265. Zhou, Y., Ahn, S., Wang, M., Hoogendoorn, S.: Stabilizing mixed vehicular platoons with connected automated vehicles: An H-infinity approach. Transportation Research Procedia **38**, 441–461 (2019)

Index

G. Orosz and T. G. Molnár, *Dynamics and Control of Connected Vehicles*, Surveys and Tutorials in the Applied Mathematical Sciences 15, https://doi.org/10.1007/978-3-031-94555-7

The manufacturer's authorised representative in the EU is Springer Nature Customer Service Centre GmbH, Europaplatz 3, 69115 Heidelberg, Germany. If you have any concerns regarding our products, please contact ProductSafety@springernature.com

Printed and bound by CPI Group (UK) Ltd, Croydon, CR0 4YY
05/10/2026
02228127-0005